PRAISE FOR
Crowded with Genius

"James Buchan has written a hugely readable and comprehensive review of this volatile period in the city's life. . . . He brings a natural story-teller's relish to his subject. . . . An absolute joy to read."
—Irvine Welsh, *The Guardian*

"Rich. . . . Illuminating. . . . *Crowded with Genius* is inherently one of those appealing works that combine group biography with intellectual history. . . . Buchan knows his material thoroughly and approaches it as a Scotsman."   —*Washington Post Book World*

"A sparkling and cleverly written book. . . . Edinburgh was the vortex of the intellectual currents that flowed through the Scottish Enlightenment. . . . James Buchan opens a fascinating portal through which to watch it happen."   —Arthur Herman, *The Scotsman*

"Persuasive. . . . Buchan [is] eminently qualified to write a history of Edinburgh in the eighteenth century, for the city was full of polymaths like himself. . . . Buchan writes brilliantly."
—*The New York Review of Books*

"A vivid, gripping account. James Buchan never tangles or drops the many threads on his loom. He is an expert weaver of this involving tale of cerebral passion and humanist achievement."   —Edmund White

"James Buchan tells the extraordinary story with a novelist's narrative zip and brilliant flashes of detail. . . . [A] marvelous book."
—*The Sunday Times*

"It is an extraordinary story—an entire city lifting itself up by its own dung-stained bootstraps and propelling itself towards civic greatness—and it is lovingly narrated and superbly depicted by Buchan in this elegant, authoritative work."   —*The Observer*

"Buchan brings this remarkable era to life. . . . Throughout, Buchan writes well and does a fine job arguing the case for Edinburgh's disproportionately large impact on eighteenth-century intellectual history."
—*Publishers Weekly*

"A lively portrait of the city once called 'the Athens of the north.' . . . Buchan makes a good argument for its having been a great place to be."
—*Kirkus Reviews*

"A vigorous and entertaining book. . . . When is improvement achieved without penalty? Buchan's book is a delightful threnody on its splendours and miseries."   —Paul Johnson, *The Sunday Telegraph*

"[An] elegant portrait. . . . [Buchan] moves gracefully among many miscellaneous topics without any grinding of gears or loss of authority."
—*The Times Literary Supplement*

"A spellbinding chronicle. . . . Alive with personalities, rich in ideas. . . . An impressively sophisticated and multilayered cultural history."
—*Booklist* (starred review)

"Buchan gives us a novelist's evocative account. . . . The book is a triumph of fact-based, imaginatively expressed writing."
—Magnus Magnusson, *The New Statesman*

"Buchan writes with verve. . . . This book is free of sentimental jingoism and offers real insight to adventurous readers."   —*Charlotte Observer*

"Beautifully written and enormously satisfying. . . . Brilliant and very readable."   —*Washington Times*

"Remarkable. . . . Vivid. . . . This is a gem of a book. . . . A bubbling social and intellectual scene meets its lively match in Buchan's prose. The combination makes this book a must-read."
—*St. Louis Post-Dispatch*

About the Author

JAMES BUCHAN is a novelist and critic. He is the author of *The Persian Bride*, a *New York Times* Notable Book, as well as *Frozen Desire*, an examination of money that received the Duff Cooper Prize. He has also won the Whitbread First Novel Award and the Guardian Fiction Prize. Buchan is a contributor to *The New York Times Book Review* and *The New York Observer*, and a former foreign correspondent for the *Financial Times*. He lives in Norfolk, England.

Other books by the author

Crowded with Genius

The Scottish Enlightenment:
Edinburgh's Moment of the Mind

JAMES BUCHAN

Perennial

An Imprint of HarperCollins*Publishers*

The Library of Congress has catalogued the hardcover edition as follows:
Buchan, James
    Crowded with genius : the Scottish enlightenment : Edinburgh's moment of the mind / James Buchan.—1st ed.
        p. cm.
    Includes index.
    ISBN 0-06-055888-1
        1. Enlightenment—Scotland—Edinburgh.   2. Edinburgh (Scotland)—History.
    3. Edinburgh (Scotland)—Intellectual life—18th century.   4. Scotland—Intellectual life—18th century.   I. Title.
    DA890.E2B83 2003
    941.3'407—dc22   20030624540

    ISBN 0-06-055889-X (pbk.)

04 05 06 07 08 ❖/RRD 10 9 8 7 6 5 4 3 2 1

For Elizabeth

Contents

Illustrations

19. The Hill of Arthur Seat, and Town of Edinburgh, from the South-west, 1774 (John Clerk of Eldin)
20. The General Assembly of the Kirk of Scotland, 1783 (David Allan)
21. Mrs Agnes McLehose, 1788 (John Miers)

The authors and publishers would like to thank the following for permission to reproduce illustrations: Plate 1, Royal Bank of Scotland, Edinburgh; 2, Bridgeman Art Library, London/Agnew & Sons, London; 3, 6, 11, 13 and 21, Scottish National Portrait Gallery; 4, Private Scottish Collection; 5, 7, 12, 16, 17 and 20, Edinburgh City Libraries; 8, National Portrait Gallery, London; 9 and 15, Bridgeman Art Library, London/ Scottish National Portrait Gallery; 10, The National Gallery of Ireland; 18, Hulton Archive; 19, British Library. Maps reproduced courtesy of Edinburgh City Libraries.

Acknowledgements

The author would like to thank the staffs of the British Library, the National Library of Scotland, the National Archives of Scotland, Edinburgh University Library, the Town Council Archives and the Edinburgh Room of the Central Public Library, Edinburgh. He is grateful for the assistance of Caroline Dawnay, David MacMillan, Caroline Knox, John Murray, Deborah Stewartby, Allan Massie, Constantine Normanby, Liz Robinson, Matthew Rice, Caroline Westmore and, last but foremost, James and Anna Buxton.

Prologue

Every age hath its consolations, as well as its sufferings.
Adam Ferguson[1]

For a period of nearly half a century, from about the time of the
Highland rebellion of 1745 until the French Revolution of 1789,
the small city of Edinburgh ruled the Western intellect. For near
fifty years, a city that had for centuries been a byword for poverty,
religious bigotry, violence and squalor laid the mental foundations
for the modern world.

The battle of Culloden Moor in 1746, which the writer John
Buchan called 'the last fight of the Middle Ages',[2] was also the
beginning of the modern age for Scotland's ancient capital. The
town, which had sat little changed on its rock until then, inconvenient, dirty, old-fashioned, alcoholic, quarrelsome and poor,
began to alter, first slowly, then in a convulsion. Lochs were
drained, ravines spanned by bridges, streets and squares thrown
out into the stony fields. Ale gave way to tea and port and whisky.
People dined at one o'clock, then two, then three, and then four.

Men discovered there were ways of charming women this side

of abduction. They ceased to bring their pistols to table, or to share the same cup. They read newspapers, became Freemasons, danced, burst into tears. Societies sprang up for encouraging manufactures, abolishing Scottish pronunciation and spreading the Gospel. Rents, profits and the cost of living all doubled, and the poor old Scots soldier in Smollett's *Humphry Clinker* vowed in disgust to return to his old life among the Miami Indians of America.[3] A new theory of progress, based on good laws, international commerce and the companionship of men and women, displaced the antique world of valour, loyalty, religion, and the dagger. 'Edinburgh, the Sink of Abomination' became 'Edinburgh, the Athens of Great Britain'.[4]

David Hume, Adam Smith, William Robertson, Adam Ferguson and Hugh Blair were the first intellectual celebrities of the modern world, as famous for their mental boldness as for their bizarre habits and spotless moral characters. They taught Europe and America how to think and talk about the new mental areas opening to the eighteenth-century view: consciousness, the purposes of civil government, the forces that shape and distinguish society, the composition of physical matter, time and space, right actions, what binds and what divides the two sexes. They could view with a dry eye a world where God was dead or had withdrawn into Himself, where hierarchy had disintegrated and luxury was super-abundant.

The American patriot Benjamin Franklin, who first visited Edinburgh with his son in 1759, remembered his stay as 'the *densest* happiness' he had ever experienced.[5] The famous *Encyclopédie* of the French philosophers had devoted a single contemptuous paragraph to *Écosse* in 1755,[6] but by 1762 Voltaire was writing, with more than a touch of malice, 'today it is from Scotland that we get rules of taste in all the arts, from epic poetry to gardening'.[7] The Russian Princess Romanovna Dashkova, who came from St Petersburg in the late 1770s so her son could study the classics at the College, was captivated by the genius, sociability and modesty of '*l'immortel Robertson, Blair, Smith et Ferguson*'.[8] The Prussian officer J.W. von Archenholz, who travelled in England and

Scotland in the 1780s, told his German readers that 'more true learning is to be found in Edinburgh than in Oxford and Cambridge taken together'.[9]

As for Edinburgh, it was intoxicated with its own brilliance. 'In the history of every polished nation,' a correspondent wrote to *The Scots Magazine* in July 1763, 'there is always one period at least to be found, which is crouded with men of genius in every art and science. This was the case of Greece after the overthrow of Xerxes, of Rome during the reign of Augustus, of France in the time of Lewis XIV, and of England in the time of Q. Anne ... For my part, I pour out my heart with the utmost gratitude to Providence for giving me a being in this illustrious period; and I have great reason to congratulate the present generation of my countrymen for enjoying the same blessing.'[10]

How did this come about? How did Scotland, bullied or ignored by its neighbours for centuries, find and keep its place in the sun? How did a city that consisted of a single long street and fewer than forty thousand inhabitants come for a season to rival Paris? What was this 'sudden burst of genius ... sprung up in this country by a sort of enchantment, soon after the Rebellion of 1745'?[11]

Nothing in history is sudden or enchanted. Edinburgh's moment had its origins in the Scottish past and still reverberates in the present. It was a belated reaction to a series of injuries to Scotland's sense of itself in politics, in religion, in morality. A mental crisis was surmounted in a physical city of stone, ordure, taverns, turnpike stairs, towering apartment blocks, dancing assemblies, icy churches and torrents of wind. From its resolution came not simply modern Scotland but modernity itself.

1

Auld Reekie

Edinburgh in the warm September of 1745 was a handsome, cramped and discontented provincial town of approximately 40,000 people,[1] just embarking on modernity. As a capital city, it was nothing much. It had lost its royal court to London in 1603, when King James VI succeeded to the English throne, and its nobility followed at the amalgamation of the Scottish and English parliaments in 1707. Edinburgh had no manufacturing, and its trade was a set of pettifogging monopolies, down to who had the right to rent out the pall at burials or run coaches to the port of Leith.[2] The town lived off lawyers attending on the Court of Session and clergymen coming to the General Assembly of the Church of Scotland and gentry sending their children up to school and spending the winter in town. In the first age of millionaires, an Edinburgh family was rich with £1,000 a year.[3]

There were nine Presbyterian churches, each with two ministers, and two more outside the walls;[4] two banks (which survive) and a couple of general merchants that could discount commercial bills; two thrice-weekly newspapers (one Whig, one Jacobite) and *The Scots Magazine*, founded in 1739 and full of trials, poetry, bills of mortality, and a narrative of Scots and world affairs; four printing-works to garble Bibles and law papers;[5] offices of the Friendly and Sun Fire Insurance schemes; a fund for the widows of ministers of the Kirk; a few brewers between the Cowgate and the walls; and three mail coaches to London a week: though there were men alive to tell Sir Walter Scott that once the return mail brought just a single letter for the whole of Scotland.[6] A stagecoach ran monthly to London, spending at least ten days on the road,[7] though a private chaise could do the journey faster. It was not until the time of Robert Burns's visit in 1787 that the journey was cut to sixty hours.

The parliamentary Union with England in 1707 and the abolition of the Scottish Privy Council a year later had demolished the formal administration of Scotland. In as much as the country was ruled at all during the long ascendancy in London of Sir Robert Walpole, it was controlled by the Duke of Argyll and a clutch of law officers. The Edinburgh Town Council, whose constitution had been violently disputed but altered very little since the time of James VI, was a permanent oligarchy that nominated its own successor from candidates submitted by restrictive merchant and craft guilds and even elected the town's MP.[8] Advocates and clergymen, being unincorporated, had no say in either election.

The Council met in a building in Parliament Close hard against the south-west corner of the high church of St Giles and did its drinking at Lucky Wilson's tavern in Writers' Court. Its ordinary membership was twenty-five, which could be expanded under precise and obscure conditions to thirty-three. In the words of a reforming pamphlet of 1746, 'Is it a Small Matter, with you, that the Gentlemen in the Administration of this great City, who should represent near Forty Thousand, do at no time represent Forty of the Inhabitants.'[9] 'Omnipotent, corrupt, impenetrable,' as a witness wrote of the nineteenth-century councillors, 'they might have been sitting in Venice.'[10] They controlled the trade of the town and of the Port of Leith, the street-lights and weights and measures and water supply, and named the ministers to the kirks, the doctors to the High School and the under-janitors to the College. As for the College professors, the magistrates might arrive in a body unannounced to hear a new appointment lecture. The purpose of the Council was to maintain peace between the guilds and, in alliance with the Kirk-sessions and the Presbytery, an atmosphere of unctuous piety.

From a distance the town was a palisade of towers rising, in the phrase of Robert Chambers, 'from a palace on the plain to a castle in the air'.[11] Between Castle Hill and Holyrood ran what Daniel Defoe called 'the most spacious, the longest, and best inhabited Street in *Europe*'.[12] It was called in its upper section the Lawnmarket; then lower down the High Street, which was closed at the bottom by the gate called the Netherbow Port; and, at the bottom, the Canongate. In parts the street was so broad that five carriages could have moved abreast, but so high-cambered that four of them would have overturned.[13]

Confined by its site, the Lawnmarket and High Street made a sort of antique Manhattan. With nowhere else to go, the pressure of population had squeezed the stone apartment blocks or 'lands' upwards. Those at the back of Parliament Close towered twelve storeys above the Cowgate.[14] Seen from the shores of the Firth of Forth, the garlands of wood- and peat-smoke round these pinnacles had given rise to a nickname for the town: Auld Reekie.

Between the lands, the wynds and closes ran steeply down ravines to the waters of the North Loch, or to the Cowgate. In those filthy lanes, between sagging houses showing their gables to the street and pigs rooting in the gutters, every condition mingled. As a young medical student named Oliver Goldsmith wrote in 1753 or 1754, 'you might see a well-dressed duchess issuing from a dirty close'.[15] Indeed, Jane Maxwell, who in the second half of the eighteenth century became Duchess of Gordon and the leader of Edinburgh society, was once seen riding up the High Street on a sow which her sister drove on with a stick.[16]

The lands themselves accommodated dancing-masters and Lords of Session and all sorts in between.[17] The dark scale-stairs were upright streets, a thoroughfare of Musselburgh fishwives, sweeps or coal-porters and barefoot housemaids. Sir Walter Scott, who had lost six siblings to the bad air of College Wynd, remembered in the next century that each inhabitable space 'was crowded like the underdeck of a ship. Sickness had no nook of quiet, affliction no retreat for solitary indulgence.'[18] Burns, as usual, was chiefly interested in sexual privacy, and showed a touching sympathy for his widowed landlady, obliged to listen to some 'Daughters of Belial ... gandygoin' with their men visitors in the apartment upstairs.[19]

The nineteenth-century publisher Robert Chambers described from this period a gentleman lawyer's apartment in Forrester's Wynd, 'a region of profound darkness and mystery, now no more', comprising a kitchen and three rooms: 'my lady's room', which was a sort of parlour, the gentleman's consulting room or study, and their bedroom. The children, with their nurse, had beds laid down for them at night in their father's room, the housemaid slept under the dresser, and the manservant made shift outside. Rent for this opulence was just £15 a year sterling.[20] The diet of such a family would have been fish, occasional butchers' meat (but only in summer), milk and oatmeal. They would have eaten off pewter 'often ill-cleaned'. At the Orphan Hospital, the children who worked weaving wool or lint from six a.m. to eight p.m. breakfasted and supped on oatmeal pottage and home-brewed small beer, and dined as follows:

Monday	Kail without flesh, and eggs when cheap, and in place thereof cheese and a mutchkin of ale to every servant, and half a mutchkin to the children, the bread as regulated by the statutes.
Tuesday	Kail with flesh, and no ale.
Wendesday	Kail without flesh and eggs when cheap.
Thursday	Kail with flesh, no ale.
Friday	The same as Thursday.
Saturday	Kail without flesh, and butter or eggs when cheap, and ale.
Sabbath	Bread and butter and ale as regulated.[21]

There was still little bread eaten. Once a week Duncan Forbes of Culloden, Lord President of the supreme civil court known as the Court of Session, Dr John Clerk of Listonshiels, who was President of the Royal College of Physicians, the mathematician Colin MacLaurin and the anatomy professor Alexander Monro *primus* supped together in the Lord President's house in the Cowgate, 'each Man', as Monro remembered, 'having his particular Dish. Dr Clerk had some Bread and Whey, P[rofessor] M[onro]'s Dish was boil'd spinage ... the President had a small fish ... and before Mr MacLaurin a Stake or a Chop was placed.'[22] Turnips, a new-fangled fodder crop being tried on improving estates, appeared on Edinburgh tables as a dessert.[23]

There were very few public buildings, so that all legal, town and church business was done in dark taverns. Indeed John Coutts, the general merchant who treated with the Highland rebels and gave his name to a long-lived bank, became in 1742 the first Lord Provost to entertain at home (President's Close) rather than in a tavern.[24] By the next decade, there were six hundred tavern licences within the Royalty, or bounds, of Edinburgh.[25] The drinks were ale at a penny an English quart, claret, whisky, West India punch and even, in 1743, the first speculative ship-load of port.[26] The town was a byword for insobriety. Even at the annual meeting of the directors of the Royal Bank, the drinking was such as to cause the old Provost George Drummond – who had a cold and thus the privilege of taking just sack and water – to ask God to 'pardon me the guilt of others'.[27] At their mother's funeral in 1716, the future President of the Session Duncan Forbes and his brother had been so drunk on the way to the kirkyard that they left the body behind.[28] Sir John Clerk, a Baron of the Exchequer Court, thought 'no man lookt so poor so contemptible and detestable as a drunken judge'.[29] James Boswell barely drank until he passed advocate and started pleading in the Edinburgh courts; thenceforth, he was rarely sober.[30]

Dress was both gaudy and slovenly.[31] There was no such thing in town as a haberdasher or a perfumer and the few hairdressers either refused or did not dare to dress hair on the Sabbath.[32] Women covered head and shoulders in long plaids that came down to the waist, scarlet or crimson in colour or, more often, check or tartan. That costume, coupled with their high spirits and good carriage, greatly impressed English visitors.[33] The older women still wore patches on their faces. Some of the older men still wore swords, though the practice of duelling in the street had abated.

The heart of Edinburgh lay round the old cathedral of St Giles, which since the abolition of bishops at the Reformation had been divided into four parish churches: the New Church in the choir, the Old Church in the nave, Haddow's Hole in the north-west corner and the Tolbooth in the west. In St Giles, every May, the supreme court of the church known as the General Assembly held its annual meeting. In a city without a royal court or a parliament, this ecclesiastical plenary was a high day in the calendar.

Facing St Giles in the middle of the High Street and separated from it by a cramped lane called the Krames was a row of four-storey houses with shops on the ground floor, known as the Luckenbooths. These were demolished in 1817, but earlier the best mercers and clothiers were found there: 'lucken' derives from a dialect word for locked or shut up, and these booths or shops could be locked to protect their valuable merchandise. At the east end, looking down the High Street and way out into east Lothian, was a flat where the poet Allan Ramsay senior had in 1725 established the first lending or 'circulating' library in Scotland. Two generations later, on the ground floor, William Creech, publisher of Burns, entertained the *literati* with a minimum of expense. The west end of the Luckenbooths faced up and out onto a black-looking, antique, turreted prison known as the Old Tolbooth or, in a bitter joke, the Heart of Midlothian. Further up the Lawnmarket was an open space, the Bowhead, with an ungainly old building called the Weigh-house where butter and cheese were sold. From here the West Bow, the most fanatical (and haunted) street in Edinburgh, ran down in 'sanctified bends' amid the incessant banging of the whitesmiths, to the Grassmarket and the West Port.

It was down these same bends that, prompt at midnight, the ghost of the warlock Major Weir, whose body was hanged and burned on 14 April 1670, would thunder, in a carriage driven by the Prince of Darkness and drawn by four headless chargers. The house he had lived in still stood, empty and damned, at the first bend in the street: an outpost of the world of darkness.[34]

At the back of St Giles was the Parliament House, where the Estates or Scots Parliament had sat from 1639 till their suspension at the Union in 1707. Each weekday except Monday from 12 June until 11 August, and then from 12 November to 11 March with a short recess for Christmas, sat here the two divisions of the Court of Session, the highest civil court of an energetically litigious society: either a single judge, sitting on the old Sovereign's throne under the hammer-beam roof of what was now known as the Outer House or, if his sole judgment was unacceptable to one of the parties, the 'haill fifteen' (all fifteen) Lords of Session, sitting in an apartment called the Inner House.

It was an informal age in Scotland, and while the southern part of the Outer House teemed with advocates and petitioners and judges in their gowns of purple cloth with crimson velvet facings, their full-bottomed wigs and white cravats, at the north end, fenced off by a half-partition, were the flimsy shops of stationers, cutlers, toy-makers and jewellers.

Each morning at nine, men such as Alexander Wedderburn (who later became Lord Chancellor) or James Boswell walked to the Outer House to while away their time amid old-fashioned and even deranged figures such as were captured in the next generation by the caricaturist John Kay: Andrew Nichol, 'Muck Andrew', who litigated thirty years over a middenstead (dungheap) and died destitute in Cupar gaol, or John Skene, 'the Heckler', who worked nights as a flax-dresser then changed into a suit of black clothes and curled and powdered his hair, so that he could attend the Outer House: for how else could the causes (cases) proceed in orderly fashion?[35] The court rose at noon, and Lord President Forbes, the only judge to use a carriage, would take two Lords of Session with him for a short airing in The Meadows, a little park on the southern outskirts of the town, and an early dinner.[36]

Sir Walter Scott, the great novelist of the early nineteenth century, was fascinated by the Edinburgh of his parents' time and sought to recover its atmosphere in a series of 'historical' novels. Bred to the law, he passed many hours of reverie in the Outer House and reproduced the splendours and miseries of Scots litigation as nobody before or since in the character of Peter Peebles in *Redgauntlet*:

> It's very true that it is grandeur upon earth to hear ane's name thunnered out along the long-arched roof of the Outer-House – '*Poor Peter Peebles against Plainstanes, et per contra*'; a' the best lawyers in the house fleeing like eagles to the prey; some because they are in the cause and some because they want to be thought engaged ... – to see the reporters mending their pens to take down the debate – the Lords themselves pooin' in their chairs, like folk sitting down to a gude dinner, and crying on the clerks for parts and pendicles of the process, who, puir bodies, can do little mair than cry on their closet-keepers to help them. To see a' this ... and to ken that naething will be said or dune amang a' thae grand folk, for maybe the feck of three hours, saving what concerns you and your business – O, man, nae wonder that ye judge this to be earthly glory! – And yet, neighbour, as I was saying, there be unco drawbacks – I whiles think of my bit house, where dinner, and supper, and breakfast used to come without the crying for, just as if fairies had brought it – and the gude bed at e'en – and the needfu' penny in the pouch – And then to se a' ane's warldly substance capering in the air in a pair of weigh-bauks, now up, now down, as the breath of judge or counsel inclines it for pursuer or defender – troth, man, there are times I rue having ever begun the plea wark, though maybe when you consider the renown and credit I have by it, ye will hardly believe what I am saying.[37]

To the west, in a room in the Council House on the site of what is now the lobby of the Signet Library, six of the Lords of Session under the Lord Justice-Clerk constituted the High Court of Justiciary, Scotland's highest criminal court, from which there was no appeal. A row of tenements to the south and east, rebuilt after a fire in 1700, formed with the Parliament House and St Giles a square known as the Parliament Close.

Out in the High Street, half-way between St Giles and the Tron Kirk, was an octagonal building surmounted by a pillar bearing the Scottish unicorn. This was the ancient burgh mercat or market Cross, where merchants gathered to do what little business there was to be done in Edinburgh before taking their meridian dram. It was a great resort also for general masculine gossip, the first port of call for visitors to Edinburgh, the place to pick up a chair or a caddy – odd-job man or street messenger – and a haunt of beggars, some of them licensed by the King.[38]

The grandeur of the High Street was brought to earth by the homely commerce of its salt-sellers, candy men, fish-wives screaming oysters and partans (crabs), sellers of peat, coals, and yellow sand for strewing on floors, cobblers, knife- and scissor-grinders, hawkers of spunk (fire-lighters), broadsheet sellers.[39] At night the street was lined along much of its length with barrows lit by paper lanterns. James Ballantine, a painter and poet of the nineteenth century, remembered an 'array of penny shows, ballad singers, speech criers, baskets o' laces, combs, caps, shoe-ties, an' twopenny mirrors, wi' hurleys fu' o' cherry-cheekit apples an' brown speldings'.[40]

At ten p.m. by the Tron church, as English travellers recalled with a fascinated horror, the tenements were cleared of ordure. With a shout of 'Gardy Loo', each household poured its waste down into the street, where it lay all night still scavengers came at first light to collect it. Citizens burned sheets of brown paper to neutralise the smell. The cry is thought to be a corruption of *gardez l'eau* or *gare de l'eau* – look out for slops! – but Mrs Jenkins, the good-natured maidservant in Tobias Smollett's *Humphry Clinker*, translated it as 'Lord have mercy upon you!'[41] For Edward Burt, an English surveyor sent north in the 1720s to build military roads to pacify the rebellious Highlands, a night in Edinburgh was best forgotten. At the inn, 'the Cook was too filthy an Object to be described, only another English Gentleman whispered me and said, he believed, if the Fellow was to be thrown against the Wall, he would stick to it.'[42] Trying to sleep, Burt was 'forced to hide my Head between the Sheets; for the Smell of the

Filth, thrown out by the Neighbours on the Back-Side of the House, come pouring into the Room.'[43]

Yet while Edinburgh looked and smelled like a medieval city, it had the makings of modern institutions. For all its parsimony and dirt, Edinburgh had made more history, as John Buchan later put it, than any town its size but Athens, Jerusalem and Rome.[44] The religious disputes and civil violence of the seventeenth century had caused the Lowland Scots to think about the purposes of political government. Important questions of religious and civil liberty had been posed. Latin was still the universal learned language and Scottish students and scholars moved without any hindrance but poverty to and from universities in the Netherlands and France, as they had since the early Middle Ages. 'The constant flow of information and liberality from abroad', according to Dugald Stewart, later Professor of Moral Philosophy at the College, helped account for the sudden 'burst of genius'.[45]

By the beginning of the eighteenth century, Edinburgh pamphleteers had passed beyond sectarian subjects. John Law of Lauriston and Andrew Fletcher of Saltoun made outstanding contributions to the chief secular debates in Europe and North America: what was money, and what were the political consequences of standing armies? They wrote as happily in French and Italian as in English. That both men were homicides, if not actual murderers, shows that Edinburgh philosophy still had something of the old Scots 'killing affray' about it.

John Law, the son of a goldsmith, condemned to death in London for killing his man in a duel, was forced to flee Scotland at the Union. A visionary financier and theorist of credit, he charmed the bankrupt French monarchy into establishing an ambitious Indies trading company financed through a stupendous Europe-wide stock bubble, which burst in 1719–20. A brilliant mind matched by a peculiarly beautiful personality, Law died at Venice in 1729. Andrew Fletcher, who killed a man in a brawl over a horse during Monmouth's rebellion of 1685, dreamed of establishing in Scotland a military society based on the aristocratic virtues of ancient Sparta. A high-minded patriot and republican, Fletcher would have felt at home in revolutionary America.

By the time of the Union of the Parliaments, Edinburgh had strong intellectual institutions of long foundation – the courts, the Kirk, the College – which were guaranteed under Acts of Security associated with the Treaty. It was through these well-established Protestant institutions that energies released by Scotland's national and religious crises were channelled towards intellectual and cultural innovation.

Pre-eminent was the university. The 'tounis colledge' had been founded after the Reformation as a Presbyterian institution to rival the ancient Catholic foundations of Aberdeen, Glasgow and St Andrews. It received its charter from James VI in 1583. It was sited in the extreme south of the town, hard by the Potterrow Port, at the site of the old Kirk o' Field where James VI's father Henry, Lord Darnley, had been blown up with gunpowder in 1567, and approached from the Cowgate through a steep and foul alley, College Wynd.

'The College', as it was known, was no ivory tower but an indispensable piece in the pattern of influence, patronage and corporate jealousy that constituted Edinburgh life. The place was modest in the extreme, with just three courtyards and no more than a couple of hundred students. From the College Yards, not so long before, Mrs Agnes Anderson, widow of the royal Master Printer, had claimed a monopoly of Bibles, catechisms and schoolbooks and vigorously defended it in the courts for forty years, though her Bibles were notorious for their blasphemous illiteracy.[46] Her successors were little better. As Principal Robertson wrote in an unsuccessful appeal for funds in 1768, 'A stranger, when conducted to view the University of Edinburgh, might, on seeing such courts and buildings, naturally enough imagine them to be almshouses for the reception of the poor.'[47] An American student called the place 'a most miserable musty pile scarce fit for stables'.[48] The students boarded in the town: Alexander Adam, who became rector of the High School, roomed in the late 1750s in Restalrig for four pence a week and dined on a penny roll in the Meadows (or, in bad weather, on an alley stair).[49]

In the early years of the eighteenth century the College had for its principal one of the most able ecclesiastical administrators of his time, William Castares, who attracted the patronage of William III and Queen Anne. In his campaign to keep Scots students at home, Principal Castares, who had spent some time in Dutch exile before the 1688 Revolution, in 1708 introduced from Leyden and Utrecht the system of specialised professors. This division of academic labour replaced the old Scots practice of regenting, in which a single teacher took a class through its entire curriculum, as in a modern primary school.[50] New chairs were created in arts and law. In 1726 a faculty of medicine was established, the first in Britain.

In printing, the Jacobite scholar-printers Thomas and Walter Ruddiman produced pioneering editions of the Scots Renaissance Latinists George Buchanan and Gavin Douglas before the Fifteen, and of Allan Ramsay senior's verses in the 1720s.[51] Under Thomas's keepership the Advocates' Library in the Parliament House, founded in 1680, had expanded to over twenty thousand volumes and but for its damp and dismal accommodation could challenge Oxford. An academy of painting had been opened in 1729, with Allan Ramsay junior, son of the poet, among its members, but closed two years later. A medical society founded in 1731 was relaunched eight years later as the Philosophical Society.

The High School, with some four hundred pupils, was among the largest grammar schools in Britain. Puritan enthusiasm for organising the poor had given Edinburgh no fewer than four 'Hospitals' or boarding schools for indigent or fatherless children: for boys, George Heriot's foundation of 1659, and George Watson's Hospital in Lauriston; and for girls, two long-established corporation schools, the Merchant Maiden Hospital and the Trades Maiden. A children's workhouse known as the Orphan Hospital, north of the Trinity Hospital and the Physick Garden, designed by William Adam and begun in 1734, was built with voluntary labour and contributions that included a striking clock, thirty pairs of shoes for the boys and twenty-five dozen baps.[52] The four-storey Charity Workhouse, built in 1743 in the space between Greyfriars Kirkyard and the town wall, accommodated up to six hundred paupers by means of donations, church-door collections (the residue of bad copper money rejected in the markets), a two per cent poor rate, and fees from the dancing Assembly established in 1723 on the pattern made fashion by Beau Nash in Bath. Edinburgh's pride and joy was George Drummond's Royal Infirmary, designed by William Adam on two acres looking north over the Cowgate and completed in 1741–2, in good time to take the wounded redcoats from Prestonpans.

Domestically there was a breath of change, with new styles of living and new professions to serve them. Of fashionable new housing, the eight storeys of James's Court, where both Hume and Boswell later lived, were completed in the 1720s; while Argyle Square (now engulfed by the University) was laid out 'after the fashion of London, every house being designed for only one family'.[53] In 1720, according to the *Edinburgh Courant*, a native teacher of French set up in the town.[54] By mid-century Edinburgh had its first coachworks, two upholsterers in the High Street, and a wallpaper maker.[55]

The town was even gaining some polish. In one of the etiquette manuals of the time, dedicated to the Lord Provost and Baillies, Adam Petrie told his readers how to drink from a shared glass: 'Be sure to wipe your Mouth before you drink, and when you drink hold in your Breath till you have done. I have seen some colour the Glass with their Breath, which is certainly very loathsome to the Company.'[56] The fashion for tea-parties after 1720 gave women a social occasion that need not degenerate into a debauch. They received in their bedrooms, as depicted in Scott's *Redgauntlet*:[57] so the Countess of Balcarres in the High Street, while her servant John leaned against the bed-post, handled the tea-kettle and contributed conversation.[58] Henry Mackenzie, who was born in 1745, claimed to remember from his childhood fifty species of tea-bread, and the women singing old songs unaccompanied where they sat.[59]

The lighter tone to conversation which the *Spectator* and *Tatler* propagated in the London of Queen Anne's reign spread to the north. As the memoirist John Ramsay of Ochtertyre wrote, 'These periodical papers had a prodigious run all over the three kingdoms, having done more to diffuse true taste than all the writers, sprightly or serious, that had gone before them.' Philosophy was 'brought down to the level of common sense, the cobwebs of metaphysics being carefully kept out of sight.'[60]

There were horse races each August on Leith sands. There was archery (for the Jacobites) at Musselburgh and golf (for the Whigs) at Leith Links where, among others, Forbes of Culloden spent his Saturdays –

> Yea, here, great Forbes, patron of the just,
> The dread of villains and the good man's trust,
> When spent with toil in serving humankind,
> His body recreates, and unbends his mind[61]

– and the surgeon John Rattray in 1745 won the silver club of the Honourable Company of Edinburgh Golfers for the second year running. For those who preferred to play in town, it was a par-six from the window of the poet Allan Ramsay's house in the Luckenbooths to the top of Arthur's Seat, the first stroke to the Cross and the second to the middle of the Canongate.[62] Despite the morose sermonising of ministers, there was the dancing Assembly, and a musical society with a weekly concert at St Mary's Chapel in the Cowgate. As Lady Panmure wrote to her exiled husband in Paris on 24 January 1723, 'so att last you may imagin old Reeky will grow polit with the rest of the world. I wish you were here to see it.'[63]

Politically, the city was divided in two. In the majority were the Whigs, who held that authority in both ecclesiastical and temporal affairs derived not from the Crown but from the public. Staunchly Protestant since John Knox imported John Calvin's creed and doctrines from Geneva to St Giles in the 1560s,

Edinburgh had been at enmity with the royal House of Stuart since the time of Mary, Queen of Scots. Knox's Presbyterians, who could not tolerate any intermediation between the worshipper and God, were revolted both by Roman Catholic Mary and by the English Protestants and their hierarchy of bishops with the sovereign at its head. As the historian and philosopher David Hume wrote of Edinburgh at the time of James VI, 'the same lofty pretensions, which attended them in their familiar addresses to their Maker, of whom they believed themselves the peculiar favourites, induced them to use the utmost freedoms with their earthly sovereign.'[64] Infallible Kirk confronted divine-right monarchy in a war to the death.

Between 1638 and 1649, in what amounted to a second Reformation, the Edinburgh Presbyterians rejected an attempt by James's son Charles I to impose from London a prayer book and government by bishops, defeated the Royalist armies under the Marquis of Montrose, and ensured that the Lowlands would remain Presbyterian. Montrose was confined to the Tolbooth and led out to execution. Charles himself was captured by a Scots army and then sold off to his English enemies, who struck off both crown and head in 1649. The restoration of his son Charles II in 1660 was also a restoration of the bishops' rule, known as Episcopalianism or, to the Presbyterians, Prelacy. The firm Presbyterians, or Whigs, as they were nicknamed, took to the moors and suffered two decades of persecution which ended only with the expulsion of James II of England and VII of Scotland and the import of a Protestant king, William of Orange, in the so-called Revolution of 1688.

Thus Edinburgh's sense of itself was founded in hostility to divine-right monarchy and the Stuarts. Indeed, the very topography of the town was insurrectionary. Here in St Giles was the pulpit from which Knox fulminated against Roman Catholic Mary, here the Kirk o' Field where her husband was murdered, here the High Street along which she was dragged in chains. Here, too, was the spot in St Giles where Jenny Geddes, on 23 July 1637, threw her stool at the bishop who read from the infidel service book,[65] here the Greyfriars churchyard where the National Covenant was signed eight months later, here the Tolbooth prison where the royal general Montrose spent his last night on earth, here the Grassmarket where the persecuted Covenanters met their death with psalms. Presbyterianism, which to outsiders seemed obsessed with minor matters of church government, actually stood for the rights of commoners, and most notably of burgesses, against monarchy.

These religious Whigs extended, by way of a new generation of reforming clergy, to a modern class of what Hume called 'political Whigs', as embodied in Lord President Forbes: pragmatic individuals whose 'chief regard to particular Princes and Families, is founded in a Regard to the publick Good'.[66]

Ranged against them were the men and women known as Jacobites – from the Latin for James – who held that the Stuarts had a sacramental and eternal right to rule Scotland. Religious schism and political union with England exacerbated their grievances and swelled their ranks. At the 'Revolution' that ended Stuart government in 1688, the Episcopalian church was disestablished and many Episcopalians became Jacobite. The Union of 1707 took Edinburgh unawares, and crippled her. Daniel Defoe, who worked tirelessly as a government agent in Edinburgh to promote the Union of the Parliaments, watched in disgust as a common loathing of Union united the squabbling sects: "Twas the most monstrous Sight in the World, to see the Jacobite and the Presbyterian, the persecuting Prelatic Non-juror and the Cameronian, the Papist and the Reformed Protestant, Parle together, Joyn Interest, and Concert Measures together.'[67] When the secret articles of Union were read out in October 1706, there was a riot. According to Hugo Arnot, cadaverous and asthmatic historian of the town at mid-century, three regiments of foot were deployed in Edinburgh, with a battalion at Holyrood Abbey and horse-guards to attend the government's commissioner, the Duke of Queensberry.[68] George Lockhart, a contemporary witness, said His Grace hurried from the Parliament House to his coach at the Cross 'through two Lanes of Musqueteers' and raced to his lodgings in Holyrood at top gallop, under a hail of curses and stones.[69]

Instead of the 145 nobles and 160 commoners who had gone to Edinburgh for the Parliament and spent their rents among the city's tradesmen, sixteen peers and forty-five members now went to London, where they made little impression.[70] The old trade with France was destroyed. Poor harvests aggravated the depression. It was said grass grew round the market crosses.[71] The Canongate, the suburbs of the nobility, was a forlorn place.

All this was grist to the Jacobite court in exile. According to Lockhart, people became daily more persuaded that 'nothing but the Restoration of the Royal Family, and that by the means of *Scotsmen*, could restore them to their Rights.'[72] Even Andrew Fletcher of Saltoun, that notorious foe of monarchy, was so bitter at England and the Union that 'in revenge to them, he would have sided with the *Royal Family*'.[73]

The city itself, which in the years before 1707 had become accustomed to low taxes and dilatory collectors, had to pay new impositions on beer, salt, linen, and soap. Those imposts were supposed to defray part of the cost of the Scottish administration, but were viewed in the town as a levy for the House of Hanover's Continental wars and national debt. The decision by Walpole's government in 1725 to impose a malt tax, which must perforce drive up the price of ale, provoked reaction of a ferocity not seen again in Scotland till the poll tax riots of the 1980s. The conflict, defused by Forbes of Culloden and by Walpole's agent in Scotland, Argyll's brother the Earl of Ilay, erupted again in 1736 when a crowd of several thousand lynched John Porteous, Captain of the Town Guard, on a dyer's post in the Grassmarket where, five months earlier, his men had fired into the crowd that had gathered to protest against the execution of a popular smuggler. Sentenced to death at the Court of Justiciary in Edinburgh, Porteous was granted a stay of execution by the government in London. Suspecting they were to be cheated of revenge, on the night of 7 September an angry mob took Captain Porteous from his cell at the Tolbooth. The discipline and single-mindedness of

the mob – they left money on the counter of the shop in the Bow
from which they took the lynching-rope – demoralised London
and was no doubt observed with interest at the Jacobite court at
St-Germain-en-Laye near Paris.[74] Porteous's Affair, as the
riot came to be known, exposed for an instant not only the dis-
content among the unfranchised smithies and shopkeepers of the
West Bow but the government's complete impotence. George
Drummond, a former Lord Provost now out of favour with the
Earl of Ilay and anyway passing through a profound religious
crisis, saw the riot as a direct consequence of a city government
stuffed with placemen by Ilay's lieutenant, the Lord Justice-Clerk,
Lord Milton. (This man, though nephew of the truculent Scots
patriot Andrew Fletcher of Saltoun, was a loyal government servant.) 'The administration in this countrey', George Drummond
wrote on 16 September 1736 in a manuscript diary now in the
University Library in Edinburgh, 'is supportd by fear, not love,
and the tools who are employed, are hated and contemnd, by
almost all people high and low.'[75] Drummond recognised that the
Council had to be 'new constituted'; but at a public meeting on 7
April 1737, where as Old Provost he was called first, he was too
timid to speak up.[76]

The government, determined to punish Edinburgh, summoned
Lord Provost Wilson to London and clapped him in gaol, and
ordered the demolition of the Netherbow Port and the disbanding
of the Town Guard. Forbes, Argyll and the intricacies of committee drew the sting. In the end, the ministry contented itself with
barring Wilson from office, imposing a fine of £2,000 sterling to
support Porteous's widow, and bullying ministers of the gospel to
establish public order. The first Sunday of each month that
autumn, between the lecture and the sermon, ministers were to
read the Porteous Act,[77] demanding that Porteous's killers give
themselves up and threatening those who sheltered them with
severe penalties. This tyrannical and futile order struck at the root
of the Act of Security protecting the Kirk's autonomy and had the
perverse effect, as Sir Walter Scott put it in *Heart of Midlothian*,
of bringing the most scrupulous Presbyterians to the side of

Porteous's killers and the town: 'since to the General Assembly alone, as representing the invisible head of the kirk, belonged the sole and exclusive right of regulating whatever belonged to public worship'.[78] According to Alexander Carlyle, son of the minister at Prestonpans and himself bred to the ministry, at least half the clergy disobeyed the order and none was deposed;[79] not one of Porteous's assassins is known by name to posterity. There were further disturbances in the winter of 1740–1 over food prices and, the next year, over the corpse-snatchers known as the Resurrection Men.

Unlike Glasgow, Edinburgh had no merchant class that could profit from the colonial trade made legal by the Union, no surplus capital to invest in the securities issued to fund the growing National Debt, known as 'the funds'. Advocates, attorneys – known in Edinburgh as 'writers' – the *petite noblesse*, all saw a greater role for themselves if the Union could be broken. James Hepburn of Keith actually rejected divine right and condemned the later Stuart government; but was reported as saying 'that the Union had made a Scotch gentleman of small fortune nobody, and that he would die a thousand times rather than submit to it'. A contemporary wrote: 'Wrapt up in these notions, he kept himself in constant readiness to take arms, and was the first person who joined [Prince] Charles at Edinburgh.'[80] Like one of Walter Scott's crazed Jacobite lairds, Hepburn came out in both rebellions and, as Alexander Carlyle put it, 'had there been a 3rd ... would have joined it also.'[81]

Carlyle wrote that a third of the men of Edinburgh were enemies of the government, and two-thirds of the women. In the country as a whole, 'the commons in general ... had no aversion to the family of Stuart; and could their religion have been secured, would have been very glad to see them on the throne again.'[82] Yet the Jacobite/Whig divide in Edinburgh does not appear to have been a source of true rancour. From this distance, it appears the two parties had more in common with each other than with strangers. Social contact between them continued throughout the Forty-five, and after the defeat of the government forces at

Prestonpans, Alexander Carlyle was to be found happily lodging in town with the Jacobite Seton family. As the marquis d'Éguilles, French envoy to the rebel army, noted in a despatch to the Court at Versailles, even Forbes of Culloden always spoke of the young Prince with respect.[83]

David Hume taught that the casual in history must not be mistaken for the causal. Yet from a twenty-first-century vantage, it seems that Edinburgh could prosper only with the *political* defeat of the Jacobites and the *religious* defeat of the Whigs. Only when Edinburgh had abandoned both its theocratic fantasies and its yearnings for a romantic independence could it at last enter the eighteenth century. This story therefore opens with the Forty-five and continues with the ecclesiastical crises of the 1750s.

2

Charlie's Year

Princes Street, which runs from east to west above the railway lines into Edinburgh, is as fine a street as any. Its distinction has little to do with its shops and hotels, which are standard examples of such things, and nothing at all to do with its climate. 'To none but those who have themselves suffered the thing in the body', the novelist Robert Louis Stevenson wrote in agony, 'can the gloom and depression of our Edinburgh winter be brought home ... The passengers flee along Prince's Street before the galloping squalls.'[1]

The beauty of Princes Street is its setting. By an Act of 1816 the British Parliament in London outlawed any new building on the south side of Princes Street, so that laden shoppers leaving Jenners department store or Marks & Spencer come up against an impregnable Castle and the high tenements of a grim, old-fashioned Scots town. Tourists arriving by rail through Princes Street Gardens must feel that the capital of Scotland is reverting to fields and woods.[2]

It is necessary to obliterate Princes Street,[3] the railway station and the Waverley Market shopping mall; uproot the floral clock and silence its mechanical cuckoo, overturn the monument to Sir Walter Scott, the regimental memorials and the picture galleries

on The Mound, and the Balmoral Hotel; refill the old North Loch
and strew its ancient garbage about; and, in place of what is now
Princes Street, lay down a narrow lane between two dry-stone
walls, called Lang Dykes,[4] down which on Monday, 16 September
1745, at between three and four of the afternoon, the citizens of
Edinburgh watching from the Castle esplanade saw two British
regiments fleeing for their lives towards Leith. Hurrying back to
their business (or, rather, to shut up their shops and hide their coin
and banknotes), those people heard the two regiments' baggage
and women rumbling up the High Street towards the security of
the castle. That is the opening of the story: when the people of
Edinburgh saw that, for their purposes, the government of Great
Britain no longer existed and they must confront not merely a
Highland army but the conundrum of Scots history.

Ever since hostility between France and Britain's ruling House of
Hanover had broken out in pitched battle at the village of
Dettingen near Frankfurt in 1743, there had been reports from
Paris that Prince Charles Edward Stuart, grandson of the deposed
James VII of Scotland and II of England and living under the pro-
tection of the French court, intended to attempt a landing to place
his father James on the throne of England, Scotland and Ireland.
A French invasion force was scattered by a storm in the spring of
1744, but that was unlikely to deter the young man and the exiled
Jacobites.

   On 8 August 1745 Lord President Forbes received a letter from
MacLeod of MacLeod in the West, reporting that a French vessel
carrying the 'Pretended Prince of Wales' had put in at South Uist
and Barra and was now 'hovering on parts of the Coast of the
main Land'.[5] Passing on the news to the Secretary of State for
Scotland in London, the Marquis of Tweeddale, Forbes was careful neither to reveal the source of his intelligence nor to cause
undue alarm. He wrote: 'I have resolved to make my journey to
the North Country earlier this season than usual ... I propose to
set out tomorrow Morning.'[6]

Lord President Forbes, who owned a fine stone house at Culloden near the Highland line at Inverness, had no illusions about the loyalty of the Highlands to the House of Stuart, driven into exile following the events of 1688. Highland insurrections of varying severity had been defeated in 1689, 1708, 1715 and 1719. Forbes himself had proposed in 1738 that the government should recruit its own Highland regiments and deploy their courage and hardihood against the enemies of the House of Hanover on the Continent and overseas. His suggestion was slow to be taken up, though the 42nd, The Royal Highland Regiment or Black Watch, formed in 1739, had passed through Edinburgh on its way abroad in 1743.

Early on the morning of 9 August, dressed for his journey, Forbes called on Sir John Cope, commander-in-chief of the government forces in Scotland. Sir John had commanded the second line at Dettingen on 27 June 1743, in the sight of George II, and had been decorated on the battlefield. He was resolved to march his guns and two thousand men into the Highlands to confront the rebels before they could organise or arm themselves. 'A little, dressy, finical man,' in the opinion of Sir John Clerk of Penicuik, 2nd Bt (formerly one of the Commissioners of Union and now in his sixties), Cope 'had already devoured the Rebels in his imagination'.[7] Forbes himself planned to go due north to Culloden and try to hold the Highland clans, and particularly the slippery Lord Lovat and his Frasers, to their loyalty to the House of Hanover in London.[8] He left in a deep depression. Crossing the Forth, he spoke his mind to a fellow passenger, a young nobleman returning to his family castle in Fife. Forbes, wrote David, Lord Elcho, 'was greatly distressed because the Prince would only light a straw fire which would soon be put out by General Cope and it would all end with the ruin of many very fine gentlemen whose fate he mourned.'[9] No doubt Forbes knew Elcho to be a Jacobite agent.

On 19 August, the very day that Charles raised the standard of his father as James VIII at Glenfinnan at the head of Loch Shiel, Cope set off. He left two less than warlike regiments of light dragoons in the Lowlands. One, the 13th Horse, under the command of Colonel James Gardiner, was to defend the crossings of the river Forth at Stirling. (Gardiner was visibly ailing, and had been summoned in the emergency from a cure at Scarborough in Yorkshire.) The 14th or Hamilton's Horse was to camp on Leith Links to protect the capital. For the next two weeks there was no news from the north, the weekly post from Inverness having to pass through the Jacobite lowlands of the north-east. But on the 31st a messenger arrived from Perth with a thunderclap. Cope had declined battle against the poorly armed Highlanders on the 26th and was now heading *northwards* to join Lord President Forbes at Inverness.

News of this manoeuvre, which was compared acidly to a figure in a country dance, plunged the capital into turmoil. John Home, a probationer for the ministry who became famous as a playwright, remembered: 'Till that change of position took place ... the insurrection of the Highlanders was looked upon as a sort of riot, which would easily be quelled by the King's troops, who were thought be be the only men in the Kingdom that knew how to fight ... The affair began to be deemed somewhat serious.'[10]

The Prince's army made a forced march down the new military roads built in the 1720s and 1730s to pacify the Highlands and descended the Braes of Atholl to the Episcopalian lowlands. They took first Perth, then Dundee, without firing a shot. Outflanked by the Prince and his military commander, Lord George Murray, Cope resolved to try to ship his two thousand redcoats back down from Aberdeen. Messengers reached Edinburgh with his demand that transports be sent up from Leith. Those vessels sailed on 10 September. On the 14th came the news that the rebels had crossed the Forth the day before, unopposed and beyond the range of the guns of Stirling Castle. It was said the Prince had been first into the water.[11] That night John Campbell, a Highlander who was chief cashier of the Royal Bank of Scotland, packed up the bank's ledgers and its cash, securities and plate, amounting in value to some £100,000 sterling, and transported them to the Castle.[12] The older Bank of Scotland, long thought to be Jacobite in its sympathies, had moved three iron chests to the Castle on the 13th.[13]

Anxious to avoid any taint of treason, the bank had been drawing in its notes to deny credit to the Prince.

Where on earth or in hell was Cope? Up in Inverness, Forbes wrote confidently to Lovat that Cope had sailed from Aberdeen on Sunday night, and Monday's fair wind should have 'brought him safe into Leith Road'.[14] But in Edinburgh, as John Home later remembered, people glanced every moment at the church weather-vanes, for fear that the winds would turn westerly and delay the flotilla's entry into the Firth of Forth.[15]

> Hey, Johnnie Cope, are ye waukin yet?
> Or are your drums a-beating yet?[16]

The town of Edinburgh, as the English journalist and government agent Daniel Defoe once wrote, had been built to withstand sur-prise attack.[17] From the extinct volcano on which the Castle sits, the town runs eastwards down a broad ridge, with steep ravines on each side. After a catastrophic Scots defeat at English hands in 1513, the so-called Flodden wall was thrown up to protect the two principal streets, the High Street and the Cowgate. The wall was enlarged a century later to take in what is still the grandest build-ing in Edinburgh, the orphanage and school known as Heriot's Hospital. Outside it were the old royal Palace of Holyrood and two areas of suburbs: Canongate, where the nobility maintained their town houses, and the more commercial Portsburgh, below the Castle to the west by what is now the financial district.

The wall, between ten and twenty feet high, lacked embrasures for cannon and was more akin to a park or garden wall than a city defence, John Home thought.[18] It was pierced by six gates known by the old-fashioned word 'Port'. While the ministry in London had been told the wall 'cannot be forced',[19] Sir John Clerk, engaged in Edinburgh affairs since the Union, knew that it was 'good for noth-ing'.[20] The Prince's Secretary, John Murray of Broughton, had spent the winter in Edinburgh preparing the town Jacobites for the insur-rection and had had leisure to reconnoitre. Murray, a down-at-heel

Tweeddale laird with no great scruple and a sensationally pretty wife, thought that to defend the town was a 'Don Quixote fancy'.[21]

On the north side there was no fortification, though the filthy North Loch, which was fordable, lapped at the foot of the Castle. On the south side the wall had bastions to provide flanking fire but was, according to Murray of Broughton, 'thine [thin] and in very bad repair'.[22] The Potterrow Port was overlooked by a tall house to the south. On the west of the town, from the Cowgate Port to the Netherbow Port, a row of houses in the lane known as St Mary's Wynd comprised the whole fortification. North of the Netherbow, at the northern end of Leith Wynd, the Trinity Hospital (now beneath Waverley Station) made another section of wall, but Trinity College Church provided cover to attackers.

As Murray wrote later, the Prince's commanders intended to run a sap from the house overlooking the Potterrow to blow a breach in the wall and stage other strong diversions at the North Loch sluice and the Trinity Hospital while launching the principal attack by firing the houses on the west side of St Mary's Wynd.[23]

In the gathering storm, the Scots loyalists felt abandoned. Walpole's Scottish system – 'mercenary, peaceable, corrupting', in the words of *The Scots Magazine* – had disintegrated with his fall from power in 1742. 'The country was entirely left to itself,' Sir John Clerk wrote bitterly, 'for no doubt some of the ministry wanted that we in Scotland should worry one another.'[24] Responsibility for the defence fell on the shoulders of the Lord Provost, Archibald Stewart, who also carried the grand dignities of Lord-Lieutenant, High Sheriff, Captain of the City Guard and Admiral of the Firth of Forth. A wine merchant with vaults in Leith and a warren of a house in the West Bow and member of the British Parliament for the city, Stewart also happened to be the leader of the Jacobite party on the Town Council. Worse, the Council was embarked on the lengthy and medieval process of electing itself, which required the burgesses and trade incorporations to select candidates and present them to the retiring Council.[25] These elections both clouded and illuminated the extraordinary events that followed.

On 27 August Provost Stewart had convened a public meeting in the New Church, one of the four parts into which the ancient cathedral of St Giles had been divided at the Reformation. At the meeting in the choir it was proposed to repair breaches in the town walls, and raise a regiment of a thousand foot at town expense. Edinburgh being Edinburgh, there were lawyers present, and the lawyers raised an objection: it was illegal to raise troops except by royal authority, and the Lord Advocate must first send south for King George's permission. Precious time was lost before the royal licence arrived on 9 September.

The town's forces consisted of a militia known as the Train Bands, not mustered since the Revolution in 1688 but possessed – for the purpose of celebrating the King's Birthday – of twelve hundred muskets with ammunition (which, according to a pamphlet attributed to David Hume, men liked to discharge on parade to impress their wives or mistresses[26]); the City Guards, a regiment of some 126 superannuated Highland men armed with Lochaber axes, housed in a shabby hut right in the middle of the High Street and commanded not by soldiers but by 'decayed' (failed) merchants; Hamilton's regiment, which had now been joined by Gardiner's, contemptuously described as 'Irishmen'; a few volunteer companies of College boys and apprentices; and a small garrison in the Castle under General Guest. He, as Sir John Clerk of Penicuik wrote, had 'in his time had been an Active, diligent Souldier but, being a man of above 86 years of age, he cou'd scarcely stir out of his room.' Guest was assisted by his predecessor, General Preston, who was the same age and in a Bath-chair.[27]

Whether the city wanted to be defended by this makeshift force was another matter. On 2 September, according to a diary kept by the Professor of Mathematics at the College, Colin MacLaurin, 'above twenty Gentlemen of known good affection to his Majesty and the Government mett at Mrs Clarks', a tavern in Fleshmarket Close, and agreed 'to apply to the Lord Provost that he would give Orders for putting the Town in as good a state of Defence as possible with all expedition.'[28] MacLaurin – appointed to the Edinburgh chair on the advice of Sir Isaac Newton himself, *ipso*

Newtono suadente[29] – would have remembered that Archimedes, the most famous of all antique mathematicians and natural philosophers, had fortified Syracuse against the Romans, converting a siege into a blockade.

The next day, MacLaurin persuaded Lord Provost Stewart to order the sluice of the North Loch to be shut 'that it might fill up', and some bullets to be cast, the Edinburgh gunsmiths having been mysteriously cleaned out of bullet-moulds by the Jacobite women.[30] But MacLaurin found the Lord Provost at best lukewarm and at worst defeatist: 'He said that if 1000 men had in mind to get into this Town he could not see how they could hinder them.'[31] To the argument that something must be done 'to save the reputation of the Town to Devert the Enemy from coming this way and to raise a Spirit in the Country', His Lordship replied that 'to pretend to do when we could do little was to expose us to Redicule'.[32] When MacLaurin showed the Town Council a plan of the most urgent works on 7 September, Provost Stewart queried the expense. Still, MacLaurin began work on 8 September, clearing the parapets, which had been blocked with rubble to deter smuggling, erecting scaffolds where they were too narrow for men to stand, and placing ships' cannon on the gates and on flanks to rake the exposed portions of the walls.

He demanded that the tall house commanding the wall by the Potterrow should be possessed, but nothing was done.[33] Meanwhile, the masons were occupied with their craft elections for the Council, and MacLaurin could find no labourers. Sometimes, he complained to his diary, he had less than two dozen men working.[34]

The defence party was led by the former Lord Provost, George Drummond, now in the throes of relaunching his magisterial career. A Highlander and a Freemason, he had been the first to bring the government news of the Jacobite landing under the Earl of Mar in 1715, had raised a company of volunteers, and fought for the government at Sheriffmuir. As Commissioner of Customs he had been responsible for collecting the tax on malt so bitterly resented by the Scots. Unlike Archibald Stewart, there was not a flicker of doubt about Drummond's loyalty to the House of Hanover. His world was a triangle of money, Edinburgh and his own soul. As courageous in battle as he was timid in Council, Drummond was an implacable opponent of both the Lord Provost and the royal house of Stuart. The great consequence of the Forty-five for Edinburgh, clear now but obscure at the time, was to discredit the old administration, and return to power in a popular election this same George Drummond, six times Lord Provost, pioneer of the New Town, and 'founder of modern Edinburgh'.[35]

> Hey, Johnnie Cope, are ye waukin yet?
> Or are your drums a-beating yet?
> If ye were wauking I would wait
>   To gang to the coals i' the morning.[36]

Alexander Carlyle spent that Saturday, 14 September 1745, in the College Yards, drilling with his fellow students John Home, William Wilkie and William Robertson in the corps of 418 volunteers. Stands of arms consisting of a firelock, bayonet and cartridge-box had been sent down from the Castle two days before and grizzled sergeants instructed the boys in their use. Hand-grenades could not be spared, and the twenty-three that had been stored in a chest in the Town Armoury since the rebellion of Fifteen were (wisely) left undisturbed.[37]

In the intervals of parading Carlyle called on his old mathematics professor MacLaurin as he was making good the ruined walls to the south of the town. He found the great Newtonian erecting cannon near the Potterrow Port. Though the trades election had been held that Friday, MacLaurin was still grumbling about the shortage of labour. Still, the cannon had been proved and shot made ready.

The next morning, the Sabbath, news arrived that the Chevalier (as the Prince was known) was at Linlithgow. The six volunteeer companies were hurriedly paraded at 10 a.m. in the College Yards to hear their company commanders tell them that they were to march out against the rebels and, as Carlyle noted years later with traces of anxiety, 'to expose our lives in defence of the capital of Scotland, and the security of our country's laws and liberties.'[38] By liberties, Carlyle had chiefly in mind the Established Kirk, which might be threatened by a Catholic James VIII or his Episcopalian friends.

Lord Provost Stewart had named no overall commanding officer, and the companies were under independent command. The College boys had elected to serve under Drummond himself, whose fifty-eight years had not dimmed his military ardour. Having fought at Sheriffmuir, Drummond no doubt had a rational respect for rather than irrational fear of the Highlanders. He now told the boys that the two companies of dragoons were to make a stand at Corstorphine but needed infantry support, which meant some two hundred and thirty volunteers, along with fifty of the Town Guard. If any volunteer preferred to man the walls rather than march out, well, he would not be blamed. Drummond's speech was drowned in applause. For the first time, the young men loaded their pieces.

Suddenly, the town's fire-bell rang out. That was the chosen signal for the volunteers to muster in the Lawnmarket, but it had an unexpected effect: men and women crowded from Kirk services in panic.[39] The College Company was halted in the Lawnmarket for an hour so the other companies could catch up. As Hamilton's dragoons marched past on their way to join Gardiner's at Corstorphine, the boys cheered them with huzzas. But not all was well.

> In one house on the south side of the street there was a row of windows, full of ladies, who appeared to enjoy our march to danger with much levity and mirth. Some of our warm Volunteers observed them, and threatened to fire into the windows if they were not instantly let down, which was immediately complied with.[40]

At last, the other companies marched up the hill. Their officers, coming up to Carlyle and his friends in the street, said quietly that most of the men had been unwilling to march. Negotiating the tight curves of the West Bow, remembering perhaps that John

Graham, 'Bonnie Dundee', had come down that way to launch the Jacobite enterprise in Scotland in 1689, or that the Grassmarket was the usual place of execution,[41] a glance up at the Old Assembly Rooms perhaps inspiring maudlin thoughts of girls in mourning, the young men began to lose their poise:

> All the spectators were in tears, and uttering loud lamentations; insomuch that Mr Kinloch, a probationer [candidate for the Ministry], the son of Mr Kinloch, one of the High Church ministers, who was in the second rank just behind Hew Ballantine, said to him in a melancholy tone, 'Mr Hew, Mr Hew, does not this remind you of a passage in Livy, when the Gens Fabii marched out of Rome to prevent the Gauls entering the city, and the whole [all the] matrons and virgins of Rome were wringing their hands, and loudly lamenting the certain danger to which that generous tribe was going to be exposed?' 'Hold your tongue,' says Ballantine, 'otherwise I shall complain to the officer, for you'll discourage the men.' 'You must recollect the end, Mr Hew, *omnes and unum perieri* [they perished to a man].'[42] This occasioned a hearty laugh among those who heard it, which being over, Ballantine half whispered Kinloch, 'Robin, if you are afraid, you had better steal off when you can find an opportunity: I shall not tell that you are gone till we are too far off to recover you.'[43]

Arriving at the West Port, Drummond turned about and 'he and his company found themselves alone'.[44] They were forty-two in number. An officer sent back up to the Lawnmarket found a scene of indescribable confusion. Meanwhile, it being past noon by the clock, the College boys were brought bread and cheese, ale and brandy by the Grassmarket brewers. While they were waiting for reinforcements, a party of clergy – divine service having been abandoned at the sounding of the fire-bell – fell upon them. The body included the high-flying and harder-drinking minister of the Tolbooth Kirk, Dr Alexander Webster, the reformist Dr Robert Wallace, and Dr William Wishart, the College Principal, 'who called upon us in a most pathetic speech to desist from this rash enterprise, which he said was exposing the flower of the youth of Edinburgh, and the hope of the next generation, to the danger of being cut off, or made prisoners and maltreated, without any just or adequate object.'[45] The ministers acted like 'prists',[46] wrote the author of a Whig manuscript that was found at Woodhouselee in Midlothian and published at the beginning of the twentieth century. A few young men remonstrated, but Carlyle himself now saw the 'impropriety of sending us out'.[47] Drummond, who a moment before had been for 'leaving his body on the city walls',[48] sent to Lord Provost Stewart, who replied that 'he was very much against the proposal of marching the volunteers out of the town'.[49] Though now more volunteers and part of the regiment raised at the Town's expense had at last come down from the Lawnmarket, Drummond abandoned the expedition and led his company back along the Cowgate to the College Yards.[50]

Some of the volunteers believed Drummond had merely made 'a parade of courage and zeal' with an eye to the Town Council elections; and indeed, as the only man untainted by the fiasco he was re-elected as Lord Provost the next year. In his history of the Rebellion, John Home left no doubt that in his view Drummond was standing for office. Yet Drummond had nothing to prove in point of military (as opposed to civil) courage. More probably, as Carlyle wrote, he 'did not think he could well be answerable for exposing so many young men of condition to certain danger and uncertain victory'.[51]

Meanwhile, on the walls, MacLaurin had the armaments ready. With the masons at last turning up for work, he carried cannon out to the flanks and placed three pieces to rake St Mary's Wynd. At 6 p.m. MacLaurin and the chief bombardier sent to the Lord Provost for permission to load up with small shot, but were kept waiting till eight, and even then Stewart 'desired another to sign the order for him'.[52] At the Bristo Port, the cannon could not be loaded until 1 a.m. for want of a sentry to guard the loaded piece.

Carlyle and the other students had repaired to a tavern – Mrs Turnbull's, next door to the Tron Church in the High Street – where after some quarrelling and heroics, they resolved to make their way with any other volunteers they could find to join Sir John

Cope's army when it came ashore in the Firth. Then they reassembled to keep watch at the Trinity Hospital in Leith Wynd, by the Netherbow Port, one of the weakest points in the town's defences.[53] At one in the morning, the Lord Provost made an inspection: 'Did you not see', Home whispered to Carlyle, 'how pale the traitor looked, when he found us so vigilant?' 'No,' Carlyle replied, 'I thought he looked and behaved perfectly well, and it was the light from the lantern that made him appear pale.'[54]

Out to the west, the two regiments of dragoons, numbering some six hundred men, were encamped that Monday, 16 September, at a crossing of the Water of Leith known as Coltbridge. The author of the Woodhouselee manuscript[55] was friendly with Colonel Gardiner. 'I was with Collonel Gardener about 3 afternoon,' he wrote, 'when one of the Scowts came in and said that 400 of the Highland advance gaird was on the north east poynt of Corstorphin hill. I took leave of Gardener and returned cross the fields and saw the dragowns mownt. They made 3 lowd huzzas and rod off to the northward and thane twrned east, and it is said they did not draw brydle till they came to Muselburgh.'[56] Other witnesses stated that Brigadier Fowkes, newly arrived from London to take command, had cast a professional eye over the forces at his disposal and opted for prudence; in John Home's account, the regiments fled at the arrival of a rebel skirmishing party.[57]

This retreat, visible from the town and soon christened the Canter of Coltbrigg, utterly demoralised the Whigs. 'The clamour arose that it would be madness to think of defending the town as the dragoons had fled,' Carlyle wrote.[58] The strong Whigs feared that Lord Provost Stewart now had his excuse to give up the city. Walter Grossett of Logie was an exciseman employed by Lord Justice-Clerk Milton as a secret agent. In a narrative accompanying an indent, now in the Public Records Office, for £3,709 sterling in expenses incurred in government service during the rebellion, he wrote 'that the Dragoons having soon after this upon the motion of the Rebells towards them quit their Post at Coltbridge and retired in some haste by the North side of the

Town about 3 that afternoon, without sending the Party of Dragoons into the Town as had been conserted in the morning of that Day, and Lord Justice Clarke observing that this might give a Handle for justifying the Provost to give up the town to the Rebells, he sent Mr Grosett to the Provost, to press the Defence of the Town, and to assure him, that as many of the Dragoons as he pleased to Desire should forthwith be sent in, to assist in the Defence thereof, till Sir John Cope, who was then hourly expected by sea from Aberdeen, should come with the troops to their Relief.'[59]

The fire-bell was rung again, this time for a meeting of citizens and magistrates at the Goldsmiths' Hall in Parliament Close, but 'when the crowd increased'[60] it was adjourned a few yards to the nave of the New Church, one of the four divisions of St Giles. While the volunteers mustered again in the Lawnmarket and sent in for orders, a man on horseback cantered up from the Bow crying that the Highlanders were at hand, sixteen thousand strong. 'Ye Fire bell rang in a most dismal manner till five at night,' Magdalen Pringle, the young daughter of a Roxburghshire laird, wrote to her cousin two days later, 'and everybody was in Terror for their friends the Volunteers imagining that ye Town would resist.'[61]

Under the high pillars of the New Church aisle, the Lord Provost was at his wits' end. He sent out for the officers of state and law – the Justice-Clerk, Advocate, Solicitor – only to be informed that those gentlemen had left town. There followed a scene such as only Edinburgh could stage. A letter was handed in that Deacon Orrock opened and began to read:

> Whereas we are now ready to enter the beloved metropolis of our ancient kingdom of Scotland...[62]

There was pandemonium. Lord Provost Stewart demanded to know the signature. It was 'Charles P[rince] R[egent]'. Refusing to have it read, he repaired to the Goldsmiths' Hall. There he sought a ruling by the Town Assessor, Mr Haldane, on the propriety of his reading the letter; but that functionary replied it was 'a matter too high for him to give his opinion on'.[63] 'Good God!' cried the Lord Provost. 'I am deserted by my arms and my assessors.' The letter was read, including the sentence 'If any opposition be made to us, we cannot answer for the consequences.' It ended all thought of resistance.[64] The magistrates resolved that a deputation under the bookseller Baillie Gavin Hamilton should go out to the Chevalier, encamped three miles away at Gray's Mill on the Water of Leith, to ask for time to consider the demand.

By then, the rumour had reached the West Port walls that the meeting had decided to capitulate. MacLaurin sent a volunteer to ask what was to be done with the loaded cannon but 'His Lordship had no time to speak to him'.[65] Drummond had already marched the volunteers up to the Castle, and as the autumn sun dipped to the west, they stacked their arms. He himself set off to join Cope at Dunbar, while Carlyle and the other boys dispersed, 'not a little asham'd and afflict'd at our inglorious campaign'.[66] With his brother he set out by moonlight along the sands to their father's manse at Prestonpans, passing the dragoons, demoralised and in abject panic.[67]

Barely had Baillie Hamilton and three deputies set out at 8 p.m. when news arrived at the Goldsmiths' Hall that Cope was off Dunbar, the westerly winds keeping him from Leith. It was too late to recall the deputies, who found at Gray's Mill not sixteen thousand men, but probably not much more than two thousand. These were the Lochaber clans that had first come to the standard at Glenfinnan: Clan Donald, the Camerons under their chief Donald 'Gentle' Lochiel, the Appin Stewarts, some of the MacGregors. The Prince was desperately short of money, and had just drawn on David, Lord Elcho, son of the Jacobite Earl of Wemyss, for 1,500 guineas in coin.[68] Even so, he sent the town deputation packing: Edinburgh had, he said, until 2 a.m. to open its gates and accept him as his father's Regent. The Lord Provost continued to buy time, sending out a second delegation under John Coutts to ask for a delay until 9 a.m., and a public meeting. But out on the walls, MacLaurin listened in vain for the 'All-is-Well' to go round and concluded that 'the Toun seem'd quite [quit] of its defence'.[69] MacLaurin lay low in Edinburgh for a while before slipping away to Newcastle, and then York.

Though Edinburgh's walls had been abandoned, the gates were shut. Charles Edward, recognising that the city was trying to stall him, sent a force of Camerons, Donalds and Stewarts of Appin, under Lochiel, to try to force entry. He ordered them to avoid excessive force, not to touch strong drink, and to pay for whatever they took. He promised two shillings a man in the event of success. They were guided by John Murray of Broughton, who had organised and prepared the Edinburgh Jacobites the previous winter, had met the Prince's little landing-party at Kinloch-Moidart, and been appointed his secretary on 25 August.

The small force marched in silence past the southern outskirts of the town and took position along St Mary's Wynd and opposite the Netherbow Port, the gate that divides the High Street from the Canongate. Just before daylight on the Tuesday, 17 September, the gate was opened to allow the coach that had brought Old Provost Coutts back from his embassy to Gray's Mill to return to stables in the Canongate. Lochiel's men rushed the gate. They marched quickly up the street, forced their way into the ramshackle City Guardhouse, placed sentries on all the gates and drew up in the Parliament Close. When daylight came, and as the first barefoot housemaids prepared to take down the slops,[70] Edinburgh was in rebel hands.

Meanwhile the main body of the Highland army, augmented by the Perthshire contingents, made its way through what is now Colinton and Morningside towards Holyrood. Every now and then they came in sight of the Castle, which fired three rounds, causing them to halt. Resuming their march, they passed under the park wall of Grange House, round Arthur's Seat by the south and across the King's Park to Holyrood. In the course of the march, the Prince received news that Cope had landed. Years later, embittered by exile and long since broken with Charles Edward, Lord Elcho still remembered that bright morning:

When the Army Came near town it was mett by vast Multidudes of people, who by their repeated Shouts & huzzas express'd a great deal of joy to see the Prince. When they Came into the Suburbs the Croud was prodigious and all wishing the Prince prosperity; in Short, nobody doubted but that he would be joined by 10,000 men at Edinburgh if he Could Arm them. The Army took the road to Dediston, Lord Strathallan marching first at the head of the horse, The Prince next on horseback with the Duke of Perth on his right and Lord Elcho on his left, then Lord George Murray on foot at the head of the Colum of Infantry. From Dediston the Army entr'd the King's park at a breach made in the wall. Lord George halted somewhere in the Park, but afterwards march'd the foot to Dediston, and the Prince Continued on horseback always followed by the Croud, who were happy if they could touch his boots or his horse furniture. In the Steepest part of the park Going down to the Abby he was oblidged to Alight and walk, but the Mob out of Curiosity, and some out of fondness to touch him or kiss his hand, were like to throw him down, so, as soon as he was down the hill, he mounted his hourse and road through St Anes yards into Holyroodhouse Amidst the cries of 60000 people who fill'd the Air with their Acclamations of Joy. He dismounted in the inner court and went up Stairs into the Gallery, and from thence into the Duke of Hamiltons apartment, which he occupied all the time he was at Edinbourgh.

Then it is as if Elcho shakes off his reverie, and reverts to his habitual bitterness:

He was joind upon his Entring the Abby by the Earl of Kelly, Lord Balmerino, Mr Hepburn of Keith, Mr Lockart younger of Carnwarth … and several other Gentlemen of distinction, but not one of the Mob who were so fond of seeing him Ever ask'd to Enlist in his Service, and when he marched to fight Cope he had not one of them in his Army.[71]

John Home was in the crowd, and he recorded an exact picture of the Prince:

The Park was full of people, (amongst whom was the Author of this History), all of them impatient to see this extraordinary person. The figure and presence of Charles Stuart were not ill suited to his lofty pretensions. He was in the prime of youth, tall and handsome, of a fair complexion; he had a light-coloured periwig with his own hair combed over the front: he wore the Highland dress, that is, a tartan short coat without the plaid, a blue bonnet on his head, and on his breast the star of the order of St Andrew ... He rode well, and looked graceful on horseback.

But something was amiss:

The Jacobites were charmed with his appearance: they compared him to Robert the Bruce, whom he resembled (they said) in his figure as in his fortune. The Whigs looked upon him with other eyes. They acknowledged that he was a goodly person; but they observed, that even in that triumphant hour, when he was about to enter the palace of his fathers, the air of his countenance was languid and melancholy.[72]

Home counted the rebel army at 1,900 men.[73]

Meanwhile, in the High Street another ceremony was taking place. Madie Pringle, writing to her sister Isabella, 'Tib', at Kelso in the Borders gives the best account:

A little before twelve a' clock seven hundred or thereabouts of ye Highlanders that had taken possession of ye Town surrounded ye Cross. This I saw myself ym marched three in a line with a Piper to every company. They surrounded ye Cross and at one o'clock five Heralds and a Trumpet with some Gentlemen, amongst them Jamie Hepburn [of Keith] ascended ye Cross and read two Manifestos in ye name of James eight King of Great Britain &c. at ye end of every one they threw up yr hats and huzza'd in which acclamation of joy they were joyn'd by all ye crowd which was so great I incline almost to call it the whole Town. Ye windows were full of Ladys who threw up their handkerchiefs and clap'd their hands and show'd great loyalty to ye Bonny Prince.

Then Miss Pringle, too, remembered herself and her family's Whiggishness: 'Don't imagine I was one of those Ladies. I assure you I was not.' She added, not without some sadness: 'All ye Ladies are to kiss ye Prince's [hand] – I've an inclination to see him but I can't be intro—' – at which point the letter breaks off.[74]

From a window on the north side of the High Street, the author of the Woodhouselee manuscript looked down in Whig contempt on a 'commick fars or tragic commody'.

All these mountan officers with there troupes in rank and fyle in order marched from Parliament Closs down to surrownd the Cross, and with there bagpipes and loosie [lousy] crew they maid a large circle from the end of Luickenbooths to half way below the Cross to the Cowrt of Gaird [presumably, the Town Guard house] and non but the officers and speciall favowrits and one lady in dress were admitted within the ranges.[75]

The lady was Margaret, wife of Murray of Broughton, and the memory of her, on horseback, sword in hand, her dress and bridle fluttering with white ribbons, became one of the Jacobite consolations.[76]

I observed there armes, they were guns of different syses, and some of innormowows length, some with butts tured up lick a heren [like a heron], some tyed with puck threed to the stock, some withowt locks and some matchlocks, some had swords over ther showlder instead of guns, one or two had pitchforks, and some bits of sythes upon poles with a cleek, some old Lochaber axes. The pipes plaid pibrowghs when they were making ther circle thus they stood rownd 5 or six men deep ... The Crosse to the east was covered with a larg fine Persian carpet. The Lyon Heralds in these formalities, coats on, and bleasons displayed, came attended but with one trumpet to the theatur or to the Cross. They were five in number, Ereskin, Lyon Clerk, on his left, Roderick Chalmers, pursevant and herald panter, the others were Clerkson, pursevant, Gray and one I knew not. All the streat and the windows and forstairs were crowded and sylence being made the manefesto was read in the name of James 8 of Scotland England France and Irland King[77]

The proclamations read by Chalmers appealed to anti-Union sentiment in Edinburgh, while reassuring the Presbyterians about the free practice of their religion. The first, signed by James VIII and dated from Rome on 23 December 1743, appointed Charles as Prince Regent of England, Scotland and Ireland, but spoke to peculiar Scottish grievances, such as the increase in taxes since the Union and the construction of forts and the disarming of the Highlanders in the years after the Fifteen. 'We see a nation,' the manifesto accompanying the commission of regency said, 'always famous for valour ... reduced to the condition of a province, under the specious pretence of an union with a more powerful neighbour.'[78] It promised a general pardon for those who had served the Elector of Hanover, free parliaments, and protection for 'all our Protestant subjects in the free exercise of their religion'.[79] The second, signed by Charles and dated from Paris on 16 May 1745, proclaimed that 'we are now come to execute his Majesty's will and pleasure'.[80]

'Thus the winds blew from Rome and Paris to work owr thraldome,' wrote the author of the Woodhouselee manuscript. 'I could hear at my distance distinctly, and many much further, for there was profownd silence after all these military dismissed with bagpipes playing and a fashon of streamers over ther showlders and the chime of bells from the High Church steaple gave musicall tunes all the whill.'[81]

He was struck by the good order of the mountain men, as was Miss Pringle, who wrote to her sister:

> After all this the Crowd dispersed and ye Highlanders march'd with Lord Elcho and John Murray Broughton on their Head back to ye Parliament Closs where they stood a while and then dispersed, they are as quiet as lambs, civil to everybody and takes nothing but what ye pay for.[82]

Still, there were bound to be mishaps among young men unused to the firearms handed out from the city armoury. The next morning, Wednesday, 18 September,

> An ugly accident happen'd ... to poor Madie Nairn who was looking over Lady Keith's window along with Katie Hepburn. On ye other

side of ye street there was a Highland Man and a Boy standing with a Gun in his hand which Gun went off and shot in at ye Window and ye Bullet went in at Mady Nairn's head. Luckily the strength of ye ball had been spent by its Grazing on ye wall so that it stuck and did not go through her skull or she must have Died instantly. Mr Ratray has taken out ye Ball and sow'd up her wound he thinks her safe if she keeps free from a Fever. The Prince has sent several messages to inquire after her which has help'd not a little to support her spirits under ye Pain of her sore wound.[83]

The Caledonian Mercury, which like the *Edinburgh Evening Courant* came out thrice weekly, had referred in its issue of Monday the 16th to 'rebels'. In its issue of Wednesday the 18th it showed the Jacobite colours of its proprietor, the very learned Thomas Ruddiman:

> Edinburgh, September 17: Affairs in this city and neighbourhood have taken the most surprising turn since yesterday without the least bloodshed or opposition, so that we have now in our streets Highlanders and Bagpipes, in place of Dragoons and Drums, of which we will be allowed to give the following narrative of facts, as far as we have been able to collect them. On Monday last the Highland army stood under arms about Corstorphine...[84]

The following day, 19 September, the Prince received reliable intelligence that Cope was setting out from Dunbar and was to encamp at Haddington. Anxious that Cope should not slip past him again, Charles Edward spent the night with his army at the village of Duddingston, to the west of the King's Park. He had first arranged for surgeons to come out from the town, and coaches and chaises for the wounded. All the leading medical men, Whig and Jacobite, came out: John Rattray, last seen tending Miss Nairne's wound, and even Dr Monro, Professor of Anatomy at the College, a staunch Whig and cousin to Duncan Forbes.

Stepping out of the cottage at Duddingston, where he had spent the night, at nine in the morning of the 20th Charles Edward made a little speech. As the *Caledonian Mercury* reported in its edition of the 23rd, 'the Chevalier putt himself att the head of his small army, drawing his sword, said with a very determined Countenance, Gentlemen, I have flung away the Scabbard, with Gods assistance I don't doubt of making you a free and happy people.'[85]

Even now, the words 'free' and 'happy' have the capacity to startle. They bring to this little village in Scotland on a fine morning the breath of a Continental ferment. Freedom was an old Scottish interest, but happiness was the great invention of the eighteenth century. It is as if in his stuffy palace in the Piazza dei Sant'Apostoli, in between his hunting expeditions, the Prince had been poring over Continental pamphlets on public happiness, *du bonheur publique, della pubblica felicità*. The Jacobite rebellion was not just an exercise in divine-right monarchy and feudal military organisation, but touched modernity in some of its most urgent interests. Then the Prince came back to earth: 'Mr Cope shall not escape us as he did in the Highlands.' They began their march westwards, along Carberry Hill.

Cope had drawn up his four thousand men on ground harvested of its corn just the evening before. The place was a fortress, with a deep ditch to the south, the village of Preston to the west, a small morass on the east, and the broad Firth to the north. Cavalry and infantry had at last effected a junction, though Colonel Gardiner, for one, was not sanguine. He had told Carlyle in confidence in the garden of the Dunbar Manse on the 19th that he had 'not above ten men in my regiment whom I am certain will follow me. But we must give them battle now, and God's will be done!'[86] Around noon the rebels appeared on the high ground to the south-west, gave a shout, and had to be restrained by their officers from falling at once on the government army.

The sun went down on another beautiful evening, and as the Highland army seemed to settle for the night, Carlyle called again on Colonel Gardiner and found him 'grave, but serene and resigned; and he concluded by praying God to bless me, and that he could not wish for a better night to lie on the field, and then called for his cloak and other conveniences for lying down, as he said they would be awaked early enough in the morning.'[87] Returning to his father's manse, Carlyle found it packed with clergymen and military tourists. At supper he was so exhausted that 'no sooner had I cut up the cold surloin which my mother had provided, than I fell fast asleep.'[88] Retiring to bed at a neighbour's house, Carlyle instructed the maid to wake him the moment the battle began, which she did; but though he sprang into his clothes, he was too late. Running to a high point in the Manse garden, in the faint, misty light he made out a scene of horror:

> The whole prospect was falled with runaways, and Highlanders pursuing them. Many had their coats turned as prisoners, but were still trying to reach the town in the hopes of escaping. The pursuing Highlanders, when they could not overtake, fired at them, and I saw two fall in the glebe. By-and-by a Highland officer whom I knew to be Lord Elcho passed with his train, and had an air of savage ferocity that disgusted and alarmed. He inquired fiercely of me where a public-house was to be found; I answered him very meekly, not doubting but that, if I had displeased him with my tone, his reply would have been a pistol bullet.[89]

The battle, according to one Jacobite officer, lasted a full three minutes.[90] At 3 a.m., the Highlanders were guided over the morass and drawn up in line. They fired once, then threw down down their muskets and charged with their broadswords at the government forces. As Gardiner had feared, the dragoons fled the battlefield, leaving the foot to be crushed up against the park walls of Preston House. Cope's fortress was also a prison. The doggerel poet Dougal Graham was in the Highland army, and reported:

> The poor foot, left here, paid for all,
> Not in fair battle, with powder and ball;
> But horrid swords, of dreadful length,
> So fast came on, with spite and strength,

Lochaber axes, and rusty scythes,
Durks and daggers prick'd their thighs…
From 'bove Cow-caney to Preston-dyke,
About a mile or near the like,
They were beat backward by the clans
Along the crofts 'bove Preston-pans,
Till the high dyke held them agen,
Where many taken were and slain,
Although they did for quarters cry
The vulgar clans made this reply,
'Quarters! You curst soldiers, mad,
It is o'er soon to go to bed.'[91]

Gardiner himself, abandoned by his men, was cut down by a scythe in sight of his own house. It was said he prayed for death, 'for his state of health was bad, & his heart was broken with the behaviour of the Irish dogs whom he commanded.'[92] Drummond, sturdily mounted on the old dragoon horse he had bought for four pounds to draw his cart, was more fortunate. Intending to fight with his friend, he was swept away in the rout of Hamilton's troop and ended up in Berwickshire.[93] Cope was there already. An officer noted acidly that he was one of the very few commanders to bring tidings of his own defeat.

When Johnnie Cope to Dunbar came
They speer'd at [asked] him, 'Where's a' your men?'
'The deil confound me gin I ken
For I left them a' this morning.'[94]

Lord George Murray, writing to his brother at Blair on the 24th, gave the government dead at 600 with as many wounded, and 1,200 uninjured prisoners, including 80 officers. On the Prince's side, the losses were 36 killed and 50 wounded.[95] Carlyle seems to have put the worst construction on Lord Elcho's demand for a public house, but it was not that he was thirsty: he needed brandy for the wounded, and the redcoat wounded at that. The account of the battle Walter Scott gives in *Waverley*, in which the hero rides about trying to save the lives of Hanoverian officers, may not be so far-fetched. As Carlyle stood on his mound,

> The crowd of wounded and dying now approached with all their followers, but their groans and agonies were nothing compared with the howlings, and cries, and lamentations of the women, which suppressed manhood and created despondency.[96]

The jaunt was over, and with it the white cockades and the hand-kissing. There would be retribution for the redcoats bleeding amid the stubble, and Edinburgh would never be the same again. The Prince spent much of the day in the field, attending to the government wounded, and passed the night at Pinkie House. Returning to Holyrood, he forbade any display of 'publick Joy' at the death of his subjects in the battle.[97] Though he remained in the Palace for a further six weeks, people complained that his court was dull and sad. There was no divine service in the Kirks.[98]

Prestonpans transformed the political game. As Lord President Forbes wrote: 'All Jacobites, how prudent soever, became mad; all doubtfull people became Jacobites; and all Bankrupts became heroes, & talk'd nothing but hereditary rights and victory; & what was more grievous to men of gallantry, & if you will believe me, much more mischievous to the publick, all the fine Ladys, if you will except one or two, became passionately fond of the young Adventurer, and used all their Arts, & industry for him, in the most intemperate manner.'[99] Among those who joined the Prince's cause was the lawyer Sir James Steuart, who became the only Jacobite political economist. Meanwhile Sir John Clerk, who had retreated with his wife to England, found Newcastle in a 'terrible consternation. A most terrible pannick had possessed all the people to that degree that many rich people about Newcastle, Durham and York, had sent off a great deal of their Effects to Holand and Hamburgh, and all their silver plate, jewels, money and such like domestic necessaries were hidden under ground.'[100] In London, government securities tumbled in value through fear that a restored House of Stuart would repudiate the National Debt.

These were frustrating days for the Prince. He was eager to march on London for, as Elcho had it, he 'said often he would have the three kingdoms or nothing at all'.[101] But it was fantasy to think of such an expedition, when so many Highlanders had drifted away to store their booty from the battle and there was no money to pay them. The Prince's Council had imposed a levy on Edinburgh at the rate of half-a-crown per pound sterling of rental (12.5 per cent), while James Hay of Restalrig, a Writer to the Signet of Jacobite leanings, was despatched to extort what he could from the Whig merchants of Glasgow.

With no artillery heavier than the field guns captured from Cope, the Prince could not expect even to reduce Edinburgh Castle; instead, the men posted at the Weigh House and the head of the Lawnmarket began to harass the Castle gate. On Sunday 29 September General Preston sent to the Lord Provost demanding that communication between town and Castle be kept open. A truce was agreed, but broke down in the afternoon of Tuesday, 1 October, when the house that Allan Ramsay senior had built on the Castle Hill, known as the Guse-pie from its octagonal shape, was occupied by Cameron of Lochiel and damaged by shot. Neither the poet father nor the painter son was at home. On Wednesday the 2nd the Prince issued a proclamation forbidding any intercourse with the Castle, on pain of death.[102] General Preston responded by turning his ordnance on the High Street.

There followed scenes unexampled in the modern history of these islands. A round aimed at the Tolbooth gate passed through two churches. Cartridge-shot from the Half-Moon Battery on Castle Hill raked the streets. At its peak, that 'damned angry bitch',[103] as the Highlanders called the Castle, was firing as many as sixty rounds a day. ''Tis not safe being in Lawn or Grassmarkets,' wrote the author of the Woodhouselee manuscript. 'I saw a musket ball was battered upon the stons in Grassmarket and a gentleman missed it narrowly.'[104] In the midst of the cannonade, at 9 a.m. on Thursday, 3 October, a bizarre civilian delegation carrying a white flag could be seen at the Castle drawbridge: it was John Campbell, chief cashier of the Royal

Bank, three directors, the accountant and the teller, come to cash the banknotes brought back from Glasgow by Hay of Restalrig.

Banknotes were current throughout the Scottish Lowlands, but the Highland rank-and-file wanted their sixpence a day in coin. Once in England, too, the army would need gold and silver to buy supplies. On 1 October Murray had demanded payment in coin within forty-eight hours on £857 in Royal Bank notes to pay the army. The next day, while dining at Lucky Clark's in Fleshmarket Close, Campbell had been by presented by Murray with a further £2,307 in notes. It was a delicate matter for Campbell and for the five bank directors who were still in town. The Royal Bank was a Whig house, founded in 1727 in the Argyll interest. Under its deputy governor, Lord Justice-Clerk Milton, it was the main channel for government payments in Scotland. Whatever his private sympathies, which remain obscure, publicly Campbell had thrown in his lot with the House of Hanover. To be set against that was the promise-to-pay on the Bank's notes.

Campbell had at first played for time, claiming the pass he had been issued, which expired at 10 p.m. that night, did not give him long enough to count out so much coin. It was the sort of Scots banker's tactic that became notorious in the so-called 'banknote wars' with the new Glasgow banks[105] in the 1750s, but Murray was having none of it. One of his deputies, Peter Smith, warned Campbell 'that a gentleman, who understood the business of banking, was with the Prince, when the pass was a granting, who said there was no difficulty in the thing.'[106] The 'gentleman' was, presumably, Sir James Steuart.

Arriving at the governor's lodgings at the Castle, Campbell airily told Guest and Preston that 'our errand in general was to get into the R. Bank's repositories to do some business.'[107] That night, over a single bottle of wine in the quiet and secrecy of Lucky Clark's, he paid over £3,076 in coin to Murray's deputy secretary, Andrew Lumsden. In the course of the month Campbell was to pay over to Murray of Broughton a further £3,600 in gold, which went to finance the march south. This treasonous accommodation, though it continues to cause some faint qualms to the officers of the Royal Bank today, was wholly in the character of the Edinburgh of that period.

Still the cannonade continued: on the worst day, Saturday, 5 October, the author of the Woodhouselee manuscript noted 'all shops shut, evry body scared off the streats, except here and ther, one skulking and runing, the Cowget full of cartes with plenishing and so at Nether Bow and all down the Cannonget.'[108] At the head of Liberton's Wynd, he came on a tradesman 'in a blew frock' lying, 'his brains dashed owt and in his blood'.[109] That night, under the threadbare cover of a proclamation deploring 'the many Murthers which are committed upon the innocent inhabitants of this City by the inhumane Commanders and Garrison of the Castle of Edinburgh',[110] the Prince called off the blockade. No doubt he was advised that there would be little hope of raising his half-crown levy on Edinburgh rentals amid such pandemonium.

By the following Wednesday, he was reviewing his men at Duddingston. 'O lass such a fine show as I saw on Wednesday last,' wrote Miss Pringle to her sister.

He was sitting in his tent when I came first to ye field. The Ladies made a circle round ye Tent and after we had Gaz'd our fill at him he came out of the Tent with a grace and Majesty that is unexpressible. He saluted all ye Circle with an air of grandeur and affability capable of Charming ye most ostinate Whig and mounting his Horse which was in ye middle of ye circle he rode off to view ye men. As ye circle ws narrow and ye Horse very Gentle we were all extremely near to him when he mounted and in all my Life I never saw so noble nor so Graceful an appearance as His Highness made, he was in great spirits and very cheerful; which I have never seen him before. He was dressed in a Blue Grogrum Coat trimmed with Gold lace and a lac'd Red wastcoat and Breeches. On his left shoulder and side were the Star & Garter and over his right shoulder a very rich Broad Sword Belt. His sword had ye finest wrought Basket hilt ever I beheld all Silver. His hat had a white Feather in it and a white cockade and was trimmed with an open gold lace. His horse was black and finely bred (it had been poor Gardners).[111]

For all the Prince's new cheer, and his success at Prestonpans, recruitment to his cause was slow.

The initiative seemed to be slipping away. On 11 October Lord George Murray wrote to his brother: 'Every thing is in great confusion in England, particularly in London, where credite is at a stand; the greatest Banquiers have stopt payment; all would go our wish if we could but march immediatly.'[112] But on the 14th Charles Edward's hopes revived. That day there arrived at Holyrood one of the most attractive of all the actors of this period, Jean-Baptiste de Boyer, marquis d'Éguilles, secret envoy of the King of France. A lawyer of Aix-en-Province, highly intelligent, handsome, enthusiastic and adventurous, Éguilles had slipped into Montrose on a Dunkirk privateer on 7 October with instructions from the marquis d'Argenson, Louis XV's Minister of Foreign Affairs, to report on the Prince's intentions and strength but to do nothing to commit the French court in his support. Éguilles's own feelings about his mission can be judged from his letter to the Duke of Atholl on October 15: '*Me voicy enfin arrivé à Edimbourg où je compte voir incessament votre adorable Prince, qui fait les délices de ses peuples, et qui fera bientôt l'admiration de l'Europe* [Here I am at last at Edinburgh, where I hope to feast my eyes on your lovely Prince, already the darling of his subjects and soon to be the toast of Europe].'[113]

On the two days following Éguilles had two audiences of the Prince in the drawing-room at Holyrood, during which the Prince announced he was leaving in a week to march direct to London, and pleaded for a diversion from France. 'Marquis,' he concluded in exasperation, 'am I not to count on an early landing?'[114] Éguilles suggested he should perhaps delay his expedition until the news of Prestonpans had reached the French court. Meanwhile, behind the scenes Éguilles was also encouraging a 'Scottish solution', in which the Union would be broken, favoured by at least some of the Council.[115] For whatever reason, the Prince resolved to send his own envoy across the water to try to negotiate full-scale assistance from the French. He selected Sir James Steuart for the mission, condemning the philosopher to two decades of exile.

Prince Charles Edward left Edinburgh, never to return, on 31 October. Very few of the Edinburgh Jacobites left with him. He pressed into service forty-four surgeons and physicians. On 3 November he set off for England, on foot at the rear of his army, 'with his target [shield] over his shoulder'.[116] On 12 November the Hanoverian government, in the form of Lord Justice-Clerk Milton and the other law officers, crept back into Edinburgh. As is well known, at Derby the Highland army lost its hopes of taking London. Retreating in good order to Scotland, a portion was cut to pieces near Forbes's house at Culloden on Wednesday, 16 April 1746. Demoralised, Charles abandoned his expedition and his followers and, after five months of hair-raising escapes and adventures, re-embarked for France on 19 September.

The victor of Culloden, George II's ruthless and able third son, William Augustus, Duke of Cumberland, was determined to punish Edinburgh for its Jacobite Saturnalia. The fourteen rebel standards captured in the battle were paraded through the High Street by the common hangman and thirteen chimney-sweeps, then burned at the Cross. When Forbes of Culloden urged restraint, Cumberland was overheard describing him as 'that old woman who talked to me about humanity'.[117] Forbes did manage to have the surgeon John Rattray spared, but received scant thanks for having held much of the Highlands for the Crown and was £1,500 out of pocket. There seems to have been some agitation in favour of using Cumberland's patronage to reform the Town Council that had so lamentably failed Edinburgh in the crisis. Drummond, who had once favoured reform, quite changed his mind when, on 26 November 1746, he was elected by a vote of burgesses to his second two-year term as Lord Provost.[118]

But before that, for the second time in half a generation a Lord Provost of Edinburgh was clapped into a London gaol. In the end, Old Provost Stewart was tried for neglect of office in Edinburgh on 24 March 1747 and, after many delays, acquitted on 2 November. Robert Drummond, an elderly and infirm printer who had issued a Jacobite squib on the Provost's trial and the character of the Duke of Cumberland, was pilloried, banished for a year and deprived of his business.[119] The parliament of 1747 took several measures designed to demolish the ancient society of the Highlands, passing Acts to deprive the Highlanders of their arms, their dress and the hereditable jurisdiction of their chiefs. Lord President Forbes was barely consulted. He went to his grave later that year, in part as a result of his exertions in the crisis. MacLaurin was already dead, on his way home from York.

The Jacobite Rebellion transformed Edinburgh, and accounts for much of what went on in the city during the next half-century: authoritarian church and town politics, frantic expressions of loyalty to the House of Hanover, attempts to reform Scottish pronunciation. For David Hume and Adam Smith, for lawyers such as Alexander Wedderburn and for the younger ministers, the best way forward was to forget the past, shed any distinctive Scottishness, unlearn the Scots language, re-forge links with the Continent grown rusty with Jacobite intrigue, and reveal the innate superiority of Scotland by out-Englishing the English.[120]

The Forty-five, while it confirmed the older clergy in their Whiggishness, entrenched the younger in their Unionism. The chaplain of the Black Watch, Adam Ferguson, told the First Highland Regiment of Foot in a sermon given in Gaelic on 18 December 1745 that they had a 'league' with society and must protect their religion and liberty from despotic government and Popish corruption.[121] Another rising clerical star, Hugh Blair, in a sermon in St Giles at the opening of the first General Assembly after Culloden, on 18 May 1746, called the Rebellion providential, sent to 'work a cure' of such evils as luxury, corruption of manners and religious apathy.[122] The tone of his encomium of Cumberland speaks for itself:

> When the proper season was come for God to assert his own cause, then he rais'd up an illustrious deliverer, whom, for a blessing to his country, he had prepared against the time of need. HIM, he crowned with the graces of his right-hand; to the conspicuous bravery of early youth, he added that conduct and wisdom, which, in others, is the fruit only of long experience; and distinguish'd him with those qual-

ities which render the Man, amiable; as well, as the HERO, great: He
sent him forth to be the terror of his foes, the confidence and love of
his friends; and in the day of danger and death, commanded the
shields of angels to be spread around him!

The General Assembly's congratulatory letter to the Duke of the
20th is in the same vein:

> As for some months past, the many fatigues you endured, and the
> alarming dangers you ran, in pursuing an ungrateful and rebellious
> crew, filled our minds with the greatest pain; so the complete victory
> now obtained over them by the bravery of your Royal Father's troops,
> led by your wise conduct and animated by your heroic example, gives
> us the highest joy.[123]

Yet the Rebellion left a deep injury, which found its expression
in romantic nostalgia and philosophical pessimism. Arriving in
Edinburgh for the first time in 1747, Ramsay of Ochtertyre
observed that nine-tenths of the women were wearing tartan
plaids.[124] Two years later, John Campbell sat to William Mosman
for the magnificent portrait that now hangs in the Royal Bank's
boardroom, leaning against a tray of coin and banknotes, armed
to the teeth and swathed in crimson-and-black tartan.[125] In contemplating the events of September 1745 for their models of universal history, the Edinburgh philosophers concluded that men
were so weakened by their commercial existence that they no
longer knew how to fight, or to die. Even the most good-natured
of them, David Hume, saw 'men fallen into a more civilized Life,
entirely unfit for the Use of Arms'.[126]

3

The Disease of the Learned

On 23 December 1696 Thomas Aikenhead, the eighteen-year-old son of a late surgeon-apothecary of Edinburgh and a student at the College, was indicted in the Court of Justiciary on a charge of blasphemy.

The Lord Advocate, Sir James Stewart of Goodtrees, grandfather of the Jacobite philosopher who was the Prince's secret envoy, told the court that Aikenhead had on several occasions in conversation with his fellow-students said the doctrine of Christian theology was a 'rapsodie of faigned and ill-invented nonsense',[1] called the Old Testament 'Ezra's Fables' and the New 'the History of the Impostor Christ', rejected the Trinity and Incarnation and Redemption, preferred 'Mahomet to the Blessed Christ' and committed various other offences against the laws of 'well-governed Christian realmes'.[2] Sir James asked for the death sentence, 'to the example and terror of others to committ the lyke in tyme coming'.[3]

No counsel appeared for the prisoner, nor was any defence presented. Lord Advocate Stewart examined five witnesses, one a Writer or lawyer and the others students at the College, all aged between eighteen and twenty-one. In their fragmentary testimony, words and names – 'faigned', 'Ezra', 'Mahomet' – keep repeating themselves as in a nightmare, and a haggard gleam falls for a moment on the student life of that era. Patrick Midletoune or Midletone, aged twenty, testified that 'about the middle of August last, about eight o'clock at night, goeing by the Tron kirk, he hard him (being cold) say that he wished to be in the place Ezra called hell, to warme himself there.'[4] The witnesses confirmed the charges and added one more: that the prisoner said he was confident Christianity 'would be utterly extirpat' by the year 1800.[5]

Aikenhead pleaded for mercy on the ground of his youth, said he was merely repeating phrases from atheistical books supplied by one of the witnesses, confirmed his belief in the Trinity and the scriptures, expressed deep contrition, and solemnly promised to make amends. As John Buchan commented more than two hundred years later, that would have saved his life at the hands of the Inquisition.[6]

This was Presbyterian Edinburgh. On Christmas Eve the jury unanimously found the prisoner guilty of railing against God and Christ. He was condemned to be taken to the 'Galowlee betwixt Leith and Edinburgh', between two and four of the afternoon of 8 January 1697, and hanged till dead, his body to be interred at the foot of the said gallows. On the scaffold, Aikenhead read from a paper that he had come to doubt the objectivity of good and evil, and to believe that moral laws were the work of governments or men.

The Aikenhead case haunted the next century, for it seemed to inaugurate a battle for the soul of Edinburgh between a rigid Calvinism which saw any deviation in doctrine or conduct as a mortal threat to the whole community, and a new conviction of the privacy and variety of conscience. This contest ran alongside the Whig–Jacobite dispute, and outlasted it. The century ended not quite as Aikenhead had prophesied, but in 1800 his old college had more professors and students of medicine than of theology, and the chief debating topics in the clubs were economic or geological. Men and women were coming to suspect that knowledge acquired through scepticism might be more useful in this world below than knowledge 'revealed' by scripture.

This revolution in belief occurred also in Paris and London and Philadelphia; Edinburgh was exemplary because of the peculiar importance of religion in old Scotland. In a country without formal political life, religion will replace politics: the dispute about religion in Scotland was also a dispute about the state.

The rise of a church reform movement that ecclesiastical historians called Moderatism, and with it the 'abandonment for good and all of the fantastic theocratic dreams of the previous century',[7] was also an accommodation with the secular power in London and its Scottish agents. It drew on the same spirit that was abandoning Jacobite dreams of French alliance or separation from England and embracing scientific agriculture and commercial opportunity. By the 1760s, Edinburgh clergymen had become powers even in England.

The pitched battles of this ecclesiastical war took place in the years immediately after the Forty-five. The first broke out over an attempt by the 'High-Flying' or 'Popular' party in the Kirk to excommunicate the philosophers David Hume and Henry Home at the General Assemblies of 1755 and 1756 on the grounds of 'infidelity', or unbelief. The second was the battle over John Home's stage play *Douglas, A Tragedy*, of the following year.

Had the High-Flyers succeeded against the philosophers, Henry Home, who had been raised to the Court of Session as Lord Kames in February 1752, could no longer have served as a judge, and both men would have left Scotland. It is conceivable, in this realm of hypothetics, that their Continental admirers would have portrayed them as martyrs of a clerical reaction. The Scottish School would have had a quite different and, no doubt, much more radical character. As for the battle over *Douglas*, it was the launch of a secular society in Edinburgh. These two contests forced Edinburgh to develop and clarify its intellectual position, with results that are felt today.

At the Revolution of 1688–9 which re-established the Presbyterian church in Scotland, the new King William III made a plea for tolerance. 'We never could be of the mind that violence was suited to the advancing of true religion,' he wrote to the General Assembly of the Kirk on 17 October, 1690. 'Moderation is what religion enjoins, neighbouring churches expect from you, and we recommend to you.'[8] His appeal fell on ears made deaf by decades of religious warfare. The late seventeenth-century Kirk in Scotland, like the contemporary Puritan sects in New England, did not believe in privacy of belief, for belief was not private from the Lord or the Devil. Any doctrine of Grace holding that only a certain community will be saved begins in intolerance and continues in persecution. Edinburgh knew in its heart it was wicked, that its citizens were 'Heirs of Wrath', as the hagiographer Patrick Walker put it, and doomed but for Christ's interposition to eternal flames. To walk in Grace, it must observe literally the precepts of scripture without any attempt at interpretation. To be tolerant was to be lukewarm.

In this grim cosmogony, the world was a public arena in which colossal unseen forces of good and evil were fighting to the death. Here, in the damned houses of the West Bow and Mary King's Close, were the very outworks of the nether world. Robert Wodrow, a Presbyterian historian of the early eighteenth century, reported with awe the titanic struggles of a Reverend Mr How in the persecution times, 'a most mighty, importunate wrestler in prayer' who wrestled so fervently in the course of a service that the 'sweet [sweat] haled doun'; at which Mrs How 'stepped to him gently, took off his wigg, and with her napkin dryed the sweet and put on his wigg again. This she was obldiged to do twice, if not thrice, and Mr Hou seemed not to knou what was done to him.'[9] In private, as recorded in the diaries of Thomas Boston or Lord Provost Drummond, men attributed to the Holy Spirit or the Devil mental states that would now be ascribed to illness or hangover, and interpreted events in the High Street or Council Chamber as God's *signs* indicating a particular line of conduct.

The blood of persecuted Covenanters shed during the years of Episcopalian domination under the later Stuarts had to be purged: not merely by driving Episcopalians out of parishes and university

offices, but by expunging all the cursing, Sabbath-breaking, forni-
cating, adultery, drunkennness and blasphemy that had supposedly been rife under Strumpet Prelacy.

In the vacuum of temporal power in Scotland, the Kirk had
many civil and judicial functions. At the lowest level, the Kirk-session consisting of the minister and elected lay-elders organised
poor relief, supervised the parish school and acted as a court of
first instance for many civil crimes as well as what are now called
misdemeanours. These Kirk-sessions were gathered by district
into Presbyteries. In Edinburgh, additionally, the Town Council
had for centuries enjoyed authority to make 'acts, statutes and
ordinances for the good government of the town'. Together they
displayed, in Arnot's phrase, a 'gloomy and morose contempt for
the social pleasures'.[10]

The Sabbath was a day in which every moment was under sur-
veillance. Work, the smallest recreation, 'vaguing' in the High
Street or on the Castle esplanade, all invited punishment from
Kirk-session or a fine from the magistrates. 'Seizers' patrolled the
streets during sermon-time to ensure attendance in the icy
churches (even in the West Kirk, the black cowl of the minister was
sometimes covered in a 'thin glaister o' sifted snaw'[11]); or patrolled
the streets in the evening, pursuing suspicious passers-by down the
steep, dark wynds.

Sir James Stewart of Goodtrees, Aikenhead's persecutor, was
still famous a century later for the unremitting severity of his
Sabbath devotions. His younger son told Elizabeth Mure of
Caldwell, who collected observations on the the century's chang-
ing manners, that

> After prayers by the Chaplin at nine o'clock; all went regularly [in
> good order] to church at ten, the women in high dress. He was
> employed by his father to give the collection from the family, which
> was a crown. Half after twelve, they came home; at one had prayers
> again by the Chaplin, after which they had a bit of cold meat or an
> ege, and returned to Church at two; was out again by four, when
> everybody returned to their private devotions, except the Children

and servants, who were conveened by the Chaplin and examined. This
continued till five, when supper was served up, or rather dinner. A few
men friends generally partaked of this meal, and sat till eight; after
which singing, reading and prayers was performed by the old gentleman himself; after which they all retired.[12]

Visitors from England were crushed by the gloom of the
Edinburgh Sunday. Sir Richard Steele, founder of *The Tatler*, who
visited Edinburgh in 1717, nicknamed the Reverend Andrew Hart
'the hangman of the Gospel' because he seemed to take such
pleasure in preaching 'the *terrors* of the Lord'.[13] As late as 1775,
Captain Edward Topham, an English traveller, said that during
Sunday service it was as if 'some epidemic disorder had depopu-
lated the whole City'.[14]

Weekdays were only a little less sombre. Kirk elders went into
the taverns at 10 p.m. by the church clock to send the occupants
home: hence the Edinburgh phrase 'ten o'clock man' for a person
fond of good cheer. The first dancing Assembly and touring theatrical impresarios and companies were visited with anathemas from
Kirk-sessions, as being injurious to religion and youth. On one
occasion – but reported only in a single source – zealots pierced the
door of the Assembly-room in the West Bow with red-hot spits.[15]
Even the circulating library started in the Luckenbooths by that
one-man literary revival Allan Ramsay senior in 1725 was raided
by the magistrates.[16] As the Reverend George Anderson told the
congregation at the Tron Church: 'A Life spent in innocent
Diversions is in itself sinful ... By doing no Good you do evil.'[17]

Elders and deacons spied on their congregations. Ministers fulminated against the wedding parties known as 'penny bridals'
held by their poorer parishioners, and all 'promiscuous dancing of
men with women'. Fornicators – except those of the laird class –
were brought to stand on a raised platform or stool directly in
front of the church pulpit, known as 'the pillory', in a cloak of
sackcloth, facing the congregation and rebuked by the minister
above, for as long as half a year of Sundays. Fear of the ordeal bred
terrible crimes. There were twenty-one convictions for child-

murder in Edinburgh between 1700 and 1706, including four on a single day.[18]

Yet as the years passed, the Kirk came to be confronted with the challenge to all religious revolutions that legislate for eternity but subsist in time: the rising generation ceases to care about the blood of the martyrs. 'The witty lown-warm Air of Edinburgh' had its effect, as Patrick Walker reported, and men found 'the Tables better covered, the Chambers warmer, and the Beds Softer than the cold Hills and Glens of Carrick, and Galloway.'[19] An act of the Edinburgh Presbytery of 29 April 1719 complained that:

A great number take an unaccountable liberty in despising and pro-faning the [Lord's day] idly and wickedly, by standing in companies in the streets, misspending their time in idle discourse, vain and use-less communications ... withdrawing from the city ... to take their recreations in walking through the fields, parks, links, meadows ... And by entering into taverns, ale-houses, milk-houses, gardens, or other places, to drink, tiple, or otherwise misspend any part thereof; by giving or receiving civil visits ... and by idly gazing out of windows ... Yea, some have arrived at the height of impiety, as not to be ashamed of washing in waters, and swimming in rivers upon the holy Sabbath ... as they would not bring down the wrath of God upon themselves and the land, that they forthwith henceforward seriously repent.[20]

The sin of 'promiscuous dancing' had become the craze for 'danc-ing promisky', and Allan Ramsay senior's Katy and her laird were able to take the air:

> Since ye're out of your mither's sight,
> Let's take a wauk up to the hill.
> O *Katy* wiltu gang wi' me,
> And leave the dinsom town a while?[21]

The Church of Scotland, which had prided itself on doctrinal purity and administrative unity, was racked by a dispute over who was or was not saved, and began to splinter into sects over the proper relation between Church and State. The Marrow Brethren, who took their nickname from a forgotten devotional book, *The Marrow of Modern Divinity*, 'rediscovered' in an Ettrick cottage by Thomas Boston in 1715, restated the old Calvinist doctrine of saving Grace: that no action by a member of the Elect of the Kirk could cause him to forfeit salvation. (The doctrine was parodied by Burns in 'Holy Willie's Prayer' and, in the next century, and with all sorts of brilliant Gothick effects, in the *Confessions of a Justified Sinner*, attributed to James Hogg.) On the other wing of the church John Simson, Professor of Divinity at Glasgow University, appeared to argue the opposite case: that virtue was indispensable to the Christian.[22] He based his theology as much on experience and observation as on the Confession of Faith, remarking: 'Our knowledge of divinity has not yet ... arrived at perfection.'[23] Simson was a red rag to the Brethren and also to James Webster, the ultra-orthodox minister of the Tolbooth Kirk in St Giles, who was his chief pursuer. Simson was tried over and over again for heresy at the General Assembly, but defended himself with spirit and, it seems, an insufferable consciousness of his intellectual superiority. He was finally suspended from teaching in 1729.[24]

Those who disputed over doctrine also disputed over organisation. In 1712 Queen Anne's ministers restored to lairds and the Crown the right they had enjoyed up to 1649, that of appointing ministers to vacant parishes. The Patronage Act knocked out one of the keystones of Scottish Presbyterianism: that the church congregation selects its own minister. As Walpole's agent the Earl of Ilay began to enforce the Act and landlords to invoke their rights, there were outbreaks of resistance or churches stayed vacant. The General Assembly was obliged after 1729 to appoint 'riding committees' to travel to parishes and induct presentees obnoxious to their congregations. It was said that when the philosopher John Reid was presented in 1737 by King's College, Aberdeen to the living of New Machar – a country parish some ten miles from the town – he was set upon by men dressed in women's clothes and ducked in a pond; and that, on the Sunday he first preached, a relation stood on the pulpit stair with a drawn sword.[25]

Appeals to the King for relief from the 'grievance' of patron-age[26] could not save the Kirk from schism. The occasion was a compromise offered by the General Assembly of 1732, giving local landlords known as heritors a voice in the appointment of ministers. 'I can find no warrant from the word of God,' said Ebenezer Erskine of Stirling, who with Boston of Ettrick had been one of the Marrow Brethren, in a sermon at Perth that October, 'to confirm the spiritual privileges of his house upon the rich beyond the poor.'[27] Together with three other ministers Erskine founded the Associate Presbytery in 1733. Further secessions weakened the Kirk by drawing off many of its most vigorous, pop-ular and conscientious members, leaving the General Assembly in the control of men more polite in manners and liberal in doctrine, but also more submissive to government authority.

More and more ministers were chosen because they were acceptable to the government, or were able to secure the favour of private patrons. Old Provost Drummond came away dispirited from an audience with the Earl of Ilay on 20 October 1736: 'He is set upon allowing no Minr to come to Edinbr who is of the warm stamp. Were this presbytery, once, all of a piece, I am afraid we shall not then be able to discern Christianity in our pulpit per-formances.'[28]

A shortage of suitable careers had brought into the Kirk many from the laird class, who in the next generation might have served in the army or in the East India Company. They identified with their class and despised the rustic enthusiasm of an earlier gener-ation. It is tempting to generalise Alexander Carlyle from his autobiography, but he may not have been alone in choosing divin-ity over law and medicine for the 'conveniency of a family still consisting of eight children, of whom I was the eldest'.[29] Of the natural historian John Walker, one of the younger generation of ministers, a parishioner complained that 'he spent the week hunt-ing butterflies and made the cure of the souls of his parishioners a bye-job on Sunday.'[30] In addition, the Kirk and the General Assembly attracted those of a legal and political bent who saw it as 'the only field left for Scotch oratory and statesmanship'.[31]

Meanwhile Freemasonry, a collection of quasi-religious rituals separate from the Kirk which had many devotees in the craft incorporations and the Town Council, could not but influence the ministers.

The old visionary ministers, with their vivid Latinate Scots, outlandish metaphors[32] and 'gospel walk', were joined in the General Assembly by a new generation born since the 1690s who bridged the gap between the fervent Presbyterianism of the seventeenth century and the ambitious young men of the College Company: between, as it were, grey homespun and powder and ruffles. These men, whom Robert Wodrow called 'Neu-Lights and Preachers-Legal', included Robert Wallace in Edinburgh and Francis Hutcheson in Glasgow. Among their opponents, the most remarkable was Alexander Webster, minister of the Tolbooth Kirk.

Of Wallace, who became minister at New Greyfriars in 1733, the memoirist Ramsay of Ochtertyre wrote: 'It was the ambition of this excellent man to make the learned, the rich, and the fashionable part of the community pious and devout without foregoing the pursuits of elegance and eloquence.'[33] Robert Wodrow said the same thing when he dismissed Wallace's sermon at the General Assembly of 1730 as 'borrowed from the Spectators [*The Spectator*]'.[34] Now in the Edinburgh University Library, Wallace's private essays, which he copied late in life into tiny notebooks in an all but indecipherable hand, are astonishing in their frank discussion of such subjects as women's sexual affections. At the end of his life he was said to have been working on an essay entitled 'The Art of Dancing'.

Born in Kincardine in 1697, Wallace was 'presented' by an aristocratic patron to the parish of Moffat, a small town in Dumfries where city people resorted for their health to drink the waters or 'the whey' (goats' milk). In the sensational sermon that launched his name and career, preached before the Provincial Synod of Dumfries in October 1729, Wallace argued that the Kirk should make use of reason to repel the assaults of Deists, who substituted the intellect for scripture in attempting to understand God, and sceptics. 'We live in an Age so enlightened,' he said, 'when weak arguments and bad reasonings will not pass so well as formerly.' In conclusion, he prayed that his auditors be preserved alike from 'a Deluge of Scepticism and Deism' and 'implicit Faith and blind Obedience. Amen.'[35] Translated to Edinburgh in 1733, Wallace was tarred with heresy, fell out with Ilay and the Walpole government, and refused to read the Porteous Act. With the fall of Walpole in 1742 he became moderator of the General Assembly, where he presided the next year over a pioneering life insurance scheme for the widows and orphans of clergy.[36]

Wallace himself made the expert actuarial calculations[37] under the supervision of MacLaurin at the College,[38] but the statistician was none other than Alexander Webster, an evangelical who had, in the usual style of nepotic Edinburgh, followed his father into the pulpit of the Tolbooth Kirk in 1737. As chief of the High-Flyers, Webster belied the conventional division of the eighteenth-century Kirk into hidebound evangelicals and modern-minded Moderates, and much other convention beside. A heavy drinker even by the inundated standard of mid-century Edinburgh who contracted a romantic and fashionable marriage, Webster was so magnetic a preacher that it was said 'it was easier to get a seat in the Kingdom of Heaven than in the Tolbooth Kirk'. Drummond attended his induction before the 'Tolbooth Whigs' on 5 June 1737 and heard him preach on Romans 15, 30–32. 'A finer discourse, a more pathetick, a more gracefully delivered sermon ... I never heard in my whole life, a fuller church I never saw us have, and folks of all denominations – even some deists.'[39] As for his drinking, the Scots Magazine's obituary was unusual for a clergyman: 'He had a constitutional strength against intoxication, which made it dangerous in most men to attempt bringing him into such a state; often, when they were unfit for sitting at table, he remained clear, regular and unaffected.'[40] His wife was Boswell's mother's sister, but even Boswell was disgusted by Webster's late hours, as on 8 November 1774, when after the Synod of Lothian and Tweeddale he kept people sitting till 2 a.m. making 'coarse merriment' of other men's sermons.[41] Webster always had his own bottle of claret at the table of Boswell's noto-riously stingy father, Lord Auchinleck.

It was Webster who brought to Edinburgh its great popular celebrity of the mid-century, the English evangelist George Whitefield. Methodism in England and Wales and the religious revivals in New England had their counterpart in Scotland in a return to the kind of field preaching last seen in the persecution years of the later Stuarts. Beginning in 1741, Whitefield made fourteen visits to Edinburgh, and by 1756 was saying grace before the Commissioner of the General Assembly. The era of Moderat-ism, far from being cold and dry and Deistic, could be startlingly passionate, as witness Whitefield's farewell to the city: 'O Edinburgh, Edinburgh, surely thou wilt never be forgotten by me.'[42] ('Zealous, pig-eyed English quack', retorted the scurrilous pamphleteer James 'Claudero' Wilson.[43])

Whitefield was invited to Scotland by the secessionists Ebenezer and Ralph Erskine, but fell out with them when he refused to submit to the Solemn League and Covenant. In the summer of 1741 he was taken up by Webster, who soon had him addressing crowds said to be over fifteen thousand strong in St Giles Churchyard, the Canongate Church, the King's Park and the grounds of the Trinity Hospital. 'Religion in this sinful city revives and flourishes,' one minister recorded.[44] 'Every morning, I have a levée of wounded souls,' Whitefield wrote to his friends on 15 August 1741.[45]

The revival reached its climax at the famous 'Cambuslang Wark' in Lanark the following year. On 15 August 1742, on a green brae scattered with broom and furze and thorn-trees, Webster and Whitefield and ten other ministers preached simulta-neously to as many as forty thousand people gathered from all over the Lowlands, 'bathed in tears',[46] terrified for their own sal-vation. 'Such a Passover has not been heard of,' Whitefield wrote.[47] The Neu-Lichts sniffed at such rustic Enthusiasm while, on the other wing of the Kirk, the Erskines called a day of fasting to confront a work of the Devil. Whitefield took no notice. Back in Edinburgh he preached in the park of Heriot's Hospital, where two thousand seats had been erected, some of them under cover, and 'let out to the best advantage'.[48] Hundreds of pounds were raised in alms for Whitefield's Orphan house in Savannah, Georgia and for the Royal Infirmary being built above the Cowgate. Attempts to outlaw Whitefield from the city churches as a representative of an 'Erastian' – state-controlled – church were easily prevented by Webster.

Even while Alexander Webster was out on the town – drinking enough at the Good Town's expense, it was said, to float a man-of-war – he still managed to be at the heart of plans for the expansion of Edinburgh and to conduct the first systematic census attempted in Britain since the Roman Empire. Encouraged by Robert Dundas the elder, who had succeeded Forbes as Lord President, he reopened the correspondence with the rural clergy which had been established over the Widows' Fund in 1743 and used his prestige to persuade them to count their parishioners (including Papists). The result was 'An Account of the Number of People of Scotland in the Year 1755'.[49] Adam Smith admired Webster's statistical mind and thought him, 'of all the men I have ever known, the most skilful in political arithmetic'; but found him erratic, and even crapulous, and remained sceptical of national statistics even in *The Wealth of Nations*.[50] Still, Webster's figures, which showed a population for Scotland of 1,265,380, were the best available until the census of 1801, which indicated a population of 1,608,000. Webster remained at the Tolbooth Kirk for forty-seven years, and was drawn by John Kay in the act of preaching there to packed galleries. By then, of course, most of the congregation was asleep.

Meanwhile, at Glasgow University, whither Alexander Carlyle went in 1743, liberal sentiments had also been gaining ground under the influence of the thought and personality of Francis Hutcheson. Hutcheson, who succeeded his old teacher Gershom Carmichael in the chair of Moral Philosophy in 1729 and occupied it, turning down a call from Edinburgh, until his death in 1746, attained a European reputation for his philosophy of sensation and benevolence. Profoundly anti-authoritarian, suspicious of metaphysics and of all systems not based on observation and experience, Hutcheson sought a foundation for ethics other than the purported will of God or of the political sovereign, or conceptions of advantage.[51] He found it not in the operations of Platonic reason but in the realm of the senses; and his sensual, hedonistic philosophy appealed not merely to the young Edinburgh clergy but to the great novelists of the period, from Henry Fielding to Walter Scott. The Man of Feeling comes bashfully onto the eighteenth-century stage, which he will leave in a hurricane of tears.

The exceptional elegance of Hutcheson's thought was matched by his good looks and attractive nature, easily imagined from the portrait by Allan Ramsay junior that still hangs in the Glasgow University picture gallery. Hutcheson lectured in English as well as the Latin that was customary. As Carlyle reported, 'when the subject led him to explain and enforce the moral virtues and duties, he displayed a fervent and persuasive eloquence which was irresistible.'[52] Not only students but townspeople flocked to his free lectures on Sunday evenings. His most celebrated pupil, Adam Smith, dubbed him the 'never to be forgotten Dr Hutcheson'.[53]

Grandson of an Ayrshire minister, Francis Hutcheson was born in 1694 at Drumalig in Co. Down and in 1710 went to Glasgow University, where Simson's liberal theology appealed to him. Returning to Ireland in 1718, he was licensed as a preacher, though it is said his first sermon, on the essential goodness of God, went down badly with a congregation accustomed to a scalding with the 'gude auld comfortable doctrines of election, reprobation, original sin, and faith'.[54] Invited to Dublin in or about 1720, he moved in a circle strongly influenced by Anthony Ashley Cooper, third Earl of Shaftesbury, and the polite moralising of The Spectator.

One of the preoccupations of the eighteenth century was virtue. What is a good action, and how is it known to be good? After the assaults of the late seventeenth century, the old Christian certainties were in tatters. For Thomas Hobbes, notoriously, man was entirely selfish. The duc de La Rochefoucauld and the London wits argued that actions not motivated by self-interest were contrary to human nature. Bernard de Mandeville, a Dutch physician settled in London, inaugurating what would now be called an economic analysis of conduct, pronounced in *The Fable of the Bees* that bad actions yielded material amenities and good actions threw people out of work:

> The Root of Evil, Avarice,
> That damn'd ill-natur'd baleful Vice,
> Was Slave to Prodigality,
> That noble Sin; whilst Luxury
> Employ'd a Million of the Poor
> And odious Pride a Million more.[55]

In contrast, Hutcheson believed virtue that needed reward was not worth rewarding. Even self-satisfaction diminished it.[56] The sole virtue was distinterested benevolence. In 1725 he published *An Inquiry into the Original of our Ideas of Beauty and Virtue; in Two Treatises, in which the Principles of the late Earl of Shaftesbury are explain'd and defended, against the Author of the 'Fable of the Bees'*. The book was not only a response to de Mandeville but also an excursion into the polite philosophy proposed in *The Spectator*'s campaign to bring 'Philosophy out of Closets and Libraries, Schools and Colleges, to dwell in Clubs and Assemblies, at Tea-Tables and in Coffee-Houses'.[57] Introducing himself to the reading public, Hutcheson wrote that 'we have made Philosophy, as well as Religion ... so austere and ungainly a Form, that a Gentleman cannot easily bring himself to like it.'[58]

For Hutcheson, the mind was imprinted not just by the objects of sense – what could be seen or heard or tasted or felt – but by objects in the aesthetic and moral orders. Man possessed an internal sense that derived an involuntary pleasure from beauty, consisting (for him) of such values as 'Uniformity, Order, Arrangement, Imitation'.[59] By analogy, an equivalent and superior *moral sense* delighted in benevolent actions without any view to its own advantage. It likewise suffered pain at malevolent actions even those to its advantage. This moral sense distinguished good and evil as infallibly as the supposed aesthetic sense distinguished between the beautiful and the ugly.

He took his epigraph from Cicero's *On Duties*, which itself drew on a beautiful passage in Plato's *Phaedrus*: that if virtue were visible it would inspire in the beholder an overpowering passion.[60] For Hutcheson, good conduct was both beautiful and pleasurable.

The argument proceeded on a largely secular plan, with a strong bias towards a sort of exalted utilitarianism – 'the greatest happiness for the greatest number'[61] – and a great deal of baffling hedonistic algebra.[62] Yet embedded in Hutcheson's thought, as in that of his pupil Adam Smith, were exploded fragments of the Christian system. At the close of the second treatise, Hutcheson introduced a religious component:

> Yea, this very moral sense, implanted in rational agents to delight in, and admire whatever Actions flow from a Study of the Good of others, is one of the strongest Evidences of Goodness in the Author of Nature.[63]

To modern readers, Hutcheson's human nature is much too pretty to be true. What of differences between people in their appreciation of beauty or of right? Or contradictions? Should human feelings be ascribed to the Deity? What about self-interest? In reality, Hutcheson's system recommended itself more to the aesthetic imagination than the reasoning intellect. It was the philosophy of the sentimental novelist. Oliver Goldsmith's *Vicar of Wakefield* and Mackenzie's *Man of Feeling* – in the novels of those names – were drawn to virtue by their generous natures, stumbled through excess of feeling, but, being sound at heart, always corrected their errors. Virtue and beauty and pleasure became incarnate as Woman and Heroine of Romance, from Sophia Weston in *Tom Jones* to Edinburgh's very own Jeanie Deans.

Through the teaching of Hutcheson and his friend the Professor of Divinity, William Leechman, Carlyle wrote, 'greater liberality of sentiment'[64] was introduced into the curriculum:

A new school was formed in the westerm provinces of Scotland where the clergy till that period were narrow and bigoted, and had never ventured to range in their mind beyond the bounds of strict orthodoxy. For though neither of these professors taught any heresy, yet they opened and enlarged the minds of the students which soon gave them a turn for free inquiry.[65]

Ramsay of Ochtertyre saw their influence as more insidious. Hutcheson was 'a zealous propagator of those ethics which are supposed to have evidences and sanctions independent of revelation ... From this, it was no very violent transition to reject or explain away everything which this universal standard of reason did not understand or wished to get rid of. Those pretended discoveries, in conjunction with the spirit of the times, led a number of indiscreet young men into Deism.'[66] He added, with some penetration: 'It gave less offence to charitable, well-meaning people, that the men who were most addicted to philosophy professed the utmost veneration for the Supreme Being, while they spoke the language of virtuous philanthrophy in glowing terms.'[67]

By the late 1740s, the heroes of the College Company were settled in parishes in Edinburgh or out in farmland within an hour's ride. Alexander Carlyle was minister at Inveresk, William Robertson at Gladsmuir, John Home at Athelstaneford in East Lothian. Their friend Hugh Blair was at the Canongate Kirk and gaining a reputation for both eloquence and literary taste. John Jardine, who in 1744 had married Drummond's daughter and won him from the influence of Alexander Webster, was called to Lady Yester's off the Cowgate in 1750. From 1753 the gaunt and uncouth William Wilkie was at Ratho, where he composed epic poetry in English and ploughed his own glebe. These young men were linked not only by their Unionism but by literature and, in many cases, by marriage. All had at least begun as aristocratic presentees, and at Inveresk Carlyle ran into some local opposition on the grounds of his dancing, enjoying the company of his 'superiors', wearing his hat 'agee', and general high spirits.[68] (He was so handsome that he became known as 'Jupiter' Carlyle, after the chief of the Roman Gods.)

In their doctrine, these men sought the middle way of classical antiquity. As Hugh Blair put it later, in a sermon he preached in the High Kirk on a text from Proverbs and later printed under the title 'On Extremes in Religious and Moral Conduct', there was an ideal path through the midst of Presbyterian controversies: between the primacy of faith against the primacy of work, justice against benevolence, severity against over-accommodation, business against retirement.[69] Profoundly shaken by the Forty-five, they expressed their Presbyterianism not in a 'Gospel walk' but in classical learning, polite accomplishments, Continental contacts, a passion for law and order, and an identification with the regime in London. Having seen off the Jacobite challenge, these ministers now turned to deal with Enthusiasm and indiscipline. Their battles won obscure Scots country parishes a temporary celebrity.

The first skirmish occurred in 1751, in the matter of the parish of Torphichen, near Linlithgow, which had rejected Lord Torphichen's candidate, one James Watson, and received no censure from its Presbytery. As Carlyle recounts, 'a select company of fifteen' met on the eve of the General Assembly to consult over what was to be done. Those present – who formed the nucleus of what became the Select Society – included alongside the young ministers George Drummond, now very much in the Moderate camp, and lawyers such as Gilbert Elliot of Minto, a Borders laird and friend of Hume who was about to launch a political career (and dynasty) in London. The lawyers confirmed 'that it was necessary to use every means in our power to restore the authority of the Church, otherwise her government would be degraded, and everything depending on her authority would fall into confusion.'[70] Watson was duly inducted by riding committee on 30 May 1751, but the mild rebuke meted out by the Assembly to the Presbytery incensed Carlyle and his friends.

Meanwhile, the elders of Inverkeithing, a country parish across the Queensferry from Edinburgh, had refused to accept as minister one Andrew Richardson: not, it appears from the reports, for any failure of piety or probity in the man, but because he had accepted appointment or 'presentation' from a laird. The Presbytery of Dunfermline refused to enforce the appointment. A Commission of the General Assembly decided in March 1752 against censuring the Presbytery. The Moderates, led by Robertson, Home, Blair and Jardine, were determined to succeed this time. Robertson was emerging as the group's most skilful politician. In what became known as the Manifesto of the Moderates, *Reasons of Dissent from the judgment and resolution of the Commission, March 11, 1752, resolving to inflict no censure on the Presbytery of Dunfermline for their disobedience in relation to the settlement of Inverkeithing*, Robertson laid down the cold, authoritarian principles that he was to impose on the Kirk for the next generation:

> There can be no society where there is no subordination; and therefore, since miracles are now ceased, we do conceive, that no church or ecclesiastical society can exist, without obedience required from its members, and enforced by proper sanctions.[71]

The counter-blast of the High-Flyers was evidently written by John Witherspoon, who later became President of the College of New Jersey at Princeton and was instrumental in delivering Scots philosophy and education to the new Amercan republic. A descendant of John Knox, he asserted ministers' right to conscience, and reduced Robertson's arguments to Jacobitical absurdity: the Moderates, it seemed to him, would have submitted even to Stuart absolutism.[72]

At the General Assembly of 1752, which opened on 14 May, Robertson and John Home won over the delegates. The Presbytery of Dunfermline were brought to the bar and peremptorily ordered to go straight over the Firth to Inverkeithing and induct Richardson. Only three members agreed, and that was insufficient for a quorum. Among the dissenting members, the Reverend Thomas Gillespie, minister of Carnock in Dunfermline, described in the records as shy, prone to despondency but very conscientious, pleaded scruple and quoted the Assembly's own appeal to London in 1736, which had argued that the Patronage Act was an infraction of the Union settlement.

It was now 22 May. In evident discomfort, the ministers prayed to God for 'light and direction'. It was decided that an example should be set, and one of the dissenting members be ejected from his parish: the choice fell upon Gillespie. On the vote to eject him most of the delegates – 102 – abstained; but a majority of 52 was found to depose him.[73] The next day Alexander Carlyle, making his first appearance at the Assembly, argued for ejecting all the members of the Presbytery. In fact this was designed purely to terrorise the High-Flyers, and indeed 'occasioned a great alarm on the other side'.[74] The matter rested with Gillespie, who accepted his banishment with dignity and, in the time-honoured Presbyterian style, led his congregation out into the fields. In 1761 he formed with Thomas Boston an anti-Patronage church, known as the Relief Presbytery.

The conflict had turned nasty. The next campaign was over not patronage but philosophy.

David Hume is a very eighteenth-century hero. In his early years, at the College in Edinburgh and in provincial France, by the operation of pure reason he plumbed the darkness of intellectual despair that Aikenhead had known in the Court of Justiciary and the Leith Road. Rather than act the part of the heretic or the social monster or the *philosophe maudit* familiar from later and more Romantic ages, Hume withdrew from the 'forlorn solitude'[75] of his advanced philosophic positions, seeking solace in political economy, *belles-lettres*, Edinburgh scenes and the company of 'modest women'.[76] This progress, profoundly *philosophical* in the antique sense, is told in autobiographical fragments of breathtaking drama, and in the hagiographies of Adam Smith.

Hume thought his best thoughts before his twenties were out, and devoted the rest of his life to reproducing them in the fashionable form of short essays and works of history. As Georg Friedrich

Hegel put it early in the next century, Hume tackled the world not as a systematic philosopher but as 'an educated man of the world'.[77] Through his polite works Hume became well-known, and well-off by the standard of mid-century Edinburgh, but was always an object of suspicion – if also, at the end, of affection.[78]

Hume's scepticism at first terrified and always baffled the town. He received no professorship, was assaulted by minor philosophers such as James Beattie, and three years after his death was resurrected as a closet Christian by Henry Mackenzie. When on his death-bed he presented his friend Katharine Mure with a complete set of *The History of England*, she accepted with good grace but begged him to 'burn a' your wee bookies' – that is, his deathless works of sceptical philosophy – before going before his Maker.[79] Burns's printer William Smellie complained of the *Treatise* and of those philosophers 'who involve themselves in clouds of obscurity, and expect their readers to understand what they themselves cannot explain'.[80] In London, Dr Johnson distrusted his politics and hated his religion.[81]

David Hume founded no school and left no successor. Even Adam Smith, who alone fully understood him, did not support him for a professorship in Glasgow and refused to carry out his final wishes. Smith saw in his friend qualities he felt he himself lacked – urbanity, gallantry, philosophic courage – and set out to make him a hero of antique cast. Hume's death-bed was a philosophical raree-show. The Common-Sense school, with its palpable, almost granular reality, benevolent deity and disreputable prejudices, drove Hume out of the university and the drawing-room in both Britain and North America. In the nineteenth century, he was barely read as a historian. It was in the dark twentieth, as men and women rediscovered belief and emotion as capital components of their understanding of the world, that Hume was crowned king of the British philosophers; recently they have found the Humean courage to make consciousness a subject of scientific study.

David Hume was born on the south side of the Lawnmarket on 26 April 1711, the son of Joseph Home, a Berwickshire laird-cum-advocate who died a couple of years later, and Katherine Falconer, daughter of a President of the Court of Session. The family, which was Whiggish in the tradition of Lord President Duncan Forbes rather than the West Bow, was not rich,[82] and as a younger son David's portion would not anyway have been opulent. David was at first intended for the law but at the College, where he matriculated on 27 February 1723, he developed a passion for literature and philosophy. 'I found', he wrote in a model autobiography composed just before his death in 1776, 'an unsurmountable Aversion to every thing but the pursuits of Philosophy and general Learning: and while they fancyed I was pouring over Voet and Vinnius, Cicero and Virgil were the Authors I was secretly devouring.'[83] The first two were lawyers; Cicero was both lawyer and philosopher; but Virgil was an epic poet. As for his religion, as a young man he had once made a list of his vices in a manuscript book,[84] but he later suggested to Boswell that he had lost his faith from reading philosophy at college or soon after.[85]

In the course of his teens Hume surmounted a species of pathological crisis which was decisive not just for his career but for his thought. 'I found a certain Boldness of Temper, growing in me, which was not enclin'd to submit to any Authority,' he wrote.[86] It led him 'to seek out some new Medium, by which Truth might be establisht'. Gradually, the veil began to lift, and when he was about eighteen 'there seem'd to be open'd up to me a new Scene of Thought, which transported me beyond Measure, & made me, with an Ardor natural to young men, throw up every other Pleasure or Business to apply entirely to it. The Law, which was the Business I was design'd to follow, appear'd nauseous to me.'[87]

Living in the Lawnmarket, studying at an intense pitch, he began to succumb to hypochondria. In the revealing letter of 1734 just quoted, addressed to a person who from internal evidence was a physician in London, Hume gave a detailed case history of the breakdown of his equilibrium. 'About the beginning of Septr 1729, all my Ardor seem'd in a moment to be extinguisht.'[88] Fearing that he was merely lazy he redoubled his efforts, only to

exhaust himself further. His reading of the antique writers, with
their 'Reflections against Death, & Poverty, & Shame, & Pain, &
all the other Calamities of Life', made him not Stoical but morbid.
He broke out into spots on his fingers. He salivated. Very uneasy,
in the spring of 1730 he consulted a doctor, who 'laugh't at me,
& told me I was now a Brother, for that I had fairly got the Disease
of the Learned.'[89] The doctor wrote a prescription for 'a Course
of Bitters, & Anti-hysteric Pills',[90] an English pint of claret and a
long ride a day. Relieved to find that he was merely ill, Hume then
began to eat immoderately, before again falling into fatigue.
Unable to concentrate, he despaired of ever delivering his philosophy to the world. To himself, he seemed to be pining away like
some religious mystic.

Resolved to seek out a 'more active Life',[91] Hume took a position
in a merchant house in Bristol, from which he was dismissed after
a few months for correcting his master's English. (Orthography and
pronunciation were David Hume's crotchets.) Yet it seemed to have
repaired his health, for he determined to pick up his studies again,
this time in France, a natural destination for a young Scot with
aspirations to social polish and literary success. Hume stopped in
Paris, where one of the resident Scots reported him as 'full of himself',[92] but the capital was too dear for his purse and he passed on
to Rheims, though not before taking time to ponder the miraculous
cult of the tomb of the Abbé Paris.[93] In 1735 he moved to La Flèche
in Anjou, where he was able to live cheaply, read in the library of
the Jesuit college, and in 'perfect tranquillity' complete his masterpiece. In January 1739, at the age of twenty-eight, he published in
London the first two volumes of *A Treatise of Human Nature:
Being an Attempt to Introduce the Experimental Method of
Reasoning into Moral Subjects* (Book I, 'Of the Understanding'
and Book II, 'Of the Passions'). The price was ten shillings. Book
III, 'Of Morals', was printed in November 1740.

For a man ambitious of 'acquiring a name by my inventions and
discoveries',[94] the response was disappointing, and the two concluding volumes, 'Of Politics' and 'Of Criticism', were abandoned. Hume later used a bitter phrase from one of Alexander

Pope's *Satires*: 'Never literary attempt was more unfortunate than my Treatise of Human Nature. It fell *dead-born from the press* without reaching such distinction, as even to excite a murmur among the zealots.'[95] That was exaggeration. Francis Hutcheson responded with encouragement, though other critics, such as the sarcastic reviewer in the *History of the Works of the Learned*, became bogged down in the discussion of cause and effect in Book I, and found its author callow, dogmatic, conceited, paradoxical, heathen and immoral.[96]

The *Treatise*, which Hume never publicly acknowledged during his life, is indeed a young man's book. 'So vast an Undertaking,' he wrote apologetically to his friend Gilbert Elliot of Minto in 1751, 'plan'd before I was one and twenty, and compos'd before twenty-five, must necessarily be very defective. I have repented my Haste a hundred, & a hundred times.'[97] Yet the same juvenile egotisms of the *Treatise* that irritated the reviewers and haunted Hume in maturity both excite and touch many modern readers: his awed consciousness of his own mental range, his assault on decayed systems of thought, his outrageous self-assurance in presenting a new account of human nature, his raw, almost flayed, sensitivity to impression, his intellectual intimacies. No other eighteenth-century publication is as immediate and direct as the *Treatise*. Apart from certain later insights into history and political economy, it contains all Hume's chief thoughts.

What was the new 'Medium' mentioned by Hume in his letter to the London physician? What was the 'Scene of Thought' the appearance of which ruined the young man's peace of mind? What was this 'Experimental Method of Reasoning'? Only the last question can be answered with certainty. The 'Experimental Method' was none other than the approach favoured by Francis Bacon and Sir Isaac Newton which abandoned the *a priori* arguments of the Middle Ages – in which a given cause produces certain effects – in favour of the most scrupulous observation of phenomena. As MacLaurin put it in a lucid account of Newton's philosophy, the investigator should proceed from 'particular causes ... to the more general ones'.[98]

At the end of the second edition of his *Opticks*, printed in 1718, Newton had thrown down a challenge to general philosophy which the West has yet fully to answer. 'If natural Philosophy in all its parts,' he wrote, 'by pursuing this method, shall at length be perfected, the bounds of moral philosophy will also be enlarged.'[99]

By 'moral philosophy' Newton did not mean merely the discussion of what was right and what was wrong. In antiquity, the science of morals had ranged widely over human activities to extract from them what was predictable and permanent. By the eighteenth century, 'moral philosophy' had become a general heading that covered systematic thinking on marriage and the family, basic jurisprudence, primitive customs, the history of institutions, international relations, religion, aesthetics, ethics, and what came to be called political economy – all separate social sciences today.

Hume's *Treatise* was not designed as an improving lecture, in the manner of Hutcheson. When mildly rebuked by him that the third book of the *Treatise* – 'Of Morals' – lacked 'a certain Warmth in the Cause of Virtue', Hume retorted that he preferred Cicero's *On Duties* to the Presbyterian catechism, *The Whole Duty of Man*.[100] Hume's concern was to establish a science of ethics, but only once the facts of human nature had been established. Unfortunately, few eighteenth-century readers progressed beyond the first volume, 'Of the Understanding'.

What if, in the operations of the mind and the heart, there were some connecting principle? What if there were a sort of law of gravity in humanity, in history, law or political economy? What if there were a single law that plausibly explained the chaos of human phenomena for all time?[101] Hume's 'New Scene of Thought' was, it seems, precisely this search for gravity, or the law of motion – not in the phenomenal world but in the 'very capital or center of these sciences',[102] human nature itself. According to a short essay known as the 'Abstract', discovered by the economist J.M. Keynes and his brother Geoffrey in the 1930s and evidently written by Hume in 1740 in a last effort to drum up interest in his masterpiece, 'It is at least worth while to try if the science of *man* will not admit of the same accuracy, which several parts of natural philosophy [natural science] are found susceptible of.'[103]

Hume believed that the mind, unlike the phenomena of natural science, could be 'perfectly known'.[104] Modern philosophy holds the complete reverse, that the mind is not a natural scientist's laboratory but a badlands, riddled with metaphor and crossed by fugitive images, possibly non-existent, best kept out of bounds. With no knowledge of the brain's physical processes, or how they give rise to subjective experience, Hume had no companion but his own sceptical introspection. His was a desperately unsafe enterprise. What started as a search for causes ended with a denial of all causation.

There seems a strong possibility that in the Edinburgh of the 1720s Hume had glimpses of some stupendous and Promethean 'Theory of Everything'. His failure to achieve it upset his mental equilibrium until, in the benign atmosphere of small-town France, he came gently down to earth. For the rest of his life, Hume's contemporaries always smelled on him the brimstone whiff of his devilish flight over the rooftops. He was lucky to have come after Aikenhead: *rara temporum felicitas, ubi sentire quae velis; et quae sentias, dicere licet* – it is seldom that we are allowed to think what we like and say what we think (Tacitus).[105]

For reasons of prudence, Hume had 'castrated'[106] the *Treatise* of what he called its 'noblest parts', which were also the sections most likely to offend the Christians. Yet what *was* published was sceptical to a degree not seen since antiquity.

In Book I, 'Of the Understanding', Hume argued that the basis of all knowledge is causation – that is, that we know something *is* because it has become so – but that causality itself was not a fact that could be demonstrated. For example, experience demonstrates that a billiard ball when struck by another will move, but does not reveal any necessity in that succession of events; we are merely accustomed to expect that one event will follow the other. Causality is not a principle of the universe but a mental postulate projected onto it, and whatever we imagine, can be. It is experience and custom, not abstract reason, that orders life. Knowledge becomes mere belief, 'something felt by the mind', not the result of a rational process. In other words, Hume extended the operations of sentiment beyond ethics and aesthetics, where Hutcheson had been content to leave them, into the realm of matters of fact.

This picture of the mental processes, as revealed in Hume's discussion of cause and effect, is both preposterous and very powerful. Obviously it makes a nonsense of religion, with its hierarchy of cause and effect, its miracles, and its divinity whose attributes and operations may not be directly experienced. If that shocked Hume's town and age, modern readers are more apt to be troubled by the wholesale assault on reason. Reason, enthroned by Plato, was thoroughly brought to ground by Hume: "Tis not solely in poetry and music, we must follow our taste and sentiment,' he wrote, 'but likewise in philosophy ... When I give the preference to one set of arguments above another, I do nothing but decide from my feeling concerning the superiority of their influence.'[107] Had Edinburgh truly taken Hume seriously, there would have been no Watt, no Black, no Hutton: no steam engine, no latent heat, no *History of the Earth*.

One comes away from 'Of the Understanding' in a sort of daze. There is science no longer, but only the succession or concurrence of particular cases. There is no more philosophy, only faint and fainter impressions of what cannot be known. There are no laws of Nature, merely habitual conjunctions mislabelled as cause and effect. There is no more self, only a chaos of sensations. There is no purpose to existence, merely the supposition that the future will resemble the past.[108] In short, as his opponent Thomas Reid put it, the Humean system 'leaves no ground to believe any one thing rather than its contrary'.[109]

Reid, who succeeded Adam Smith as Professor of Moral Philosophy at Glasgow in 1764, seems to have admired Hume, but was alarmed by the consequences he had drawn. In the work that established his name, *An Inquiry into the Human Mind, on the Principles of Common Sense*, published in his first year at Glasgow, he worked back through the philosophy of his age, not sparing Descartes, Locke, and Berkeley, in a tone of orderly belligerence. 'It is genius,' he proclaimed, 'and not the want of it, that adulterates philosophy, and fills it with error and false theory.'[110] His 'common sense', or 'common judgement', an amalgamation of practical good sense (generally held to be uncommon) and various intuitive principles, was popularised by another Aberdonian, James Beattie, and enjoyed a great vogue; but the suspicion that common sense might be mere common prejudice has always clung to the school he founded. As for Hume's view of Reid, that can only be surmised from their correspondence. His sole judgement on the 500-page *Inquiry* was the correction of one Scotticism.[111] Such straight-faced pedantry was Hume's way of being rude.

Hume had his own antidote to scepticism. In the Luciferian passage that closes 'Of the Understanding', his sceptical monstrosity threatens to overwhelm him:

> The *intense* view of these manifold contradictions and imperfections in human reason has so wrought upon me, and heated my brain, that I am ready to reject all belief and reasoning, and can look upon no opinion even as more probable or likely than another. Where am I or what? From what causes do I derive my existence, and to what condition shall I return?[112]

Abandoned by reason, assaulted by chaotic impressions and hypochondria, the philosopher is rescued from his delirium by the Edinburgh High Street and the parlours of La Flèche:

> I dine, I play a game of backgammon, I converse, and am merry with my friends; and when after three or four hour's amusement, I wou'd return to these speculations, they appear so cold, and strain'd, and ridiculous, that I cannot find in my heart to enter into them any farther.[113]

To advocate a moderation in wisdom, as in Presbyterianism, or wine, or love affairs, is to take the antique philosophy of the golden mean to a ludicrous conclusion. In place of wisdom, Hume now prefers Addisonian urbanity. In the manifesto for his new direction, 'Of Essay Writing,' in *Essays, Moral and Political*, published in 1741/2, he set himself up as 'a kind of resident or ambassador from the dominions of learning to those of conversation'.[114] He enjoyed small parties and paid court to clever women. Soon he was finding the operation of gravity in the international balance of payments.[115]

Was this a tactical withdrawal, or what Hume called in a letter to Henry Home 'cowardice'? Did Hume, in the beautiful phrase of the modern French scholar Paul Hazard, have 'no wish to be crushed beneath the last remaining columns of the temple'? Or did his scepticism, as Hazard continues, 'fail him at the end, telling him that the illusions on which men feed are after all not of such great importance that one must always refuse to share them'?[116] Was Hume just lazy?

From the evidence not just of the letter to the physician but of clues scattered through the works, it would seem that David Hume had a sometimes frantic need for society to restore his equilibrium of mind. As he wrote in the *Treatise*, 'I find myself absolutely and necessarily determin'd to live, and talk, and act like other people in the common affairs of life.'[117]

Hume decided to be happy. Philosophy has never sounded so captivating as in his definition of it to Hugh Blair: 'Reading and sauntering and lownging and dozing, which I call thinking, is my supreme Happiness.'[118] He wanted to cook for his friends. His letters to Adam Smith display an intense affection under a formal restraint. 'Be a philosopher,' he proposed in a new version of 'Of the Understanding' published in 1748. 'But, amidst all your philosophy, be still a man.'[119]

The ban on tartan after the Forty-five was not enforced, it seems, for the men who paid the British government's bills. John Campbell, cashier of the Royal Bank of Scotland in 1749

Topographical views of Edinburgh are scarce before the end of the eighteenth century. This view of the castle from the Grassmarket is by the pioneering Yorkshire watercolourist Francis Nicholson (1753–1844)

Prince Charles Edward Stuart in a pastel done in Paris in 1748 by the French court painter Maurice-Quentin de La Tour. The Prince wears armour and the badge of the Order of the Thistle to commemorate his Scottish adventure and to reassure his supporters that he would return

The philosopher as dandy: David Hume aged forty-three in 1754, by his friend Allan Ramsay. 'I dine, I play a game of backgammon, I converse, and am merry with my friends . . .'

The High Street in the 1790s, just before the shift in social gravity to the New Town

The philosopher as yokel: James Hutton by Henry Raeburn. The curious treatment of the left arm and shoulder somehow adds to the intellectual power of the portrait

By the time Joseph Farington drew this view of Edinburgh, from the Water of Leith to the north-west in 1788, the town was well-established as capital of the sublime and the picturesque

Of the many portraits Raeburn made of Henry Mackenzie, this was the one the painter kept for himself. It shows the Man of Feeling as the very type of a modern Highland gentleman: elegant, gracious, wistful, sensitive, tough as nails

For the frontispiece of the Edinburgh edition of Burns's poems, William Creech early in 1787 commissioned a likeness of the poet by Alexander Nasmyth, an able pupil of Ramsay but with a bent towards landscape. After a couple of sittings, Nasmyth felt he had a likeness and, for fear of spoiling it, left the picture unfinished

'The happiest Marriages,' said David Hume, 'are found where Love, by long Acquaintance, is consolidated into Friendship.' Raeburn's stupendous double portrait of fifth baronet Sir John Clerk of Penicuik and Rose Mary D'Acre, Lady Clerk, might have been painted in illustration

Duncan Forbes, who wore out himself and his fortune for the Hanoverians in the Forty-five. The portrait, most likely by Jeremiah Davison, shows Forbes in the robes of a justice of the Court of Session

As official draughtsman to the Military Survey sent to Scotland after the Forty-five, Paul Sandby brought a tactical eye to this sketch of Edinburgh Castle from the east, but already there are hints of a picturesque approach

Edgar's plan of Edinburgh, engraved in 1742, shows a town still confined by its walls. It is the most accurate picture of the city on the eve of its great expansion

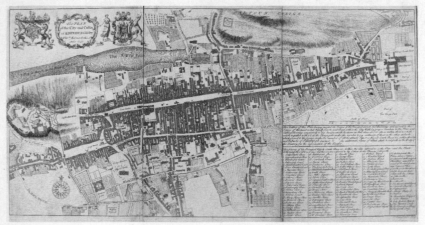

'PEACE to thy shade, thou wale o'men . . .' George Drummond, six times Lord Provost, creator of modern Edinburgh

Engraving by P. Fourdrinier of Sandby's drawing of the north front of the Royal Infirmary. The building, by William Adam, is essentially a standard Ordnance Board barracks altered to accommodate an operating theatre on the top floor plus columns, pediments and a niche for statuary to flatter the charitable subscribers to the scheme

James Craig, surrounded by his plans for the Edinburgh New Town, painted by David Allan, probably in the 1770s

Heavenly city of the philosophers or suburbs on the cheap? Engraving of the version of Craig's plan that was accepted by the Town Council on 29 July 1767

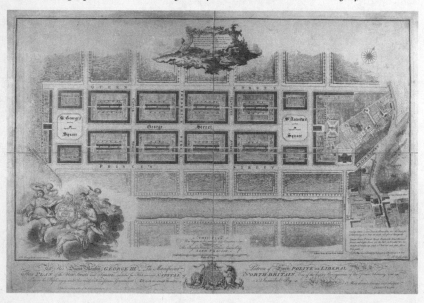

Faden and Jeffrey's 1774 plan clearly shows the new residential districts to the south of the Cowgate and the New Town creeping westwards from St Andrew's Square

The North Bridge as seen across the bed of the North Loch, before the creation of Princes Street Gardens and the arrival of the railway. The building of the bridge was a catalogue of disasters, but without it there would have been no New Town

To have lured Dr Johnson to the howling wastes of Edinburgh and
the Highlands was a spectacular literary coup for Boswell, and
Thomas Rowlandson makes the point

Edinburgh as geological laboratory. Clerk of Eldin chose a viewpoint from the
north-east to show the majestic Salisbury Crags dwarfing the capital of
Scotland into insignificance

The General Assembly meets in the High Kirk in May 1783. Among the ministers and laymen present are Principal William Robertson, Alexander Carlyle and old Alexander Webster (facing left). James Boswell is speaking

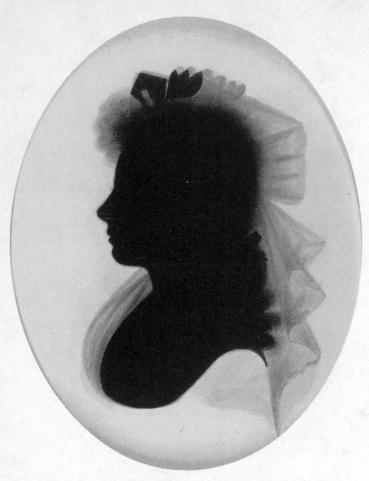

'I thank you for going to Miers . . . I want it for a breast-pin,
to wear next my heart.' Silhouette of 'Clarinda' done for
Burns in Edinburgh in February 1788

4

The Philosopher's Opera

The failure of his *Treatise* to provoke even the bigots of England and Scotland did not long trouble David Hume. 'Being naturally of a cheerful and sanguine temper,' he later wrote in the biographical sketch known as 'My Own Life', 'I very soon recovered the blow, and prosecuted with great ardour my studies in the country.'[1]

Persuaded that the book's length and philosophical ambition had told against it, he decided to follow the path blazed by Joseph Addison and Richard Steele in London in the early years of the century and try his hand at the short literary exercises known as essays. Living with his mother and brother at Ninewells, Hume was soon exchanging philosophical papers with his older neighbour and remote kinsman, the able if overbearing lawyer Henry Home. Gathered up as *Essays, Moral and Political*, printed this time in Edinburgh (by Kincaid) in two volumes over the winter of 1741/2, the 'work was favourably received, and soon made me forget my former disappointment'.[2]

Here, then, were the two unwritten books of the *Treatise*, 'Of Politics' and 'Of Criticism', in morsels. Addison he was not. The leaden gallantries in 'Of Essay Writing' or 'Of Love and Marriage'

must have fallen heavily on the ears of eighteenth-century
London: commerce between the sexes in Edinburgh was perhaps
as yet too stilted for the Addisonian manner. More happily, from
the first sentence of the first essay, 'Of the Delicacy of Taste and
Passion', Hume set out to convey the psychology of the *Treatise* as
masculine conversation. The essays, as those of Francis Bacon,
give a sense of a very powerful mind at leisure. And Hume had a
manner all his own, learned yet not academic, virtuous without
priggishness, forward-looking, balanced in both style and argument.

Above all, Hume was free from the entanglements of 'Party-
rage' or social allegiance. If he applauded the English system of
politics, with its unscientific mixture of monarchy and republicanism, it was not from complacency, chauvinism, or misty-eyed
idealism. In essays such as 'Of the Liberty of the Press', he portrayed Britain as a precarious equilibrium of often disreputable
forces – Court patronage, parliamentary corruption, a free press,
commercial competition – that were the residue of the violent
political conflicts of the seventeenth century. For all his Scottish
origins and friendships, he had no time for Whig or indeed any
ideology: there was rarely, he later wrote, any 'philosophical
origin to government'.[3] The British constitution was for Hume the
product of violence, and its form was both unintended and precarious. It was also, as might have been expected, civilian: a creation, as he also later wrote, of 'that middling rank of men, who
are the best and firmest basis of public liberty'.[4]

Having regained his literary poise, Hume now sought public
employment. When in 1744 John Pringle, Professor of Ethics and
Pneumatical (spiritual) Philosophy at the College, requested an
extension to his leave of absence serving as a physician with the
army in Flanders, in exasperation Provost Coutts asked Hume to
stand. An innocent in the ferocious party politics of Kirk and
Town Council, Hume was out of his depth. Pringle had no wish
to forego either of his salaries, while William Cleghorn, who was
now teaching Pringle's classes (and who was to march with the
College Company in 1745), stood in the way. Archibald Stewart,

elected Lord Provost on 7 November 1744, favoured Hume but his influence on the Council was under threat from the Drummond Whigs. In March 1745 Pringle at last resigned, and on 10 April 1745 the Council sought an *avisamentum*, a ruling on the candidates, from the ministers of the town churches. William Wishart, Principal of the College, bent himself to extracting unsound propositions from Books I and III of the *Treatise*, and these were soon in circulation.

At a loss, Hume had that February as a temporary measure accepted a post as tutor in the English household of the Johnstone of Annandale family, at Weldhall near St Albans in Hertfordshire. A tutor's position was not always dignified, and Hume was unfortunate in his charge: the young Marquess of Annandale was known to be eccentric, and soon began to display symptoms of depression and violence. In such degrading circumstances, and without a copy of the *Treatise* to hand, Hume sought to rebut Principal Wishart's charges in an open *Letter* published in 1745. He seemed to have lost much of his confidence in the book by this point, and showed more vigour in listing the complaints against him – 'Universal Scepticism', 'Principles leading to downright Atheism', 'Sapping the Foundations of Morality' – than in refuting them. At the close, he observed that 'the Author [himself] had better delayed the publishing of that Book', by reason not of its dangerous principles but of its imperfections.[5] In the *avisamentum* a large majority of the ministers spoke against him. The shining exception was Robert Wallace, who did not see why Hume's abilities should be overshadowed by a juvenile and anonymous work. Hume withdrew his candidacy, and on 5 June Cleghorn was named professor *ad vitam aut culpam*. *Et*, it might be added, *oblivium*.

Despite this rebuff, and new difficulties with the Annandale household over his salary, Hume was not idle. It was during this period, it seems, that he composed a new version of the first book of the *Treatise*, longer than the 'Abstract', and both impersonal and elegant, which appeared in 1748 as *Philosophical Essays Concerning the Human Understanding*. He also planned the

History of England. In the spring of 1746 he was dismissed from his tutorship and escaped into the world of action, becoming secretary to a kinsman, General James St Clair, and acting as Judge-Advocate during a chaotic attack on Port Lorient in Brittany in the dying throes of the War of the Austrian Succession. In the next two years he served as the general's *aide-de-camp* on embassies to Vienna and Turin. In the course of these uncharacteristic peregrinations Hume read more than he wrote – including the baron de Montesquieu's influential *De l'Esprit des Lois* – witnessed military and political incompetence, courage and cowardice, put on weight, discovered a taste for whist and the company of Italian women, and became, as he noted modestly, 'Master of near a thousand Pounds'.[6] On his return, he went back to Ninewells and composed a digest of the third book of the *Treatise*, now called *An Enquiry Concerning the Principles of Morals*, which was published in 1751, followed by a series of essays of a more political and economical character, published in 1752 as the *Political Discourses*.

By now, Hume could afford to live in town. 'In 1751,' he wrote in 'My Own Life', clearly with intense satisfaction, 'I removed from the Country to the Town; the true Scene for a man of Letters.'[7] That he should have chosen Edinburgh as the setting for his literary activity was not a foregone conclusion: Scots writers in English such as James Thomson in the 1720s and Tobias Smollett in the 1730s, had decamped to London. MacLaurin had died in 1746 and Forbes a year later – exhausted, it was said, by their exertions during the Rebellion – and with their passing the mantle of literary Edinburgh had fallen on Henry Home and Lord Elibank. Home, a Jacobite until the 1730s, had spent the Rebellion quietly at his estate working on his *Essays on Several Subjects concerning British Antiquities*. This included attacks on Jacobite doctrine and broke the philosophical ice in Edinburgh on its publication in 1747.[8] In 1751 Home indulged his passion for metaphysics in his *Essays on the Principles of Morality and Natural Religion*. An industrious, clever, coarse, angular and domineering man, with the beaky look of a Scots Voltaire, he was

to shape the careers of Edinburgh men and women for a generation. Patrick Murray, Lord Elibank is one of those frustrating historical figures, much-liked but diffident or idle, who appear in all
the memoirs but leave few literary memorials: in Elibank's case,
little essays on paper currency, the national debt, entails and the
Scottish peerage.

Edinburgh was not necessarily Hume's first choice. He made a
glancing attempt to secure the chair of Logic at Glasgow, though
he must have known he was too notorious for that town. Even
Adam Smith, the outgoing professor, did not support him.[9] His
consolation was old Thomas Ruddiman's post of Keeper of the
Library of the Faculty of Advocates in Parliament Close, though
it lay less, it seems, in the salary of £40 a year than in the 20,000
or more volumes he was able to browse for his *History*. It is evident that Hume wanted to create out of Ruddiman's crabbed
repository a modern or 'polite' library, and he soon quarrelled
with the curators over the books he was ordering from London.
Racy French novels and *belles-lettres* were too much for the
churchy antiquarian Sir David Dalrymple (later Lord Hailes) and
the classical purist James Burnett (later Lord Monboddo).[10]
Hume was furious, and though he stayed on as Librarian until
1757, he gave his salary to the blind poet, Blacklock.[11] (Son of a
bricklayer in Annan, Dumfriesshire, Thomas Blacklock had lost
his sight to smallpox at six months. Sent to grammar school at the
late age of twenty to learn Latin, he published his *Poems on
Several Occasions* in 1746. He later played a key role in 'discov-
ering' Burns.)

Hume lived at first in Riddel's Land, a six-storey mansion which
survives in the Lawnmarket, near the head of the West Bow, but
moved at Whit Sunday, 1753[12] to Jack's Land, described by
Chambers as a 'somewhat airier situation', in the Canongate.[13] In
this nondescript building (knocked down in the early twentieth
century) Hume spent the next nine years writing his *History*, until
in 1762 his growing prosperity took him to James's Court in the
Lawnmarket and then, in 1771, to the New Town.

The first volume of Hume's *History*, which covered the reigns of the first two Stuarts, was published by Hamilton, Balfour and Neill in 1754 as *The History of Great Britain: The Reigns of James I and Charles I*. Baillie Hamilton paid no less than £400 for the copyright. A second volume, taking the story up to the Revolution of 1688, appeared in December 1756. Two volumes on the Tudors followed in 1759. Two further volumes covering the history of England from Julius Caesar to the time of Henry VII completed the series in November 1761. Hume's bizarre scheme of composition did not go unremarked: Richard Hurd, a sharp critic of Hume's religious and political writings (who rose to be Bishop of Worcester), said that he wrote his history 'as witches use to say their prayers, backwards'.[14]

Yet there is nothing surprising about Hume's point of departure. For a philosophical historian looking out of Jack's Land, both Stuart absolutism and the combination of religious fervour and political self-assertion that destroyed it were living monuments. Hume's view of the British constitution, as not some primordial set of liberties such as Montesquieu in the *Esprit* or the English Whig historians had painted it, but a creation of the bitter civil and religious conflicts of the seventeenth century, has long been accepted.

Though the *History* eventually became as popular as any general history before Edmund Gibbon's, the first volume sold poorly. Hume wrote in 'My Own Life' that he was assailed from every side: 'English, Scotch, and Irish, Whig and Tory, churchman and sectary, freethinker and religionist, patriot and courtier, united in their rage against the man, who had presumed to shed a generous tear for the fate of Charles I.'[15] Such criticism might be taken as tribute to his impartiality. James and Charles encompassed their own downfall through obstinacy, but the 'enthusiasm' and 'fanaticism' of the Scots Reformation was worse than anything 'during the darkest night of papal superstition', according to Hume. His picture of the mental state of the enthusiastic Presbyterian is as brilliant as anything in the *Treatise*:

The mind, straining for these extraordinary raptures, reaching them by short glances, succumbing again under its own weakness, rejecting all exterior aid of pomp and ceremony, was so occupied in this inwards life, that it fled from every intercourse of society, and from every sweet or cheerful amusement which could soften or humanize the character.[16]

Nothing could so well describe the bugbear of the young Moderates. My enemy's enemy is my friend: no wonder the young clergy flocked to *le bon David*.

Hume became bolder over the years. In the *Treatise* he had attacked rational defences of religion. In the *Philosophical Essays* of 1748, now known as *An Enquiry Concerning Human Understanding*, he launched a full assault on 'revealed' or scriptural religion. In the well-known chapter 'Of Miracles', he discussed the miracles attributed to the Abbé de Paris at the time of his first visit to Paris, arguing that it was quite unreasonable to believe in those violations of natural laws, such as reports of miracles and prophecies, that were the foundations of revealed religion. (Although these 'natural laws' are, for Hume, established by experience, that does not weaken his reasoning: the argument against miracles is simply as strong as any argument from experience can be.) The chapter ends with a sentence worthy of Voltaire:

> So that, upon the whole, we may conclude, that the *Christian Religion* not only was at first attended with miracles, but even at this day cannot be believed by any reasonable person without one.[17]

In the following chapter Hume used dialogue to attack the well-known proof of God's existence, dear to Francis Hutcheson and to Colin MacLaurin in his account of Newton, from so-called Design: that the universe reveals such marks of intelligence or benevolence that it is perverse to ascribe its formation to chance or blind matter.[18] His essay of the same year, 'Of National Characters', included a quite gratuitous attack on priests.

In the course of the 1750s Hume composed most of the *Dialogues concerning Natural Religion*, his most profound and systematic thinking on faith, but prudently suppressed them for the time being; but in 1757 in his 'Natural History of Religion', the first of *Four Dissertations*, he published his sideswipe at instinctive or what he termed 'vulgar' religious belief. Hume found religion arising in idolatry and ending in superstition, but these 'spectres of false divinity' vanished before a 'manly, steddy virtue' and – beautiful phrase – the 'calm sunshine of the mind'.[19] Yet the phrase that was to become notorious in Edinburgh was the one that ended the piece: religion was not susceptible to reason but was 'a riddle, an aenigma, an inexplicable mystery'.[20]

As to Hume's sincere feelings on religion, they are clearer now than they were to his contemporaries. During the last summer of his life, in 1776, he amused himself and his friends by imagining a conversation with the infernal boatman, Charon, in which he begged for a little longer in the light so that he could have the pleasure of 'seeing the churches shut up, and the Clergy sent about their business'.[21] Hume had no sympathy for the religious attitude and not much knowledge of existing faiths. (One would not know from him, for example, that George Sale's meritorious translation of the Koran had been available for twenty years.) His tone could be scathing, crudely anti-clerical, abusive, defiant, or obviously insincere. Yet to the end – and beyond, in the *Dialogues concerning Natural Religion* published at his behest by his nephew in 1779 – Hume continued to feign a prudent theism. Even today, there are doubts about which character in the *Dialogues* most nearly reproduces their author's sincere feelings. It is possible to surmise that Hume probably accepted, with the sceptical character Philo, 'That the cause or causes of order in the universe probably bear some remote analogy to human intelligence';[22] but only because they are investigated by that intelligence.

Hume's model, it appears, was really Cicero, for whatever 'sceptical liberties' the Roman orator might have used in his writing or his philosophical conversation, he avoided in the common conduct of his life any imputation of profaneness.[23] Above all, Hume was determined to break the traditional link between moral conduct and the system of rewards and punishments established by scripture. As he wrote to his friend James Oswald in connection with the *Philosophical Essays* in October, 1747: 'I see not what bad consequences follow, in the present age, from the character [reputation] of an infidel; especially if a man's conduct be in other respects irreproachable. What is your opinion?'[24]

Nor did such attitudes prevent him seeking out the company of clergymen. According to Alexander Carlyle, Hume did not attempt to convert the young ministers, but merely wished to enjoy the only literary society on offer in Edinburgh. In his lodgings Hume, who used to boast of his 'great talent' for cookery, gave suppers of roast chicken and minced collops – eighteenth-century hamburgers – with a bowl of punch. Whenever the out-of-town ministers – John Home, William Robertson, Hew Bannatine, he himself – rode in, Carlyle said, they would send by caddies to assemble their friends in a tavern 'and a fine time it was when we could collect David Hume, Adam Smith, Adam Ferguson, Lord Elibank, and Drs Blair and Jardine, on an hour's warning.'[25] Hume used to take the apartment key from Jack's Land with him, so that he did not have to wake his lass, Peggy Irvine, when he got back at one in the morning.

At this time Hume was painted by Allan Ramsay junior, son of the poet, in a crimson cap and brown coat with flowered waistcoat. While there is a hint of dissipation about the face, it is far too intelligent to be that of the 'Turtle-eating Alderman' the youthful Irish peer Lord Charlemont encountered at Turin.[26] He looks very much like the hero of the *Treatise*, who has learned that in society he is 'in every respect more satisfied and happy, than 'tis possible for him, in his savage and solitary condition, ever to become'.[27] His peace of mind was about to be tested.

━━◆━━

In 1753 there appeared anonymously at Glasgow a little pamphlet entitled *Ecclesiastical Characteristics: Or, The Arcana of Church Policy, being an humble attempt to open up the Mystery of Moderation*. Priced at sixpence, it consisted of a set of facetious Maxims, reminiscent of La Rochefoucauld, in ironic defence of the 'Moderate Man'. The author was John Witherspoon, who had roomed with Alexander Carlyle at the College but had developed in the opposite ecclesiastical direction: in later years he abandoned Moderate Scotland for New Jersey, and signed the American Declaration of Independence.

In his pamphlet, which demonstrated that the Moderates had no monopoly of wit or sense, Witherspoon ridiculed the preaching of the Neu-Licht men, their classical learning and Hutchesonian aestheticism, their affectation of politeness and good breeding, their submission to wealthy patrons, and their unpopularity with congregations used to the 'pathetic way of raising the passions'.[28] His characters are as clear and sharp as John Kay's drawings of the next generation:

> Maxim IV: A good preacher must ... have the following special marks and signs of a talent for preaching. 1. His subjects must be confined to *social duties*. 2. He must recommend them only from *rational considerations*, viz. the beauty and comely proportions of virtue, and its advantages in the present life, without any regard to a future state of more extended self-interest. 3. His authorities must be drawn from *heathen writers*, NONE, or as few as possible, from *Scripture*. 4. He must be very *unacceptable* to the common people.[29]

More abusive was a long work that appeared later the same year from the High-Flying camp, entitled *An Estimate of the Profit and Loss of Religion, Personally and publicly stated: Illustrated with References to Essays on Morality and Natural Religion*. The chief target, as was evident from the way the title echoed that of his work, was Henry Home, under the sophistical pseudonym of 'Sopho', but 'his assistant, *David Hume, Esq.*', was not forgotten. The fact that the two men were the philosophical equivalent of chalk and cheese was quite lost on the author, the Reverend George Anderson, last met penning pamphlets against the stage in 1733, who had just retired as Master of George Watson's Hospital, a little short of eighty years old.

In his *Essays on the Principles of Morality and Natural Religion* of 1751, Henry Home had sought to rein in his protégé David Hume and refute the section 'Of Liberty and Necessity', in Book

II of the *Treatise*, in which Hume ingeniously argued that the 'free will' so prized in Christian theology was little more than constant conjunction, though in this case of action and motive rather than of two billiard balls. Henry was not a sceptic, like David, but the opposite: a Scots dogmatist of familiar type who could not rest till he had found a specious certainty. He set racing all manner of philosophical hares, all more or less unorthodox, before concluding that Free Will was a sort of delusion given mankind by a benevolent God. It was a bizarre position for an elder of the Kirk, let alone a Scottish judge (elevated to the bench as Lord Kames in 1752). His arguments did not trouble Hume, who wrote affectionately of 'our Friend Harrys Essays' to his friend Michael Ramsay on 22 June of that year. Generous as always, Hume merely feared the work would get Kames into trouble.[30]

In truth, what interested both Witherspoon and Anderson was that Kames and Hume were friends and allies of the Moderate ministers, and their soft doctrinal underbelly. Anderson unveiled his purpose at the end of his long tirade, where he called on the General Assembly of the Church to exclude such men 'from their fellowship, not only in sacred things, but likewise from all unnecessary conversation upon other subjects'.[31]

Anderson's sectarian appeal had its answer in the midst of the General Assembly of 1754. On 22 May fifteen ministers and laymen met at the Advocates' Library to form a debating club that became known as the Select Society. The instigator of this plan to bring together the best of Kirk, College and Parliament House (as well as some of the nobility) was the painter Allan Ramsay. David Hume was the host, while Adam Smith presented the proposals. Among the ministers were John Home, William Robertson, Blair, and Carlyle; the lawyers included the pious and prolific Sir David Dalrymple and Andrew Pringle; while Dr Monro and Dr John Hope, later Professor of Botany, represented the College. The man elected president or *preses* was one of the youngest members, Alexander Wedderburn, aged twenty-one, who had passed advocate only that year and was already making a name for himself in the High Street and the Outer House:

Men then wore in winter small muffs, and I flatter myself that as I paced to the Parliament House, no man of fifty could look more thoughtful or steady. My first client was a citizen whom I did not know ... I asked him, 'how he came to employ me?' The answer was, 'Why I had noticed you in the High-street going to Court – the most punctual of any as the clock struck nine, and you looked so grave and business-like, that I resolved from your appearance to have you for my advocate.'[32]

Judging from the minutes of the Society, Wedderburn and his cockiness were much indulged.[33]

The original members, limited to fifty, were to meet in the Advocates' Library every Wednesday evening during the Court Session, from six to nine o'clock, for a debate and informal conversation. No topic for debate was outlawed except 'such as regard Revealed Religion or which may give occasion to vent any Principles of Jacobitism'.[34] Topics noted in the Society's surviving records breathe the spirit of self-conscious Improvement: 'Whether Brutus did well in killing Caesar?' – 'Whether presentation by patrons, or election by the parishioners is the best mode of settling ministers?' – 'Whether the difference in national character be chiefly owing to the nature of different climates, or to moral and political causes?' – 'Whether ought we to prefer ancient or modern manners, with regard to the condition and treatment of women?' Though it is reported that David Hume was silent at the meetings, some of the chosen topics reflect the economic preoccupations of the *Political Discourses*: 'Whether the establishment of Banks in Scotland has increased wealth?' – 'Whether the bounty should be continued on the exportation of low-priced linen made in Scotland?'

Nor did Adam Smith, it seems, ever make a formal speech. Possibly the Society was always too large for these two philosophers. The membership expanded to sixty in the first summer and to over one hundred and twenty within five years. The Select Society had long outgrown the Advocates' Library, and its programme to promote correct southern English pronunciation was the laughing-stock of the town. It expired, unmourned, in 1763.

Wedderburn, an impetuous and unscrupulous man who later became Lord Chancellor, was the prime mover of the other great literary scheme of the mid 1750s, *The Edinburgh Review*. In the preface to the first issue, which came out in July 1755, Wedderburn set out a Unionist view of Scots history that has survived more or less intact to the present day. The revival of Scottish Latin literature in the sixteenth century had been a false dawn. Scottish culture had fallen prey to the dissensions of the seventeenth century. Fortunately,

At the Revolution, liberty was re-established, and property rendered secure; the uncertainty and rigor of the law were corrected and softened; but the violence of parties was scarce abated, nor had industry yet taken place. What the Revolution had begun, the Union rendered more compleat. The memory of our ancient state is not so much obliterated, but that, by comparing the past with the present, we may clearly see the superior advantages we now enjoy, and readily discern from what source they flow. The communication of trade has awakened industry; the equal administration of laws produced good manners; and the watchful care of the government, seconded by the public spirit of some individuals, has excited, promoted and encouraged, a disposition to every species of improvement in the minds of a people naturally active and intelligent. If countries have their ages with respect to improvement, *North Britain* may be considered as in a state of early youth, guided and supported by the more mature strength of her kindred country. If in any thing her advances have been such as to mark a more forward state, it is in science. The progress of knowledge depending more upon genius and application, than upon any external circumstance; where-ever these are not repressed, they will, exert themselves. The opportunities of education ... in this country ... ought to make it distinguished for letters.[35]

Scotland still lacked a 'standard of language' and a tradition of good printing, but those last obstacles were now being cleared.

In reality, the *Review* was both provincial and premature. That year 1755 saw, beside any number of ranting threepenny and six-penny pamphlets, the publication of just two Scottish books of note: Hutcheson's *System of Moral Philosophy* and, from the Episcopalian Ruddiman press, Bishop Keith's *Bishops of Scotland* (Webster's brilliant 'Account of the Number of People in Scotland in the year 1755' was not printed). Hume did not contribute to the *Review*; nor, it seems, did he even know his friends were behind the magazine. At the close of the second number Adam Smith proposed that it should also cover Continental literature; but few could have followed him there, and the first *Edinburgh Review* folded after just two issues.

None of the articles was signed, but surviving annotated copies show who wrote what. John Jardine and Hugh Blair reviewed the books of divinity, making fun of the homely style of the Popular preachers. In the review of a volume of sermons preached at the Tolbooth in March by Thomas Boston the Younger, son of the rediscoverer of the *Marrow of Modern Divinity*, Jardine quoted this passage on the Incarnation: 'He wanted a body, a *suit* of flesh and blood, such as divine nature never *wore* before; and God him-self was at the whole *cost* of making it.' The authentic breath of the West Bow was too much for Jardine: 'such vulgarisms as these, are indecent, even in conversation, but much more so in a solemn discourse from the pulpit.'[36] Meanwhile Blair praised Hutcheson, and also William Robertson's sermon in January to the Society in Scotland for Propagating Christian Knowledge, with its call for a polite crusade: 'Christianity not only sanctifies our souls, but refines our manners; and while it gives the promise of the next life, it improves and adorns the present.'[37]

The High-Flyers were also girding up. On 23 May, the second day of the General Assembly, there appeared *An Analysis of the Moral and Religious Sentiments Contained in the Writings of Sopho, and David Hume, Esq; Addressed to the consideration of the Reverend and Honourable Members of the General Assembly of the Church of Scotland*. Attributed both to Anderson and, with more probability, to the Reverend John Bonar of Cockpen, the paper excerpted eleven objectionable propositions from

Kames, and six from Hume. In the *Treatise* and the *Essays, Moral and Political*, Hume, it seems, maintained that:

1 All distinction betwixt virtue and vice is merely imaginary.
2 Justice has no foundation further than it contributes to public advantage.
3 Adultery is very lawful, but sometimes not expedient.
4 Religion and its ministers are prejudicial to mankind, and will always be found either to run into the heights of superstition or enthusiasm.
5 Christianity has no evidence of its being a divine revelation.
6 Of all modes of Christianity Popery is the best, and the reformation from thence was only the work of madmen and enthusiasts.[38]

As a charge sheet, it was not at all bad. Only 3 is manifestly unfair, and as for 6, Hume may well have felt Roman Catholic 'superstitions' to be preferable to Protestant 'enthusiasm'. It was only towards the end that Bonar, if he it was, unveiled his hidden battery. How could the members depose Gillespie for defying the Assembly's authority in the Inverkeithing affair while accepting as a member of the Assembly an Elder – Kames – who defied God's authority, and while 'some of you at least live in the greatest intimacy with one who represents the blessed Saviour as an impostor, and his religion as a cunningly devised fable.'[39]

The fate of the two philosophers was referred to the Committee of Overtures which was responsible for submitting resolutions to the General Assembly, where, thanks to Robertson, the problem expired. On 28 May the Committee proposed a bland and anonymous condemnation of 'several books published of late in this country'.[40]

That was merely a truce. As Hume wrote to Allan Ramsay in June: 'My friends ... prevailed, and my damnation is postponed for a twelvemonth. But next Assembly will surely be upon me.'[41] Meanwhile, Bonar had his answer on 6 June, four days after the session rose. 'Observations upon an Analysis of Sopho and Hume', probably written by Hugh Blair, published by the *Scots Magazine*, conceded that there were in the *Treatise* 'some principles by no means consistent with sound doctrine'; but these, Blair suggested, made it all the more unreasonable for Bonar to ascribe to Hume 'positions he does not advance'.[42] Blair's chief argument was that deployed by Robert Wallace in his sermon at Dumfries long before, in 1729, that without freedom of conscience there might have been no Church of Scotland:

> By means of free inquiry, the Church of Scotland was originally established. In this country, therefore, all attempts to infringe so valuable a privilege, in cases where the peace of society is not concerned, must ever be regarded with concern by all reasonable men. The proper objects of censure and reproof are not freedom of thought, but licentiousness of action; not erroneous speculations, but crimes pernicious to society.[43]

Naturally enough, the *Edinburgh Review* noticed not only Blair's 'observations' but also, and unkindly, Bonar's original, and a half-mad attack on Hume by one Andrew Moir, a former divinity student sent down from the College and excommunicated from the Kirk, with the gruesome title *The Deist Stretch'd on a Death-bed*.

As Edinburgh prepared for the General Assembly of 1756, there was an omen: the 'Cross at Edinburgh', which had been the chief resort for business and society since the Middle Ages, was demolished by order of Drummond's modernising Town Council. 'As soon as the workmen began, which was in the morning of March 13,' reported the *Scots Magazine*, 'some gentlemen who had spent the night over a social bottle, caused wine and glasses carried thither, mounted the ancient fabric, and solemnly drank its dirge. The beautiful pillar, which stood in the middle, fell, and broke to pieces, by one of the pullies used on that occasion, giving way.'[44]

On 20 May the General Assembly convened in an atmosphere made gloomier by news of the outbreak of hostilities with France that marked the beginning of the Seven Years' War. Members' satchels were heavy with pamphlets on the Infidel Writers: a fresh blast from Anderson, *Infidelity a Proper Object of Censure*, and various more or less injured responses to the first Stuart volume of Hume's *History*. The Moderates established their caucus for the duration of the General Assembly at the Carriers' Inn, at the bottom of the West Bow. This obscure little tavern had only three

rooms and stabling for a mere half-dozen horses, but Nelly
Douglas, the landlady, was both handsome and respectable and her
husband Thomas Nicolson, though a coarser article altogether,
had been known to extend credit. The place was given the medieval Latin term for lodging, *Diversorium*, Nicolson was ordered to
lay in twelve dozen bottles of claret at eighteen shillings a dozen,
and a party was soon under way. 'The attempt to be private', said
Carlyle with satisfaction, 'made it the more frequented.'[45]

Learning from their frustration in 1755, the High-Flyers had
altered their tactics. To avoid the indecorum of naming a supreme
court justice – Kames – they confined their assault to Hume alone,
who had openly avowed himself 'the author of books containing
the most rude and open attacks upon the glorious Gospel of
Christ'.[46] On 28 May they presented a written overture to appoint
a Committee to summon Hume before the spritual courts.
Attending as an Elder of Dunfermline, Wedderburn sprang up to
demand that Hume's name be dropped from the overture. His
maiden speech gave a taste of the insolence, ruthlessness and sarcasm he later brought to the Woolsack.

He demanded to know whether every man now preparing to
'crush Mr Hume with the censures of the Church' had read the
writings to be condemned?

> Am I to believe that the holy presbyters, trusted with the care of souls
> of which they are to give an account, instead of preaching, praying,
> and catechising, have been giving up their days and nights to Mr
> Hume's 'Treatise on the Human Understanding', and to his 'Essay on
> Miracles', and to 'Cause and Effect' – writings said to be so poison-
> ous and so pernicious – in neglect of the spiritual good of others, and
> possibly to the peril of their own principles ... Can you all tell us the
> difference between coincidence and causation?

And what was the point of condemnation?

> The opinions complained of, however erroneous, are of an abstract
> and metaphysical nature – not exciting the attention of the multitude
> – not influencing life or conduct...

Suppose the General Assembly were to pass the sentence of the
Greater Excommunication, by which Mr Hume would be
'excluded from the society of all Christians and handed over to the
evil one',

> ...this is a sentence which the civil power now refuses to recognize,
> and which will be attended with no temporal consequences. You
> may wish for the good of his soul to burn him as Calvin did Servetus;
> but you must be aware that, however desirable such a power may
> appear to the Church, you cannot touch a hair of his head, or even
> compel him against his will to do penance on the stool of repen-
> tance.
>
> Your 'libel', as we lawyers call it, is *ex facie*, inept, irrelevant, and
> null, for it begins by alleging that the defender denies and disbelieves
> Christianity, and then it seeks to proceed against him and to punish
> him as a Christian...

Wedderburn moved that the Assembly drop 'the overture anent
Mr David Hume, because it would not ... minister to edification'.[47]

Robert Wallace, in a pamphlet which like so much of his writings never saw publication, made those points less heatedly and
added to them: why go after 'calm contemplative wronghead writers' and not the 'Drunkards, revellers, whore-mongers, adulterers'
et cetera that made up so much of Scots society.[48] In the
Committee itself, Robertson wound up the debate, and the vote
was taken whether to transmit the original overture to the
Assembly or not. Once again, many of the clergy left the meeting
or abstained, and the negative was passed, by 50 votes to 17; after
which, Wedderburn's resolution was carried.

Anderson was not a man to surrender. He registered a formal
complaint in the Edinburgh Presbytery against Alexander Kincaid
and Alexander Donaldson, the printers of Kames's *Essays*, while
passing on the battle against Hume to his son. He died in
December 1756. The Presbytery dismissed the case against the
printers a month later; and with the death of this 'godly, spiteful,
pious, splenetic, charitable, unrelenting, meek, persecuting,

Christian, inhuman, peace-making, furious'[49] old man, an ecclesiastical era drew to an end.

Ramsay of Ochtertyre's judgement was that Anderson's campaign had been favourable to the philosophers, since it had had the effect 'chiefly to make their books more read and longer talked of than they would otherwise have been ... The two culprits were more caressed and admired than ever, and by none more than the moderate clergy ... In a word, this rash and feeble attempt to check the progress of free-thinking, convinced the philosophers of Edinburgh that they had no longer anything to dread from the Church courts.'[50] Hume's modern biographer E.C. Mossner contends that for the High-Flyers, success would have been a disaster. It might have made of Hume an opponent of the clergy as pertinacious and as scathing as Voltaire: a harbinger of bloody revolution, not of a sociable Toryism.[51]

As for Hume himself, one detects in his letters to Adam Smith a deep sense of injury. 'Scotland suits my fortune best,' he wrote on 28 July 1759, 'and is the Seat of my principal Friendships; but it is too narrow a Place for me, and it mortifies me that I sometimes hurt my Friends.'[52] For the rest of his life, and like many Scots before and since, he dreamed of France and sunshine.[53]

Staged performances, which had been a feature of both medieval Scotland and the Stuart court at Holyrood, were seen by the Kirk and the West Bow as the bottomless pit, but that did not deter enterprising companies. According to Edinburgh's asthmatic historian Hugo Arnot, a certain Signora Violante arrived in the city from Dublin in about 1720 and, in a room in Carrubber's Close, performed feats of strength and tumbling 'disgustful in any, but in a woman, intolerable'.[54] In 1727 the Edinburgh Presbytery instructed ministers to reclaim their congregations from 'a Company of Stage-players, who are acting Plays within the Precincts of [the town], and have begun with acting one, which is filled with horrid Swearing, Obscenity and Expressions of a double Meaning'.[55] Allan Ramsay senior, the poet, opened the first regular playhouse in Carrubber's Close in 1736 but immediately ran into difficulties. Walpole's new Theatres Act restricted the performance of plays to two licensed houses in London, Covent Garden and Drury Lane, while the Porteous Riot made the city magistrates excessively nervous about any public assembly. The Town Council frustrated all attempts to legitimise Ramsay's venture by applying for a royal patent for an Edinburgh theatre, while Anderson and other Popular preachers invoked St Paul to close it.[56] Ramsay went out of business in 1737 with his habitual good humour, addressing a verse petition for damages to Forbes of Culloden, newly elected Lord President of the Court of Session:

> I kept our troop, by pith of reason,
> Frae bawdy, atheism and treason.

And again:

> Shall London have its houses twa
> And we be doom'd to've nane awa?[57]

Like other towns in the United Kingdom, Edinburgh took to subterfuge. The advocates and the Lords of Session and their wives and daughters were generally enthusiasts of the stage. Unlicensed performances were given in the intervals of concerts, which were not covered by the Theatres Act of 1737. They were staged at the cramped old Taylors' Hall in the Cowgate under elaborate but threadbare disguise. The *Edinburgh Courant* reported on 27 January 1743: 'By desire of a Lady of Quality, for the benefit of Mrs Hamilton, on Monday next, being the 31st instant, will be performed a concert of vocal and instrumental music. After the first part of the concert, will be given, gratis, The Mourning Bride.'[58]

With the defeat of the Highlanders and a certain relaxation of political censorship, subscriptions were raised for a regular theatre in a close on the south side of the Canongate. Work began in August 1746. According to John Jackson, who made his debut at the Canongate Playhouse in 1762 and later managed the Theatre

Royal in the New Town, boxes were half-a-crown and the pit one shilling and sixpence. The house held £60 to £65, as against £40 to £45 for the Taylors' Hall.[59] In December of 1756 a play was performed there which diverted the battle between the two wings of the Kirk from philosophy, and brought it to a head. This time it was not David Hume but his homonym John Home who was the bone of contention.

Home, who served in the College Company and was captured by the Highlanders at Falkirk, had been settled since 1747 at the parish of Athelstaneford in East Lothian. He was at the heart of Edinburgh literary society, less perhaps for his brilliance than for the beauty of his personality: Walter Scott remembered someone saying that he 'came into a company like a sunbeam into a darkened room'.[60]

Home's first dramatic effort, written in the aftermath of the Forty-five, was a verse drama entitled *The Tragedy of Agis*, set in the capital of ancient manliness, Sparta, in which he addressed the Edinburgh theme of civic virtue and how it could be revived. He took his play to London late in 1747 but David Garrick, actor-manager of the Drury Lane theatre, rejected it, not unreasonably, as unstageable. Home returned to Scotland, both physically and for his setting, and – encouraged by Hugh Blair, whose eight-volume edition of Shakespeare appeared in 1753 – adopted a Shakespearean treatment. By the summer of 1754 his manuscript of a bloody tale of foundlings and mistaken identity was circulating in Edinburgh, to great enthusiasm. On a snowy morning in February 1755 Home set off on his Galloway pony Piercy to ride the four hundred miles to London, with the manuscript in one pocket of his greatcoat and his clean shirt and night-cap in the other. (Later, Carlyle and other friends bespoke him a valise and came with him two days of the journey.) Once again Garrick pronounced the work unfit for the stage, but in the course of his stay in London Home mixed with a group of London Scots who were growing in influence. It was probably Sir Gilbert Elliot, now MP for Roxburgh, who introduced him to John Stuart, third Earl of Bute, friend and mentor to the young Prince of Wales.

Back in Edinburgh, the *literati* had rallied to the play. West Digges, the handsome English actor whom Boswell admired so much he once impersonated him on an assignation,[61] was at the Canongate with an English company, and in a striking declaration of dramatic independence, David Hume and others were agitating to have the play put on in Edinburgh rather than London. In October Lord Milton, ever-cautious, took John Home to Inverary to meet his master, Bute's uncle the Duke of Argyll. The visit was a success and, according to Carlyle, 'the Duke's good opinion made Milton adhere more firmly to him, and assist in bringing on his play in the end of that season.'[62] It is a curiosity of the story that John Home and his friends, having taken such trouble to prepare the political ground for the staging of *Douglas* in Edinburgh, neglected the ecclesiastical.

The play opened on 14 December, with Digges and Mrs Sarah Ward in the lead roles. Of the atmosphere of that opening night, Robert Fergusson's elegy on the Canongate theatre gives the flavour:

> No more from box to box the basket piled
> With oranges as radiant as the spheres,
> Shall with their luscious virtues charm the sense
> Of *taste* and *smell*. No more the gaudy beau,
> With handkerchief in lavender well drench'd,
> Or *bergamot*, or *rose-watero* pure,
> With flavoriferous sweets shall chase away
> The pestilential fumes of vulgar cits,
> Who, in impatience for the curtain's rise,
> Amus'd the lingering moments, and applied
> Thirst-quenching *porter* to their parched lips.[63]

'There never was so great a run on a play in this country,' *The Scots Magazine* reported.[64] 'Persons of all ranks and professions crouded to it.' Mrs Ward, it seems, made a deep impression. When interest in the play among the quality was exhausted, Carlyle anonymously wrote a blood-curdling broadside, *A full and true History of the Bloody Tragedy of Douglas as it is now to be seen acting in the Theatre at the Canongate*, to bring in apprentices and tradesmen.[65] The play ran for two more nights.

In February 1757 Home travelled again to London, and on 14 March had the pleasure of seeing *Douglas* produced at Covent Garden by Garrick's rival John Rich, with Garrick's estranged mistress Peg Woffington in a leading role. The performance was attended by William Pitt (the Elder) and Lady Bute; probably through the agency of Lord Bute, Home was given a royal pension of £100 a year.

After the Covent Garden performance David Hume – whose literary tastes were as partial as his political judgements were impartial – wrote to Adam Smith: 'When it shall be printed (which will be soon) I am perswaded it will be esteem'd the best; and by French critics, the only Tragedy of our Language.'[66] Dr Samuel Johnson called it 'that foolish play'.[67] *Douglas* is neither of those extremes, but a derivative piece which but for its political effect would have gone the way of so many other eighteenth-century tragedies, into general oblivion. It might have made a decent opera.

The story takes place at a time of Danish invasion. Lady Randolph, while acting the part of dutiful wife to her second husband, secretly mourns the death in battle of her first husband, Lord Douglas, and the loss of their son. Glenalvon, Lord Randolph's nephew and heir, is in love with her. Lord Randolph is rescued from attack by a mysterious young man, who turns out to be none other than young Douglas, raised among shepherds. Mother and son are revealed to each other at a secret tryst, but Glenalvon insinuates to Randolph that their relationship is that of lovers. Douglas kills Glenalvon and is himself killed by Randolph. Lady Randolph throws herself from a high rock and Randolph sets off for battle with the Danes, determined to leave his body on the field.

Hume, who found Shakespeare grotesque,[68] rejoiced that John Home had shaken off his corrupting influence and was now a true disciple of Sophocles and Racine.[69] In fact the play, written in workmanlike blank verse, resounds with Shakespearean echoes: of *Hamlet*, in the supposedly *louche* relationship between mother and son (and also the Danes); *Othello*, in the jealous interloper; *Romeo and Juliet*, in the misinformed and misdirected slaughter. *Douglas* teeters on the abyss of literary bathos.

> Then must this western army march to join
> The warlike troops that guard Edena's tow'rs.[70]

How could an Edinburgh audience that remembered the Canter of Coltbrigg hear this couplet without laughing? How could John Home, volunteer in the College Company, have written it? *Douglas* is 'tragedy' in its modern guise, in which feeling replaces terror and propriety, shame. In other words, it is not a tragedy in the Shakespearean sense, but a pioneer of the romantic literature of the generation of Scott. In the person of Mrs Ward's or Peg Woffington's Lady Randolph, with her morbid grief, indomitable courage and eventual suicide, can be detected the first stirrings of that uncontrolled feminine sensibility that burst out in William Robertson's characterisation of Mary, Queen of Scots in 1759 and swept Europe and America in the next half-century.

Among the supporters of the play two non-literary impulses are distinguishable. The first is civic patriotism. Humiliated in the Forty-five, insulted by the Duke of Cumberland, Edinburgh was in desperate need of reassurance. Verse tragedy had great prestige and, in the words of Alexander Carlyle, the town was 'in an uproar of exultation that a Scotchman had written a tragedy of the first rate, and that its merit was first submitted to their judgement'.[71]

In an anonymous pamphlet rushed out early in 1757, *An Argument to prove that the Tragedy of Douglas ought to be Publickly burnt by the hands of the Hangman*, Carlyle solemnly argued that a 'cold, barren, and remote country' could not possibly produce 'a tragic poet to rival Sophocles and Euripides, Corneille and Racine, Shakespear and Otway'.[72] So as not to provoke the envy of the English, or of a government suspicious of Scottish loyalty, or of a red-faced Garrick, he demanded, 'let us industriously suppress every appearance of genius and spirit'.[73] The pamphlet ended with a brilliant topographical flourish,

Carlyle calling on the Edinburgh Presbytery to come in a body to the High Street the following Wednesday, precisely at one o'clock, there at 'the place where the cross once stood' to see *Douglas*, the abomination of abominations, consigned to the flames.[74]

The second discernible impulse was a belief among the Moderate clergy and their friends that they had come to constitute a vanguard of politeness and modernity. Carlyle made this point obliquely in his *Argument*, in his taunting of the old-fashioned clergy for sticking to controversy, experimental preaching, farming, horse-coping, enclosing their glebes with their own hand, and 'the impartial estimate of the profit and loss of religion'.[75] The same year Adam Ferguson, who had resigned his military chaplaincy in 1754 and returned to Edinburgh in hope of a professorship, argued in *The Morality of Stage-Plays Seriously Considered* that the theatre could not only teach virtue, but also encourage trade and industry.

It is said, though not on very good authority, that at the first rehearsal of *Douglas*, held at the lodgings of Mrs Ward in the Canongate, the parts were read by Hugh Blair, Alexander Carlyle, John Home, David Hume, William Robertson and Adam Ferguson; and that the audience included Lords Elibank, Kames and Milton and James Burnett, later Lord Monboddo.[76] With his habitual innocence, Hume dedicated *Four Dissertations* – the last significant philosophical publication of his lifetime, it included 'The Natural History of Religion' – to 'the Author of Douglas', only to become anxious as it went through the press that it would merely exacerbate the quarrel.

For Witherspoon and the High-Flyers, it was a scandal that a clergyman should write a play, of however moral a character. In the language of a pamphlet of 1757, *Douglas, A Tragedy, weighed in the Balances and found wanting*, the very elegance of the play told against it: it was like a 'dunghill covered over with snow'.[77] *Douglas* was a 'most unaccountable medley of impiety, profaneness, error, immorality and vice'.[78]

The author, who signed himself A.B., objected to the elements taken from pre-Christian Greek tragedy – blind fate, raging at the gods, suicide – and also had difficulty distinguishing the stage from everyday reality, criticising the characters as if they were backsliders among his flock. He censured Glenalvon for his wickedness, young Douglas for his lust for fame, and Lord Randolph for swearing ('By heaven!'); but he reserved his harshest judgement for that 'composition of errors and wickedness', the heroine.[79] Lady Randolph was far from being a model of feminine virtue. She should have accepted whatever fate Heaven had prepared for her. Instead, she blasphemed and swore and committed the ultimate crime of suicide. By praying on the stage, she brought prayer into ridicule.

On 5 January 1757 the Edinburgh Presbytery had issued an official Admonition and Exhortation, to be read from every pulpit on the last Sunday of the month, castigating the playhouse as a footling luxury in time of war (with France) and 'prejudicial to the interests of religion and morality'.[80] It had written in similar terms to the other presbyteries. Some bristled furiously at the interference. In the Haddington Presbytery, Robertson and other Moderate ministers protected John Home, who anyway was absent in London for the staging of *Douglas* at Covent Garden.

Alexander Carlyle was not so fortunate. In February he was summoned to present himself before the Presbytery of Dalkeith. At his first appearance, on 1 March, he plunged into a tactless defence of the morality of *Douglas*. He had fewer friends than he believed, and Robert Dundas the younger, son of the old Lord President and now himself Lord Advocate, was noticeably cool to his cause. On 25 March a libel or formal accusation was issued, accusing Carlyle of associating with stage players who were people of notorious bad character, of directing rehearsals, openly appearing at the Canongate playhouse, and taking over a box in disorderly fashion. (It seems he ejected some young men from the box so as to accommodate women in his party.) By 5 April Carlyle had come down to the point of apologising for his conduct, but that was not enough for his opponents. At a meeting on 19 April one Archibald Walker said: 'He is not yet humble enough, Moderator – Moderator, he is not so humble as I would have him.

Yea, Moderator, he is not so humble as I aim to make him if I can get my will.'[81] The Presbytery voted to retain the libel. Carlyle appealed to the Synod of Lothian and Tweeddale, the next rung up the Kirk's administrative hierarchy, which he might have expected to be both more liberal, and more open to aristocratic influence. Yet even here he was taunted for his friendship with Hume and the notorious close of Hume's 'Natural History of Religion' in the recently published *Four Dissertations*, in which Hume described religion as 'an aenigma, a riddle, an uncertainty'. William Robertson and Andrew Pringle spoke persuasively in Carlyle's favour, and in the end the Synod expressed its 'high displeasure' but, to the fury of Alexander Webster and other High-Flyers, rebuked the Presbytery for proceeding with a public libel rather than a private admonition. The Presbytery appealed in its turn to the General Assembly, where it was roundly defeated on 24 May by 117 votes to 39.

Three days later, the Assembly bent its attention to a general discussion of the morality of stage plays. Wedderburn attempted sincerity: 'In all the sermons produced by the united genius of the Church of Scotland, I challenge you to produce any thing more pure in morality, or more touching in eloquence, than the exclamation of Lady Randolph:

> – Sincerity!
> Thou first of virtues! Let no mortal leave
> Thy onward path, although the earth should gape,
> And from the gulph of hell destruction cry
> To take dissimulation's winding way.'[82]

The ministers were not convinced: and Dr Johnson's criticism of those lines – 'Pooh!' – is comprehensive. Robertson, though he had vowed to his father never to enter a playhouse in Scotland, supported Wedderburn in preventing the passage of an act of censure against the stage-struck clergy. What emerged instead was a most lenient compromise overture – written by Robert Dundas – asking that presbyteries 'preserve the purity and decorum of the ministerial character; and that they take care that none of the

ministers of this church, do upon any occasion, attend the theatre.'[83] Names were not named.

The double drama of Hume and Home was summed up in a
satire written by Professor MacLaurin's son John, who disliked
the Select Society and the pretensions of its members. The piece
tells the story of the wooing of Mrs Sarah Presbytery, relict of Mr
John Calvin and mother of the budding playwright Jacky, by a certain Mr Genius (*alias* Mr David Hume, Esq.). It was called *The
Philosopher's Opera*.

MR GENIUS: You have read my books then, Sir?

SATAN: Yes, Sir, with great delight.

MR GENIUS: Why then, Sir, you are convinced, I suppose, that there is
no God, no devil, no future state; that there is no connection betwixt
cause and effect; that suicide is a duty we owe to ourselves; adultery
a duty we owe to our neighbour; that the tragedy of DOUGLAS is the
best play ever was written; and that *Shakespeare* and *Otway* were a
couple of dunces. This, I think, is the sum and substance of my writings.

SATAN: It is, Sir.[84]

By 1766, what most interested the young lawyers and college
students packing the Canongate theatre was not ethics but value
for money. The Edinburgh theatre, as described in a pamphlet
written by the future publisher John Murray, was by then a world
of captive audiences, high box-office takings, and managers who
pushed their mistresses into unsuitable roles.[85] Under the management of two amateurs – John Dowson of Newcastle and David
Beatt, who had read the Rebel proclamations at the Cross in 1745
– the theatre was as riven by faction as a modern football club.
There were riots by the partisans and enemies of an itinerant actor
named Stayley and on 12 January 1767 a handbill appeared sus-
pending performances 'till they [the actors] can be assured of a
proper protection'.[86] On 24 January the house was opened again
for *Hamlet*, the stage was overrun, the City Guard was called out
and, as ever, repulsed, and the inside of the theatre was ransacked.
The next season passed quietly enough, but the Edinburgh stage

was ripe for regulation. Relicensed as the Theatre Royal, the Canongate was reopened by the London actor David Ross on 9 December 1767, with a performance of Henry Jones's twenty-year-old tragedy *The Earl of Essex* and a Prologue by James Boswell.

The season was a success, and a subscription was opened for a new theatre at the far end of the New Bridge (of which three arches were nearing completion), in what became the New Town. The new Theatre Royal (on the site of what is now the Post Office) duly opened in 1769, but the collapse of the south abutment to the bridge that year not only arrested development in the New Town but 'depressed the theatrical spirit of the audience'.[87] Across the Atlantic, Witherspoon himself allowed, or at least left unpunished, a student production of Home's *Alonzo* (1773) at the College of New Jersey. In 1784 the General Assembly went almost in a body to see the great English actress, Mrs Sarah Siddons.[88] John Buchan, who disliked the Moderates and the *literati* in equal measure, saw this as the moment when the 'giddy parson'[89] arrived.

❦

Douglas changed many lives. Wedderburn, aware that he had burned his boats, was determined to leave Edinburgh in glory. Towards the close of the summer session in 1757 he insulted opposing counsel in the Outer House, calling him a cuckold; when rebuked by Lord President Craigie from the bench, he pulled off his advocate's gown, laid it on the bar, bowed and left the Parliament House and, that night, a Scotland suddenly grown small.[90] He set up in chambers in London, took lessons in English elocution from Thomas Sheridan, father of the playwright, rose to be Lord Chancellor.

With his case still pending before the Haddington Presbytery, in June 1757 John Home resigned his parish charge. Soon afterwards he became secretary to Lord Bute and tutor to the Prince of Wales. On the death of the old king in 1760 he thus found himself at the seat of power. His pension was increased to £300 a year and, to the disgust of Londoners as varied as Dr Johnson and John Wilkes, the next five years saw patronage showered upon him and his Edinburgh friends.

By way of Gilbert Elliot and the Earl of Bute, the Moderate clergy now had the backing of the government in London as well as of Lord Milton and the Duke of Argyll in Scotland. They were able to rout their opponents in the General Assembly debates. John Jardine had already brought his father-in-law George Drummond over to the Moderate camp. As Carlyle reported, Jardine 'kept [Provost Drummond] steady, who had been bred in the bosom of the Highflyers'.[91] Such ambitious or sociable Evangelicals as Alexander Webster who 'wished to be well' with the Lord Provost came under Jardine's 'management'.[92]

By 1760 Hugh Blair was installed in the High Kirk, where his sermons were a must for fashionable visitors to Edinburgh. He was also Professor of Rhetoric and *Belles-lettres* at the College, the first such professor of literature in the country. Adam Ferguson, who had acted as tutor to Bute's sons, was appointed Professor of Natural Philosophy in 1759 (in 1764 he was translated to the more congenial chair of Pneumatics and Ethical Philosophy).

Douglas broke the literary dam in Edinburgh. The *Epigoniad*, on which William Wilkie, rough-hewn minister of Ratho in Midlothian, had worked so long – sometimes, according to David Hume, taking his Homer out into his father's wheat-fields where he stood guard against pigeons – appeared at the close of the General Assembly of 1757. A narrative *tour de force* of three thousand lines telling the story of an attack on Thebes in the generation before the Trojan War, it flopped. Hume fired off his usual circular letters of publicity, but the London critics were quite damning. Wilkie had not helped his case with a pugnacious preface in which he rushed headlong at the English canon, calling Milton's *Paradise Lost* 'altogether irregular' and complaining of 'quaintness of expression' in Pope's translations of the *Iliad* and *Odyssey*.[93]

Like *Douglas*, there is an air of hard study about the *Epigoniad*:

as the eighteenth century might have said, it 'smells of the lamp'. In a thorough and brilliant notice in the September 1751 number of the *Monthly Review*, Oliver Goldsmith – who had studied at Edinburgh – reached the heart of the matter. 'The *Epigoniad* seems to be one of those *new old* performances, a work that would no more have pleased a peripatetic of the academic grove, than it will captivate the unlettered subscriber to one of our circulating libraries.'[94] The problem was a lack of conviction: 'We assign as a reason of our disgust, our being conscious, that the Poet believes not a syllable of all he tells us.'[95] Furthermore, as a Scottish critic put it, an epic subject must concern the people for whom it is written; and what modern Briton cared a fig about events in the Aegean two or three millennia before the birth of Christ?[96] The reception of the *Epigoniad* should have warned Edinburgh that, in point of literature, it had to come down off its stilts: that Wilkie was no 'Scottish Homer', any more than blind Blacklock was a 'Scottish Pindar' or John Home a 'Scottish Shakespeare'. Wilkie himself, with that versatility characteristic of the Edinburgh *literati*, turned to his other talent, geometry, and was appointed Professor of Natural Philosophy at St Andrews in 1759, in time to befriend a real lyric talent, Robert Fergusson.

It was William Robertson who, in his calculating way, saw what was wanted. His *History of Scotland during the Reigns of Queen Mary and of King James VI* caused the eyes of literary Europe to fall on Edinburgh. Published by the London Scottish bookseller Andrew Millar in January 1759 and reprinted in April, it quite overwhelmed Hume's volume on the Tudors, which the philosopher accepted with his usual good nature. Its title notwithstanding, the main interest of the *History* lay in Robertson's dramatic account of the conflict between Mary of Scotland and Elizabeth of England. John Home's treatment of female motivation in *Douglas* had been scandalous. Robertson took a woman's story off the stage, lodged it in the grand historical process of Reformation and Counter-Reformation, and dressed it up in Latinised rhetorical periods. In the process, he made women a respectable topic for literature.

He also launched an enduring preference. Even today, it is through biography that the vast majority of English and Scots readers consume their history – to the bafflement of more analytical races. Horace Walpole, in a typically penetrating letter of 18 January 1759, congratulated Robertson for preserving 'the gravity of history without any formality'. Though clearly partial to Mary, Robertson had not violated the truth in her favour; and had he but recognised that such-and-such was not an English usage ...[97] A brilliant career surely awaited Robertson in London but he chose, like David Hume, to stay in Scotland. For his next work, a history of the Holy Roman Emperor Charles V, Robertson was paid no less a sum than £4,500 – enough, for example, to build a decent-sized country house: the equivalent of perhaps half a million pounds in 2003. In 1762 he was installed as Principal of the College in a house on the site of the Kirk o' Field. 'Mr Robertson's Administration' controlled Kirk and College until a crisis over civil rights for Scottish Catholics arose at the end of the 1770s.

Fifty years after the Union, and not a moment too soon, Edinburgh was prospering. Agriculture was improving, the linen trade was expanding by leaps and bounds, Leith was full of ships, new public companies were being formed. The first of George Drummond's civic schemes, a covered Exchange for merchants, was a-building in the High Street.

Robert Wallace, since the death of MacLaurin the best mathematician in town, computed in *Characteristics of the Present Political State of Great Britain* of 1758 that the Scottish national product had increased by £1,125,000 since the expulsion of the Stuarts.[98] Scotland was becoming more of a commercial proposition: estates that at the time of the Union had sold for fifteen or sixteen years' rent were now selling, according to Wallace, at twenty-three years' rent. In other words, investors who had demanded annual returns of 6.5 per cent from their property in Scotland were now content with less than 4 per cent.[99] As for Edinburgh, Wallace added with uncharacteristic sarcasm, Scotland might 'have lost the *name* of independence. Two or three of their *old palaces* may possibly decay', yet for Scots to mourn the processions of nobility and gentry and the brilliant Court during the sessions of the Parliament, lost to them since the Union, was plain ludicrous. 'In place of *empty titles* and an insignificant pomp, they have acquired the more solid blessings of security, liberty and riches.'[100]

In truth, the Jacobites were a spent force. 'Mr Evidence' Murray of Broughton, composing his memoirs alone and ostracised, his name scratched out from the minutes of the Canongate Kilwinning Lodge of Freemasons, let a draught of bitter air in onto the musty rodomontades of the Old Town Jacobites:

> The Castle of Edinburgh is starved before the punch Bowl is empty. The Batteries against Stirling erected anew, and the Garrison made Prisoners of War by the time it is replenish'd ... A parcel of Antiquated Attorneys, with the help of a black Gentleman in a gown and Cassock, will march us to Derby, from thence make our way straight and easy to the Capitall, render the March of the Enemy impossible, rouse the Sleeping English [Jacobites], seize the Treasury, make the two Armys under C–mb–r——d and Wade disband, their Officers sue for Pardon, and the Fleet send their Submission, erect Triumphall Arches, make the Mayor and Aldermen meet us with the Regalia of the City, which with their Charter returnd, and protection promies, compleat the Cavalcade to St James's ... On the other hand, mortifyd beyond expression when they reflect that their easy Scheme was not put in Execution; plainly discover treachery in our retreat ... Every Old Woman, Green Girl, Cock Laird and Pettefogger* being now become equally soldiers and Politicians, denouncing one a Coward, t'other Traitor, and a third a Blockhead...[101]

On 20 November 1759 any hope of a Jacobite restoration was finally sunk with the French invasion fleet in the waters of

---

* 'Green-sickness' or chlorosis was an eighteenth-century distemper of adolescent girls. A 'cock laird' was a small farmer with pretensions to gentility, a 'pettifogger' a legal hack.

Quiberon Bay. Prince Charles Edward had already retreated into a twilit exile of drink, incognito, and undignified attachments. In the next generation, Edinburgh was diverted by the daft Highland laird James Robertson of Kincraigie, who had been out in Forty-five and had 'extreme anxiety to be hanged drawn and quartered'. Though he toasted the Chevalier and 'bawled treason in the streets', nobody would help him. He managed to have himself shut up in the Tolbooth for debt and would not leave until the exasperated magistrates sent a summons to the Justiciary Court, on a sham charge of High Treason.[102]

In Edinburgh, the argument for Union had, late in the day, been won. The problem, as will become clear, was in London.

5

Smaller Joys from Less Important Causes

On July 2 1757, shortly before quitting the town, Wedderburn wrote from Edinburgh to the leading Scottish MP in London, Sir Gilbert Elliot of Minto, about the literary efforts of his friends. Having failed through lack of matter to launch a literary revival in 1755, two years later Wedderburn felt engulfed. 'The most agreable prospect in this Country arises from the Men of Letters,' he wrote. 'Robertson has almost finished a History, which will do any honour to any age & bids fair to dispute the prize wt Davd Hume – John Hume has finished the first act of Agis & applies in earnest – Ferguson is writing a very ingenious System of Eloquence or Composition in general – Wilkie's Epick poem you have certainly seen. Smith has a vast work upon the anvil, it discloses the deepest principles of philosophy.'[1]

Of those works, the last and greatest was Adam Smith's *Theory of Moral Sentiments*. When it appeared in the spring of 1759, a year of triumphant British military victories over the French in Europe and North America, it caused a sensation. Less abstruse than Hume's *Treatise*, less precious than Hutcheson's *Inquiry*, it was the work that established Scottish philosophy in the first rank.

When Smith died, after a life of intellectual adventure and social prudence, the *Caledonian Mercury* complained in its obituary of 4 August 1790 that he had converted his chair of Moral Philosophy at Glasgow University into one of trade and finance, a judgement that has lasted to the present day. His name stands as shorthand for an ideology of unrestricted commerce and blithe social optimism. Business people who know nothing of Edinburgh or the eighteenth century yet feel they possess in Smith a venerable sanction to go on doing what they are doing already (which is what other people are also doing). Some have heard of the Invisible Hand, an invention of his which pardons errors and makes good transgressions: wields, as it were, an invisible moral brush and dust-pan.

The picture of Adam Smith as the apostle of amoral modern capitalism has been under attack in Scotland for some years, and is indeed unhistorical in both its terms. Smith's interests ranged over wide horizons, from cosmology through language to morals and the increase in wealth. He wrote well on Italian verse and on garden topiary. He believed that all philosophy, including medicine and what are today called the natural and social sciences, was imaginary, hypothetical, provisional and impermanent. Like that other great intellectual David Hume, Smith had little time for the intellect: both philosophers held that humanity followed its instincts far more than its reason.

'Capitalism', a word invented as '*Kapitalismus*' by Werner Sombart at the turn of the twentieth century, did not exist as a mental entity in the eighteenth. The economy Smith depicted in *The Wealth of Nations*, with its small tradesmen and yeomen farmers, had as much in common with the world of the dialogues of Plato as with the economy of today. In comparison with the rococo financier John Law of Lauriston or the Jacobite Sir James Steuart, Smith's grasp of banking and finance was not especially strong, and his excursions into commercial anthropology were thrilling but implausible. Smith had social connections with wholesale merchants in Glasgow who were exploiting the colonial markets opened to Scots trade by the Union of the Parliaments, but had almost nothing good to say about them or their company. By the end of his life, he was expressing the most profound misgivings about the moral complexion of commercial society. In the revised edition of *The Theory of Moral Sentiments* of 1790, he sometimes sounded as Spartan as Andrew Fletcher of Saltoun.[2]

Dugald Stewart, Professor of Moral Philosophy at Edinburgh between 1785 and 1820, was Smith's first biographer. In a paper read to the Royal Society of Edinburgh in early 1793 he sketched out what might be called the 'Adam Smith problem'. Stewart made it his task to find the connection between *The Wealth of Nations* and *The Theory of Moral Sentiments*: between the great philosopher's 'system of commercial politics, and those speculations of his earlier years, in which he aimed more professedly at the advancement of human improvement and happiness'.[3] Stewart was not successful. In the next century H.T. Buckle, who thought *The Wealth of Nations* 'perhaps the most important book that has ever been written', simplified Smith out of existence: *The Theory of Moral Sentiments* was about sympathy, *The Wealth of Nations* about selfishness, and that was humanity sorted.[4]

Smith himself made the task difficult. He did not like to acknowledge his sources or display his calculations. On 16 April 1773, during one of his bouts of Scots hypochondria during which he feared he might die suddenly, he asked David Hume to suppress his unpublished work. His letter breathes the bachelor domesticity inseparable from Scots philosophy at that period. Apart from the manuscript of what became *The Wealth of Nations* (which he said he kept by him), and a thin folio in his writing desk containing the fragment of a great juvenile work in the form of a history of 'astronomical systems ... down to the time of Des Cartes', which he left to Hume's discretion, nothing was to be published: 'All the other loose papers which you will find either in that desk or within the glass folding doors of a bureau which stands in My bed room together with about eighteen thin paper folio books which you will likewise find within the same glass folding doors I desire may be destroyed without any examination.'[5] As it turned out, Hume predeceased him by fourteen years, but in 1790, when he realised he was failing, Smith sent for his friends the natural scientists Joseph Black and James Hutton, who burned sixteen volumes of his writings.

Even so, modern scholars, beginning with Edwin Cannan, have reassembled the scattered fragments of Adam Smith's mental colossus, including his correspondence with Hume, and students' notes of his lectures at the University of Glasgow. The outline of a magnificent system, covering every area of human activity, such as seems to belong in the pages of *Tristram Shandy* or *Candide*, and based not on fact but on logic, flickers into view.

Adam Smith was born on 5 June 1723 at Kirkcaldy, the son of Adam Smith and Margaret Douglas. His birth date places him right in the middle of the second cohort of Edinburgh *literati*: younger than Blair (1718), Robertson and Wilkie (1721) and Carlyle (1722), older than Adam Ferguson (by a few days), the natural scientists Hutton (1726) and Black (1728), the architect Robert Adam (1728) and the botanist John Walker (1731).

Adam Smith's father had practised as a Writer to the Signet in Edinburgh, served as private secretary to the Earl of Loudoun when he was Secretary of State at the time of the Union of Parliaments, and been appointed Comptroller of Customs at Kirkcaldy in Fife in 1714. Smith never knew his father, who died five months before his birth, but was also to end his days as an official in the Customs Service.

Kirkcaldy in 1723 was recovering from the devastation of its Continental export trade that was a consequence of the Act of Union.[6] Defoe in his *Tour* of the next year reported that the town consisted of a mile-long street hugging the shore, and spoke of 'considerable merchants' sending corn and linen to England and Holland, a shipyard, coal-pits and salt-pans.[7] One of the principal businessmen of the town was William Adam, married to William Robertson's aunt and best-known today as the architect of such Scots country houses as Sir John Clerk's Mavisbank, Hopetoun House and the Dundas family seat at Arniston,

Midlothian, and of the Edinburgh Royal Infirmary. Adam's son Robert, an architect with broader horizons than his father and born in Kirkcaldy in 1728, was a childhood friend of Smith.

In his earliest years, Dugald Stewart related, Adam was 'infirm and sickly'.[8] At the age of three, on a visit to his maternal uncle at Strathendry just inland of the town, in one of those incidents in his life that seem so incorrigibly uneconomical, Adam Smith was abducted by gypsies. The story, told first in the *Caledonian Mercury* obituary and repeated by Dugald Stewart, over time became quite circumstantial. According to John Rae, Smith's Victorian biographer, a gentleman told the frantic household that he had seen a tinker woman carrying a child 'crying piteously'. In Leslie Wood a search-party came on her, at which she threw the little boy down and fled.[9] Thus, wrote Dugald Stewart, there was preserved to the world 'a genius, which was destined not only to extend the boundaries of science, but to enlighten and reform the commercial policy of Europe'.[10] More to the point, Adam Smith was better suited to be a philosopher than a travelling man.

From the grammar school of Kirkcaldy he passed, in 1737, to the University of Glasgow, where he came under the spell of Francis Hutcheson's lectures. Oxford, in contrast, where he took up an Exhibition to Balliol College in 1740, was sunk in wine and Jacobite sloth. The university was simply not organised to accommodate the new sciences. It is said that the Fellows of Balliol had so little to do, they used to sit in Broad Street to await the arrival of the London mail coach.[11] Scots Exhibitioners were unpopular for their poverty and their Hanoverian allegiances. No doubt Smith had Oxford in mind when in *The Wealth of Nations* he wrote of ancient universities as 'sanctuaries in which exploded systems and obsolete prejudices found shelter and protection, after they had been hunted out of every other corner of the world'.[12] If so, the university returned the compliment. Even when Smith was its most famous living son, it never conferred on him a doctorate; and to this day it remains suspicious of political economy.

The founder of Smith's Exhibition had intended its beneficiaries to take orders in the episcopal church in Scotland; what Adam Smith did, according to Stewart, was to read in the broad humane direction laid down by Hutcheson in his lectures. '[To] the study of human nature in all its branches,' he said, 'more particularly of the political history of mankind ... he seems to have devoted himself almost entirely from the time of his removal to Oxford.'[13] Smith 'diversified his leisure hours by the less severe occupations of polite literature'.[14] He worked on translations from the French to improve his English style; and indeed Smith was never, like Hume and William Robertson, afraid of the bogeys of Scotticism.[15] In English lexicography, he dared to stand up to Johnson. Nor did he ever relax his early grasp of the philology and literature of Latin, Greek, French and Italian.

Like Hume, Smith seems to have worked himself sick. Writing on 29 November 1743 he told his mother: 'I am just recovered of a violent fit of laziness, which has confined me to my elbow chair these three months.'[16] In another letter of the following 2 July he reported alarming symptoms of this laziness: 'inveterate scurvy and shaking in the head'.[17] Fortunately, tar-water – which no less a philosopher than Bishop Berkeley had made fashionable at Oxford that spring – 'perfectly' cured him. According to Smith's modern editors, it was at Oxford that he laid the foundation of, if he did not actually complete, the fragment of that 'intended juvenile work' dealing with the history of 'astronomical systems' that he left it to Hume's discretion whether or not to publish[18] (and which was eventually published by Hutton and Black in the posthumous *Essays on Philosophical Subjects* of 1795).

What the full scope of this juvenile work was intended to be, nobody knows. In a letter written in his old age, Smith told the duc de La Rochefoucauld that he had two great works 'upon the anvil', of which one was a 'sort of Philosophical History of all the different branches of Literature, of Philosophy, Poetry and Eloquence';[19] he feared that it must fall prey to the 'indolence of old age'.[20] It must be assumed, as Smith's modern editors do, that the history of astronomy, together with two yet more defective essays on ancient physics and ancient logics and metaphysics that share the same preamble and were also published in *Essays on Philosophical Subjects*, were originally contributions to that Promethean enterprise.

The astronomy essay is the door into Adam Smith's thought. Its full title gives a clue to his intentions: 'The Principles which lead and direct Philosophical Enquiries; illustrated by the History of Astronomy'. The preamble shows that although he covers the four 'systems' of Ptolemy, Copernicus, Descartes and Newton with his usual precision, Smith's prime object was not in fact to provide a history of astronomy. His interest was, rather, anthropological and psychological. A more prudent man than Hume, he conveys his version of the *Natural History of Religion* by way of the ancient and medieval cosmos.

With Smith, as with Hume, it is the personality, as it absorbs impressions of the world, that is the chief actor in that world. In one of many fine passages, Smith says that he will examine the four cosmological systems not in 'their absurdity or probability, their agreement or inconsistency with truth and reality' but only in their fitness or not 'to soothe the imagination, and to render the theatre of nature a more coherent, and therefore a more magnificent spectacle, than otherwise it would have appeared to be.'[21] Smith pays his debt to Hutchesonion aestheticism and to Addison's 'pleasures of the imagination', but for him the actual origins of philosophy lie in what the ancients called panic. In passages that evoke a great Latin original – the verse cosmology of Lucretius known as *De Rerum Natura* – Nature is a chaos of inexplicable and contradictory phenomena. Philosophy, by 'representing the invisible chains which bind together all these disjointed objects, endeavours to introduce order into this chaos of jarring and discordant appearances, to allay this tumult of the imagination, and to restore it, when it surveys the great revolutions of the universe, to that tone of tranquillity and composure, which is both most agreeable in itself, and most suitable to its nature.'[22] Philosophy as tranquilliser. In his lectures, Smith was

unashamedly hedonistic: 'It gives us a pleasure to see the phaenomena which we reckoned the most unaccountable all deduced
from some principle.'[23]

The Newtonian system, according to Smith, was simpler than
its precursors, more coherent and more comprehensive, and thus
more *pleasant*, but it was no less imaginary than that of Ptolemy.

> And even we, while we have been endeavouring to represent all phil-
> osophical systems as mere inventions of the imagination, to connect
> together the otherwise disjointed and discordant phaenomena of
> nature, have insensibly been drawn in, to make use of language
> expressing the connecting principles of this one, as if they were the
> real chains which Nature makes use of to bind together her several
> operations.[24]

This assault on the objectivity of science is far more shocking to
the modern mind than Hume's attacks on miracles. Was Smith
sincere in this provocation? Did his scepticism embrace even the
divine Newton? Is gravity merely a notion of the mind?

'Yes' must be the answer to all three questions. Smith is saying
that by our very manner of talking about the world we convert the
imaginary into the certain. Even today, long after Einstein
reminded us that even the most successfully-tested scientific
theory, such as Newton's, should not be regarded as more than an
approximation of the truth, we are still reluctant to accept the
hypothetical character of natural science.[25]

As for the social sciences, the consequence of Smith's argument
would be this: we should not hurry to attribute to such literary
constructions as the 'Impartial Spectator' or the 'Invisible Hand'
or the 'Propensity to Barter and Truck' any reality in the external
world. Modern economics, which claims a precision to its theories
quite absent from the phenomena observed, would seem pre-
sumptuous to the Adam Smith of the 'Astronomy'. The 'Invisible
Hand' makes an appearance in the 'Astronomy', but not as the
commercial mechanism of *The Wealth of Nations*: it is 'the invisible hand of Jupiter' behind alarming and inexplicable natural
phenomena.[26]

In the late summer of 1746 Adam Smith left Oxford without prospect of employment. As Dugald Stewart reported, Smith had no taste for the 'ecclessiastical profession',[27] at least in Episcopalian form. If he hoped to serve as tutor to some nobleman's son, like Hume and Ferguson, his 'absent manner and bad address'[28] would surely have alarmed my lord and my lady. Instead, while a vengeful British state pursued its reprisals against the Jacobite rebels and Prince Charlie hid out in the heather, Adam Smith rode up to Kirkcaldy and rejoined the person who was his chief attachment, his mother.

On the shores of the Firth he passes out of view, to reappear in 1748, this time in print. Adam Smith's first published work was an unsigned preface to a collection of verses by, of all people, the cavalier poet William Hamilton of Bangour. A protégé of Henry Home and the laureate of his circle in the deep Edinburgh taverns of the 1720s, Bangour had strayed somewhat too far in the cause of the exiled Stuarts. In Italy for his health, as Ramsay of Ochtertyre reported, 'When sauntering one day about the Capitol, a young man laid his hand on his shoulder, saying, with a smile: "Mr Hamilton, whether do you like this prospect or the one from North Berwick Law." '[29] It was Prince Charles Edward. Bangour was captivated, later welcomed the Chevalier to Holyrood on 17 September 1745, wrote a celebratory poem on the battle of Preston under its Highland name of 'Gladsmuir' that Burns accurately criticised as 'far ower sublime', and subsequently escaped to France.[30] Presumably Henry Home and others wanted both to help 'poor Willie' and to launch Adam Smith into print.[31] *Poems on Several Occasions*, printed by the up-and-coming Foulis brothers of Glasgow, is much sought by collectors.

Henry Home gave Smith more lasting employment that year when he arranged for him to be invited to Edinburgh to deliver public lectures on rhetoric. These, attended by such ambitious men as Wedderburn and Hugh Blair, brought Smith a clear hundred pounds a year for three years and ensured his appointment, in 1751, as Professor of Logic and Rhetoric at Glasgow.[32] They also inaugurated a campaign in Edinburgh for supposedly

'correct' English which culminated in the Select Society import-
ing Englishmen to teach pronunciation. Adam Smith was a brilliant Scots mind, but what perhaps appealed more to Henry
Home was that he was a brilliant Scots mind *that had been to
Oxford University*

The *literati* were avid for distinction, but what language were
they to employ? John Knox and the Protestant reformers of the
sixteenth century, by insisting on translations of scripture into
southern English, had dealt a heavy blow both to Latin and to the
northern variant of English generally called Scots. The wars of the
seventeenth century, by disrupting all learning, did them further
injury.

The revival of Scottish Latin literature in the Jacobite circle of
Archibald Pitcairne, Robert Freebairn and Thomas Ruddiman
had ended with Ruddiman's great edition of George Buchanan's
Opera Omnia at the time of the Fifteen. Poor Ruddiman saw Latin
and Greek as 'the only sure Channels through which all useful
Learning did and should run',[33] and only 'Ignorance and
Barbarity' in their place.[34] Meanwhile, 'for more than a century,'
wrote Ramsay of Ochtertyre of the end of Queen Anne's reign,
'nothing of character had appeared in the dialect usually called
"broad Scots".' That was not true, and in his *Choice Collection
of Comic and Serious Scots Poems Both Ancient and Modern*
(1706–1711) James Watson, a patriotic printer in Ruddiman's
circle, had collected three volumes of Scots poetry of the previous
150 years. This is true: 'To render it polished and correct would
have been a Herculean labour, not likely to procure them much
renown.'[35] Fletcher of Saltoun had written in southern English
and Italian, Law of Lauriston in English and French. As for
Gaelic, to all the Edinburgh *literati* but Adam Ferguson the
ancient language of Scotland might as well have been double-
Dutch.

So, southern English it was to be. The intention of Henry Home
and his friends was not to sacrifice Scottish national culture to
southern politics but, by making it general, cosmopolitan, classical, businesslike, polite and loyal, to promote it. These were ener-

getic and ambitious Scots who wanted to act on the British and colonial stage; and for that they thought they must themselves adopt the speech of the metropolis. Henry Home, though he spoke 'pure Scots' in his social hour, when elevated to the bench in 1752 used a language that 'approached to English'.[36] David Hume's letters are peppered with warnings against purely Scottish usages, such as 'park' instead of 'enclosure' and 'a compliment' instead of 'a present'. He drew up a list of Scotticisms to be avoided which was published in the *Scots Magazine*.[37]

Over the course of the 1750s, as Great Britain increased in power and military success, the Select Society's interest in polite English became obsessive and ridiculous. In 1761 the elocutionist Thomas Sheridan, father of the playwright, who had trained Wedderburn and other Scots to perform at the bar in London, delivered sixteen lectures in St Paul's Chapel, Carrubber's Close, off the Canongate, to up to three hundred auditors, including young James Boswell, who immediately began to worship him.[38] Under the name of 'The Society for promoting the reading and speaking of the English Language in Scotland' the Select launched a fund to import English-language teachers to Edinburgh. Advertisements were taken out in the *Courant* and the *Caledonian Mercury*. A Mr Leigh was engaged.[39] Unfortunately, the project coincided with a period of vicious anti-Scots feeling in London. Subscribers fell away, and the Society expired in 1765.

Only one of Smith's Edinburgh lectures was printed.[40] His later lectures on similar topics given at Glasgow University after he joined the faculty in 1751 survive in the form of students' notes, while when Hugh Blair published his own lectures as Regius Professor of Rhetoric, he admitted they contained 'several ideas' on style he had derived from Smith.[41] 'We in this country,' Smith told his students in Glasgow (as he had no doubt told his Edinburgh auditors), 'are most of us very sensible that the perfection of language is very different from that we commonly speak in. The idea we form of a good stile is almost contrerary to that which we generally hear. Hence it is that we conceive the farther one's stile is removed from the common manner it is so much the nearer to purity and the perfection we have in view.'[42] This over-elevated sense of style was the heart of the Edinburgh problem. It was the source (alongside mere good nature) of Hume's consistent misjudgements in literature, and of the over-promotion of *Douglas* and the *Epigoniad*. In one of several brilliant passages in his lectures Smith recognised too an organic connection between the rise of commerce and the rise of prose as a medium of communication: 'No one ever made a Bargain in verse'[43] any more than anyone ever wrote a believable tragedy in a Lothian manse. But this insight was not carried through, and occasionally the reader feels the exasperation of Wordsworth, who wrote in 1815 that Smith was 'the worst critic, David Hume not excepted, that Scotland, a soil to which this weed seems natural, has produced'.[44]

In reality, Smith was less interested in giving practical advice about how to write and speak than in pursuing his love of system. As John Millar, his prize pupil at Glasgow, explained in a letter to Dugald Stewart, Adam Smith believed that the best method of 'explaining and illustrating the various powers of the human mind ... arises from the several ways of communicating our thoughts by speech, and from ... the principles of those literary compositions which contribute to persuasion or entertainment'.[45] In the one lecture that was published, as *Considerations concerning the first formation of Languages, and the different genius of original and compounded Languages*, Smith applied to the formation of language the philosophical method that Montesquieu had applied to law and government in his *De l'Esprit des Lois*, Henry Home to Scottish jurisprudence and David Hume to the history of religion. It was for this essay that Dugald Stewart coined the terms 'Theoretical or Conjectural History'.[46] They require a little explanation.

The eighteenth century had more ideas about the past than it had facts: archaeology and philology were infant sciences. (The twenty-first century has more facts than ideas.) Montesquieu's approach, which he derived from the ancient writers Tacitus and Aristotle,[47] was to apply logic in the absence of historical fact, devising a chain of cause and reasonable effect. 'In this want of direct evidence,' Dugald Stewart said in his eulogy of Smith read at the Edinburgh Royal Society in 1793, 'we are under the necessity of supplying the place of fact by conjecture; and when we are unable to ascertain how men have actually conducted themselves upon particular occasions, of considering in what manner they are likely to have proceeded, from the principles of their nature, and the circumstances of their external situation.'[48] In thrall to the prestige of Newtonian natural science, Edinburgh sought to overlook variations of age and locality and aspire to the universal: rather in the way that the law of gravity explained phenomena as diverse as an apple falling from a tree and the behaviour of the tides.

Human beings were pretty much the same over far-flung distances and epochs. What interested Edinburgh was not where man differed from man, but where men resembled one other. Deluged in comparative ethnography by the reports of travellers in North America and, later, the Pacific, the philosophers were neither parochial nor dogmatic. They preferred to point out how they resembled the Spartan or the Iroquois in thought or feeling rather than how they differed from the Spartan or the Iroquois in behaviour. Human history was no longer a degeneration from Paradise in which distinctions of race and tongue might be attributed to the Deluge and Babel; nor was it merely one damn thing after another; it was an explicable progression not so much of morality but of quality and number: from rudeness to polish and simplicity to refinement.

In the works of conjectural history we fly through time and space, gather the fruits of the earth or snare wild animals, tame and rear cattle, cultivate fields, acquire private property, promulgate laws, progress to opulence or luxury. We pass from Tartary to the halls of the ancient Germans, converse with Abraham and Lot, reside in antique Greece and Rome, hear the death-songs of the Iroquois: borrowing, as Dugald Stewart said of Montesquieu, 'lights from the most remote and unconnected quarters of the globe'.[49]

The dizzying scope of such histories and their haughty approach to matters of fact were to have a deplorable effect on both Edinburgh and the social sciences that drew their inspiration from the Edinburgh school: sociology, anthropology, political economy and the study of language. Of Lord Kames's *Historical Law Tracts* of 1758, which offered a sort of *De l'Esprit des Lois* for the lawyers of the Outer House, Hume commented with some justice: 'a man might as well think of making a fine Sauce by a mixture of Wormwood and Aloes as an agreeeable combination by joining Metaphysics and Scotch Law.'[50] Smith's account in his lecture of the origin of language, in which two cave-dwellers 'by mutual consent agree on certain signs' to distinguish their cave, tree and fountain,[51] is as fanciful as his explanation of the origins of commerce at the beginning of *The Wealth of Nations*. By the 1770s this Edinburgh approach had become the object of satire. In *Humphry Clinker*, Smollett lampooned the philosophers' passion for stories of native American hardihood and cruelty.[52] The poet Robert Fergusson invoked fashionable ethnography to describe sanitary conditions in Scotland: 'Nae Hottentot that daily lairs/'Mang tripe, or ither clarty [dirty] Wares,/Hath ever yet conceiv'd or seen,/Beyond the Line [the Equator] sic scenes unclean.'[53]

In 1752 Smith was appointed Professor of Moral Philosophy at Glasgow. John Millar, in his letter to Dugald Stewart mentioned earlier, remembered that Smith's lecture course had been divided into four parts. The first concerned natural theology, and looked at the proofs of God's being and attributes and (more to Smith's psychological taste) 'those principles of the human mind upon which religion is founded'.[54] The second part covered ethics, and consisted chiefly of the doctrines worked up in *The Theory of Moral Sentiments*, published in 1759. The third dealt with morality as codified into justice, and seems to have been based strongly on Montesquieu: an attempt to derive the principles of law and government and, as Smith himself put it in the *Theory*, to trace out 'the different revolutions they have undergone in the different ages and periods of society'.[55] The fourth concerned 'the political institutions relating to commerce, to finances, to the ecclesiastical and military establishments'.[56] These last two, for which students' lecture notes were discovered in 1890 and 1958, can be discerned as the framework supporting *The Wealth of Nations*.

Though he lived in Glasgow Smith took part in his vague way in all the projects of the Edinburgh *literati*. The Glasgow–Edinburgh stagecoach arrived in time for early-afternoon dinner.[57] He was elected in 1752 to the Philosophical Society, and two years later was a founder-member of the Select. In the first issue of the short-lived *Edinburgh Review*, he alone dared to criticise Dr Johnson's dictionary – as insufficiently 'grammatical', or analytical – in an article which may or may not have been the foundation of their vigorous and lasting enmity.[58] In the second and last issue, of January 1756, Smith attempted, prematurely, to establish the triangular intellectual trade – Paris, Edinburgh, London, Edinburgh, Paris – that was achieved seven years later with Hume's unexpected triumph in the *salons*.

In the course of the 1750s the Edinburgh circle became increasingly interested in commercial questions. Hutcheson, both in the *Short Introduction to Moral Philosophy* of 1747 and especially in Book II of his posthumously-published *System of Moral Philosophy* (1755), discussed value and price, money and interest with such enthusiasm as almost to forget the jurisprudential framework of his argument. In his *Political Discourses* of 1752 Hume included seven essays on commercial themes in which he demolished the prejudices of a hundred years: the prosperity of other nations – even France – was of benefit, not detriment, to Britain; there was no inherent conflict between trade and agriculture; precious metals were merely mediums of exchange; and so on. Hume's anxieties – about the proliferation of the public debt and the conversion of foreign markets into colonies – were not prejudices but premonitions. Lord Elibank published short essays on paper money and the national debt.

In 1755 the Select Society founded a subsidiary, 'The Edinburgh Society for encouraging Arts, Sciences, Manufactures, and Agriculture in Scotland', which was to meet on the first Monday of each month and award premiums, 'partly honorary, partly lucrative', for Scottish enterprise. As Hume wrote in great enthusiasm to Allan Ramsay junior, in the high preoccupations of the Select 'we have not neglected porter, strong ale, and wrought ruffles, even down to linen rags'.[59] Hume was optimist enough to believe, as he wrote in 'Of Refinement in the Arts' (1752), that 'the same age, which produces great philosophers and politicians, renowned generals and poets, usually abounds with skilful weavers and ship-carpenters'.[60] (Ramsay disliked this plebeian trend and withdrew from the Select Society to pursue fame in London.) With the outbreak of war with France, these commercial debates took on a new urgency: in his *Characteristics* of 1758, Wallace rebutted charges that Britain had become too indebted and 'financial' to confront France.

Smith never shone in the Edinburgh clubs. Carlyle said he heard him speak 'but once', when he opened the inaugural meeting of the Select Society in May 1754. Otherwise, he seems to have passed his social hours sunk out of sight in mental composition. 'He was the most absent man in Company that I ever saw,' Carlyle continued, 'Moving his Lips and talking to himself, and Smiling, in the midst of large Companys. If you awak'd him from his reverie, and made him attend to the Subject of Conversation, he immediately began a Harangue and never stop'd till he told you all he knew about it, with the utmost philosophical ingenuity. He knew nothing of characters.'[61] He used to shout during sermon, and revoked at whist.[62] The market-wives of the Edinburgh High Street thought he was mad.[63]

The Theory of Moral Sentiments, when it appeared in April 1759, showed precisely what Smith had been up to during his reveries. Never was there a more fascinated observer of his own mental state, and of the curiosities and customs of a provincial society. Never was small-town society – closed, alert, forensical, with not very much to do[64] – more philosophically evoked, down to its furniture,[65] tweezer-cases[66] and ear-pickers,[67] or its aldermen's wives, elbowing one another black and blue.[68] The *Theory* sought to explain not merely why we think such an action right and another wrong, but why, indeed, we do what we do. Hume, who was in

London to see the Tudor volumes of his *History* through the press, wrote on 12 April a letter which captured the spirit of Smith's book and of their friendship. The older man's gruelling disappointments are brought in to serve the triumph of the younger:

My Dear Mr Smith, have Patience: Compose yourself to Tranquillity: Show yourself a Philosopher in Practice as well as Profession: Think on the Emptiness, and Rashness, and Futility of the common Judgements of Men: how little they are regulated by Reason in any Subject, much more in philosophical Subjects, which so far exceeed the Comprehension of the Vulgar. *Non si quid improba Roma, Elevet, accedas examenque improbum in illa, Perpendas trutina, nec te quaesiveris extra* [If wicked Rome disparages you, don't seek to tinker with their false balance or look to anybody but yourself]. A wise man's Kingdom is his own Breast; Or, if he ever looks farther, it will only be to the Judgement of a select few, who are free from Prejudices, and capable of examining his Work. Nothing indeed can be a stronger Presumption of Falshood than the Approbation of the Multitude; and Phocion [an Athenian general], you know, always suspected himself of some Blunder, when he was attended with the Applauses of the Populace.

Supposing, therefore, that you have duely prepard yourself for the worst by all these Reflections; I proceed to tell you the melancholy News, that your Book has been very unfortunate: For the Public seem disposd to applaud it extremely. It was lookd for by the foolish People with some Impatience; and the Mob of Literati are beginning already to be very loud in its Praises. Three Bishops calld yesterday at Millar's Shop [in the Strand] in order to buy Copies, and to ask Questions about the Author: The Bishop of Peterborough said he had passd the Evening in a Company, where he heard it extolld above all Books in the World. You may conclude what Opinion true Philosophers will entertain of it, when these Retainers to Superstition praise it so highly … The Duke of Argyle is more decisive than he uses to be in its favour … Millar exults and brags that two thirds of the Edition are already sold, and that he is now sure of Success. You see what a Son of the Earth that is, to value Books only by the Profit they bring him. In that View, I believe it may prove a good book.[69]

Up to that point, philosophers had assumed that a man examined the morality of his own actions, and then used the results of that self-examination to judge the actions of others. Adam Smith's first innovation was to turn that on its head. He argued that we judge first of the morality of others, and then come back to judge ourselves. We put ourselves in the position of others, and by means of *sympathy* or *fellow-feeling* (which is the principle or gravity of Smith's system), we share – even for a just a moment – their joys and sufferings, pride, affections and dislikes, and from there judge their actions. In transferring those sympathetic judgements to ourselves – by means of a tribunal or judge to which Smith gave the character of an Impartial Spectator – we acquire our senses of duty or moral obligation, of dereliction and shame.

In Book III of his *Treatise* Hume conveyed, by means of an image from the infirmary, the tremendous power of sympathy:

> Were I present at any of the more terrible operations of surgery, 'tis certain, that even before it began, the preparation of the instruments, the laying of the bandages in order, the heating of the irons, with all the signs of anxiety and concern in the patient and assistants, wou'd have a great effect upon my mind, and excite the strongest sentiments of pity and terror.[70]

Smith's sympathy was altogether more serviceable. 'As we have no immediate experience of what other men feel,' he wrote at the opening of the *Theory*, 'we can form no idea of the manner in which they are affected, but by conceiving what we ourselves should feel in a like situation.'[71] This sympathy is carried through into the moral arena. We judge of the propriety of the feelings of others only by their coincidence with those we feel when we put ourselves in the same circumstances. From the propriety of feeling, we judge the merit or demerit of the actions that arise from those feelings.

From those premises, Smith conjured a bewildering array of social phenomena. For sympathy to function as a social mechanism, emotions must be calibrated. The spectator attempts as much as possible to raise the pitch of his sympathetic emotion, while the person displaying emotion attempts to restrain it to the level of the spectator. Both strive to reach what Smith, in a phrase which might be chosen to caption his age, described as a 'pitch of moderation',[72] from which various social virtues instantly appear: on the part of the sympathiser, attentiveness and human-ity; on the part of the sympathised-with, self-government and self-control.

We approve of displays of emotion only in so far as we can sympathise with them. Shrieks of pain or fright are unbecoming because the spectator cannot hope to match in sympathy the pain or terror felt by the sufferer. Ingeniously, Smith went on to argue that we sympathise more strongly with displays of social or benevolent emotions than with shows of resentment or hatred, or mere bursts of egotistical joy or sorrow. Through sym-pathy, we admire the rich and successful, follow the fashions they set, pursue ambition beyond our animal requirements, embrace distinctions of social class, rejoice to see justice done. Thus, the British find to their enduring surprise that good manners, fair-ness and knowing-one's-place are quite philosophical: that they have, as one of Molière's characters discovered, been speaking prose all along.

But what of our own actions and obligations? After all, we cannot sympathise with ourselves without going round and round, like a dog chasing its tail. Really, there was no end to Adam Smith's ingenuity. To judge the propriety of our feelings and the merit of our actions, we invoke the judgement not of conscience – such as might operate on a desert island – but of society itself. Aware of what our society thinks, we become spectators of our own appearance and behaviour, and make our judgements of both. In other words, society is 'the only looking-glass by which we can, in some measure, with the eyes of other people, scrutinize the propriety of our own conduct'.[73]

But society is ill-informed, partial and prejudiced. In one last twist to the argument, Smith suggests that we submit not only to the actual judgement of our peers but also to a higher instance, 'the supposed impartial and well-informed spectator, to that of a man within the breast'.[74] This 'man within', a collection of general rules about what is to be done and what avoided, is a balance to worldly misjudgement and our own self-delusion; or, as Burns had it, 'O wad some Pow'r the giftie gie us,/To see oursels as others see us!' as he stared at the louse on a girl's bonnet in Mauchline church.

The *Theory* entranced a public caught up in the cult of sensibility and virtue. Yet Smith's sensibility was sociable, not isolated and melancholy as in Gray's *Elegy* or *The Deserted Village*, or gothick as in Hugh Blair's lectures on literature, or suicidal as in Goethe's *Werther*. The more violent emotions, those which fascinated Romantic literature, must be toned down to yield their ethical value or, in the case of sexual love, suppressed. Smith's concern was with 'smaller joys from less important causes'. There was no place for heroism or zealotry in his picture: no Cato, but also no Presbytery.

The beauty of the *Theory* was its anti-authoritarian character, its sociability and its optimism. It was easier to believe that we draw our notion of ourselves, and of the purposes of existence, from others, rather than from some fabulous Hutchesonian moral faculty. Yet to replace conscience with society and schemes of objective morality with imaginary social judgement is to be bereft of guidance in any society which has gone off the moral rails (as, for example, SS officers who sympathised with their colleagues in the labours of gassing Jews, but not with the murdered men, women and children). Under criticism from Sir Gilbert Elliot and Hume, Smith tinkered with the Impartial Spectator in later editions, but without fully resolving the problem.

Smith himself was uneasy about a society forever gaping at the rich and fortunate. In a section that might have made a sentimental novella, he wrote of a poor man's son, 'whom heaven in its anger has visited with ambition'.[75] The boy pursues riches that are merely 'enormous and operose machines to produce a few trifling conveniences ... immense fabrics, which it requires the labour of a life to raise, which threaten every moment to overwhelm the person that dwells in them'.[76]

The way out for Smith is not moral but what would today be called economic. Riches are a mere 'deception', but none the less they keep in motion the industry of mankind.[77] Communities are held together not by affection or obligation but by 'a mercenary exchange of good offices according to an agreed valuation'.[78] The rich follow their own footling lives, but are obliged by the mechanism of money to share some of their riches with the poor.

> The rich only select from the heap what is most precious and agreeable. They consume little more than the poor, and in spite of their natural selfishness and rapacity, though they mean only their own conveniency, though the sole end which they propose from the labours of all the thousands whom they employ, be the gratification of their own vain and insatiable desires, they divide with the poor the produce of all their improvements. They are led by an invisible hand to make nearly the same distribution of the necessaries of life, which would have been made, had the earth been divided into equal portions among all its inhabitants.[79]

Smith's readers may accept the last sentence, or not. What is certain is that already, in 1759, he had the germ of the self-regulating commercial society of *The Wealth of Nations*.

The thread that runs through Smith's thought is the notion of the ancient Stoic philosophers that the world is organised in such a way that all activities and propensities, selfish and unselfish, combine for the benefit of the whole.[80] Smith was aware of the moral and economical paradoxes arising from that doctrine, and sought all his life to resolve them.

In the *Theory of Moral Sentiments*, the individual becomes so dependent on and absorbed in society that he or she somehow vanishes into that society, which becomes its own self-managing organism and has no need for an external moral authority. In *The Wealth of Nations*, the government – whether Prince or parliament – has dwindled into a mere drudge whose purpose is to undertake a few unprofitable public works. God as First Cause has all but vanished from explanations of phenomena; or, rather, survives only as a sort of phantom or vestige, the ghostly Invisible Hand. The Smithian system – whether literary, social or political-commercial – is like some great edifice that rests on decrepit foundations which are then, with great pains and skill, dug out from under it. The building trembles, then settles on a new foundation. Henceforward it will support itself. A new ideology is born.

6

The Thermometer of the Heart

Moffat, a small town in the green hills of Dumfries some fifty miles south-west of Edinburgh, was known in the eighteenth century as the 'Spa of Scotland'. Robert Wallace had been minister at Moffat in the 1720s and early 1730s, and Boswell was twice sent to take the waters and drink goats' milk for scorbutic or depressive ailments – and fell in love with his fellow-patients. The playwright John Home, a celebrity after his triumphs with *Douglas* and *Agis* in Edinburgh and London, liked to spend two or three weeks in the town each year.

One day in September 1759,[1] while walking on the Bowling Green, Home was introduced to a lanky young man named James MacPherson. MacPherson, who was just coming to his twenty-third birthday, was acting as tutor to the son of a Perthshire laird, young Graham of Balgowan, and staying with the boy's grandfather, Lord Hopetoun. MacPherson was a Highlander, and a published poet, and ambitious even by the measures of Scotland and his time.

Years later, Home told an investigating committee of the Highland Society of Scotland that he had been 'not a little pleased' to discover a Highlander who was also 'an exceeding good classical scholar'.[2] In conversation with MacPherson Home had raised the subject of ancient Gaelic poetry and, when MacPherson said he had several pieces with him, pressed him to translate a specimen into English. A day or two later, Home related, MacPherson returned with 'The Death of Oscar'.[3]

On 2 October Alexander Carlyle passed through Moffat on his way back from Dumfries. Stopping to dine with Home, he was introduced to the Highlander. Carlyle remembered MacPherson as a tall, good-looking young man wearing unfashionable long boots, evidently to hide his thick legs. MacPherson seemed both shy and touchy, and made some excuse not to join the two friends at dinner.[4] When they were back in Edinburgh, Carlyle and Home showed the 'Oscar' fragment to Hugh Blair, Professor of Rhetoric and *Belles-lettres* at the College, Adam Ferguson, Professor of Moral Philosophy, Principal William Robertson, and Lord Elibank. In high excitement, Blair summoned MacPherson to town and begged him to translate more.

Thus was born Ossian, son of Fingal, last survivor of the heroic age.

At the simplest, the works attributed by MacPherson to Ossian and allegedly translated from ancient Gaelic texts were a vulgar literary fraud. Even before Dr Johnson, who had his own crypto-Jacobite purposes in the Highlands, labelled *Fingal* 'an imposture',[5] critics in England were demanding to see the Gaelic manuscripts of the 'ancient epick poems' MacPherson claimed to have translated. London wanted to see Gaelic manuscripts of the third century AD, and MacPherson would not produce them.

A committee of the Highland Society, formed in 1797 under Henry Mackenzie and reporting in 1805, could find not a single Gaelic poem in manuscript that corresponded 'in title and tenor' to what MacPherson had published. The committee resolved, with excruciating reluctance, that MacPherson had used plots and passages from popular Gaelic ballads, taking various sentimental and polite 'liberties' with them 'and elevating what in his opinion was below the standard of good poetry'.[6] A more straightforward and vigorous treatment the same year by Malcolm Laing[7] noted so many resemblances to Virgil, Homer and Thomson as to convince the anonymous critic of *The Edinburgh Review* – who reads very much like Walter Scott – that 'the writer of Ossian's poems was habitually familiar with modern poetry'.[8]

These judgements have survived in more or less severe form to the present day, though it was not until the 1950s that a dozen Gaelic ballad sources of MacPherson's *Fingal* – and his scheme of history from the Irish historians – were finally run to earth.[9] It is now believed that stories from two Irish saga cycles, of Cù Chulainn and Finn – MacPherson's Cuchullin and Fingal – survived in both oral and manuscript versions in the Scottish Highlands up to MacPherson's time. They had intermingled, and also attracted to themselves semi-historical traditions from the much later epoch of the Viking invasions. MacPherson's innovation was to stitch them all together, remake ancient Scotland from a cultural vassal of Ireland into its equal or even overlord, appropriate the saga hero Finn and his Fenians to the Highlands – which had been a literary vacuum since Tacitus – and drape them all in the sentimental and feminist clothes of his own time.

There is now much less interest in manuscript authenticity, and modernity has come to regard all literature as fiction. From the hither bank of the Romantic Movement, MacPherson's Ossian can be seen as a historical symptom. Government reprisals against the Highland clans after Culloden cleared out not just a population and a political authority, but also a traditional culture. The Ossianic poems were an attempt to repopulate those wild spaces with something other than disaffected Jacobites and Roman Catholics. Before there could be sheep, and factory lairds in new tartan, there must be pale phantoms of boundless chivalry and sensitivity.[10]

Yet Ossian was not the creation of a single rootless Highlander: the blind Celtic bard was a fulfilment of the wishes of the Edinburgh *literati*, a figment of their rash aestheticism and their almost total ignorance of the Highlands. The chief witnesses reported that MacPherson was very reluctant to translate the poems. MacPherson himself, in a learned preface to *Fingal*, said that John Home – or, rather, 'a gentleman, who has himself made a figure in the poetical world' – suggested he should put them into English prose.[11] MacPherson was drawn ever deeper into Ossian by the encouragement first of Home and then, fatally, of the Professor of Literature at the College, Hugh Blair. It was the Edinburgh *literati* who financed MacPherson's trip north in the summer of 1760 to look for what Blair unwisely called 'our epic'.[12] That epic, which MacPherson called *Fingal*, was substantially composed in Blackfriars' Wynd, in an apartment directly below Blair's lodgings. As Blair wrote to David Hume on 29 September 1763: 'MacPherson was entreated and dragged into it.'[13] The High-Flying ministers and preachers, by contrast, always regarded Ossian with suspicion and the poems as 'vain, hurtful, lying earthly stories'.[14]

Ossian was an accident waiting to happen. That day in Moffat, John Home was looking for a new subject to bring to the stage.[15] Hugh Blair, the first and at that time only Professor of English Literature in Great Britain, needed justification for his literary aestheticism, and his passionate sentimentality. For Lord Kames, at work on his *Elements of Criticism*, mere literary standards were much less important than the sentiments of grandeur and novelty excited by a work of art. The Rousseauans were looking for noble and courageous natural humanity located nearer at hand than the Americas. The bellicose Reverends Carlyle and Ferguson sought stimulation for their martial fantasies, Adam Smith for his progressive view of human history. David Hume wanted to help a Scots poet, as he had helped poor, blind Blacklock.[16] The Edinburgh women wanted heroes who, unusually for any epoch, were chivalrous, affectionate and monogamous.[17] Everybody was mad for savagery, the primitive, historical conjecture, the sublime, obscurity, vastness, the infinite, sad and fuscous colours, clamour, suddenness, and the Joy of Grief.[18] The sentimental English poet William Shenstone – of all people – saw what was happening. They were living in an age, as he told a Scots correspondent in 1761, where Taste was in better supply than Genius. 'This Turn ... favors ye Work the Translator has to publish.'[19]

Above all, Edinburgh needed a Scots masterpiece. The *literati* were fed up with John Bull. The grievances of Coltbrigg and Culloden were compounding daily. The English-speaking campaign of the Select Society had been answered in London by Garrick's rejection of *Douglas*, the failure of Wilkie's *Epigoniad*, and scurrilous attacks on the Prince of Wales's Scottish favourite, the Earl of Bute. A campaign by Ferguson and the *literati* to allow Scotland to muster its own militia against the possibility of an attack from France had foundered on London's prejudice in the matter of Scots loyalty. The British military victories of 1759 at Quebec, Minden and Quiberon Bay twitched a curtain to show a glimpse of the future: a colossal commercial empire administered from London in which Scotland was a mere province and Scotsmen no more than colonial cannon-fodder. The Ossianic heroes were without ambivalence, effeminacy and luxury. They defended the British Isles against foreign invasion. In the words of the modern American scholar of eighteenth-century Edinburgh, Richard Sher, Ossian was 'a poetical response to a political crisis'.[20]

Edinburgh had tried sympathy, benevolence and scepticism, Quixotry, deism, clubs and claret, and none was quite satisfactory. Why not ghosts and wistful lyricism? MacPherson, the first literary Frankenstein, repaired all Edinburgh's national and military and sentimental injuries. So much for answered prayers. MacPherson left Ossian to his fate, and it was Edinburgh that had to defend the poems against the ridicule of London.

The Fragments of Ancient Poetry Collected in the Highlands of Scotland and Translated from the Galic or Erse Language is a modest enough little pamphlet. Yet coming to it without prejudice or, even better, by accident or in ignorance, the reader may feel a tremor of excitement, as of having crossed into a new world:

My love is a son of the hill. He pursues the flying deer. His grey dogs are panting around him; his bow-string sounds in the wind. Whether by the fount of the rock, or by the stream of the mountain thou liest, when the rushes are nodding with the wind, and the mist is flying over thee, let me approach my love unperceived, and see him from the rock. Lovely I saw thee first by the aged oak; thou wert returning tall from the chace; the fairest among thy friends.[21]

Boswell, who was twenty and keeping his journal, wrote: 'As to myself, I can only say ... that when the fragments of Highland poetry first came out, I was much pleased with their wild peculiarity.'[22] The English poet Thomas Gray allowed the most intense enthusiasm to drown his lively misgivings: 'I was so struck, so *extasié* with their infinite beauty.'[23] Even now, the first lines of the *Fragments* transmit an authentic thrill – like Captain James Cook's long entries from the Antarctic pack ice, say, or Goethe's *Winter Journey in the Harz*, but with a decade and a half of priority. There is a sense of assisting at the birth of a new way of looking at the world.

In *Fingal*, published in December 1761, MacPherson starts to toy with his public. To the purple passages he appends footnotes and invites comparison with equivalent scenes in Homer or Virgil or Milton. There is something wilful about his effrontery, like a thief parading in stolen finery. Look, he says, see how I outdo my epic models. By the time of *Temora*, published in 1763 with a dedication to the Earl of Bute, no less, MacPherson has become so careless or so self-confident that he has all but dispensed with Gaelic sources.[24] His truculence is such that he taunts the Irish with the cultural priority of Scotland. There is an air of unvarying and tedious solemnity. Whole passages read like *The Song of Songs* from the King James Bible. More sober Scots critics – notably David Hume – wake from their Celtic reveries.

James MacPherson, christened by Boswell 'the Sublime Savage',[25] was born at Ruthven in the upper Spey valley on 27 October 1736. The district, known as Badenoch, had come out for Prince Charles Edward in 1745 under Ewen MacPherson of Cluny, the so-called 'Cluny of the Forty-five'. As a boy James would have witnessed the burning of the government's barracks at Ruthven in February 1746 as the Highland army moved northwards towards catastrophe at Culloden Moor; and the counter-destruction of Cluny Castle by three hundred men sent up by the Earl of Loudoun after the battle. Perhaps he joined his cousin Allan in stoning the redcoats.

Badenoch was sullen under its military occupation. His estate forfeit to the Crown, Cluny vanished among his people. Despite a bounty of a thousand guineas on his head, he was able to hide out under the threat of the rope and the quartering-block for nine years before escaping to France.

By then, James MacPherson had left the district. In 1752 he attended the ancient University of Aberdeen, long the centre of Scottish episcopal civilisation and Latin study. Of the professors he would have known, Thomas Reid, who was working to refute Hume's scepticism, was his regent in the second year. Alexander Gerard, whose influential *Essay on Taste* was published in the *annus mirabilis* of 1759, had just been called to the chair of Natural Philosophy. There were two celebrated classical scholars: Thomas Blackwell, who had written an enthusiastic commentary on Homer, and William Duncan, whose translation of Caesar's *Commentaries* (1753) sounded a note of mourning for the martial spirit of Roman times.

Even at college, Ramsay of Ochtertyre reported, MacPherson 'gave indications of that harsh, overbearing spirit which make [*sic*] him so unpopular in the after-part of his life.'[26] He says MacPherson tormented a student named MacHardy for his poverty and awkwardness in 'Hudibrastic verses' till a professor intervened. That is very much the uncharming MacPherson of Boswell's *London Journal*. Ramsay of Ochtertyre claimed he was at some point destined to be an episcopal minister, but the reprisals taken against non-jurors after 1745 scarcely recommended that as a career for a Highlander.[27]

MacPherson seems also to have spent some time studying at Edinburgh. In May 1755 he entered print for the first time with a poem in the *Scots Magazine*, which ever since its launch in 1739 by the printers Murray and Cochrane had sought to encourage

new Scots verse.[28] 'To a Friend mourning the death of Miss ...'
was a conventional elegy, such as might be expected of a college
boy of that era, a feminised version of Horace's elegy for
Quintilius, *Odes*, I:24. Malcolm Laing, whose contention was
that only when MacPherson failed as a poet in English did he
adopt the Gaelic persona of Ossian, attributed several other
English poems to him.[29] Of the poems he printed, 'The Hunter'
has an authentic enough sound to it. It tells how a Highland hero,
one Donald, as a punishment for killing her faun is cursed with
ambition by a fairy princess, and goes down into a very 1750s
Edinburgh:

> Once the proud seat of royalty and state,
> Of kings, of heroes and of all that's great;
> But these are flown, and Edin's only stores
> Are fops, and scriveners, and English'd whores.[30]

Donald takes command of the Scottish lowland forces, defeats an
English army, and marries well.

In 1756 MacPherson was back in Ruthven without purpose or
profession, teaching at his old grammar school. Years later – in
1797 – his clansman Donald MacPherson remembered coming
across the poet in the fields on the way to Kingussie Kirk. 'I mind
perfectly well that ... upon a Sunday ere publick service began at
Kingussie, I met Mr MacPherson off the road near the Church
walking alone. Upon joining, I found him more morose, silent &
pensive than usual. With a serious feeling, I asked the reason. He
said, with the same humour, that he was quite wearied of teaching a Schooll, was at a loss how to acquire genteel bread.'[31]
Donald reassured him:

> I told him he was blessed with several talents, singular good memory,
> particularly poetry, did he hit right, he had no reason to perish so
> soon. He said there was little room there's o for him. I answered that
> there was a theme and meaning out for which I was very sorry
> untouched hitherto which in the general greatly pleased whoever
> heard it less or more & by some admired & by me to witt Ossean's

poems, & for instance I heard my father tell that my Grandfather John MacPherson of Benchar would different times cause my father sit down by him & write some of them from his mouth & strongly recommend to him minde to adher to some passage as a good rule for life it might do him honor which my father might still enlarge upon did he ask him. Mr MacPherson heard me without a word from him in the same humour.[32]

Donald went on to recall that they had shared a house in Edinburgh where James worked, as part of this project, on the indubitably authentic 'The Highlander', which appeared in 1758. The story is similar to that of 'The Hunter', only this time Scotland is saved – as in Home's *Douglas* – from the Danes. The poem was not a success, and though MacPherson did manage to escape from Ruthven, it was only to that equivocal position of tutor to the son of a minor laird, Graham of Balgowan.

Like MacPherson, the *literati* were converging on the Gaelic. In his *Ode on the Popular Superstitions of the Highlands of Scotland, Considered as a Subject of Poetry* of 1749 or 1750 the English poet William Collins boasted there were still to be found in the Highlands, 'Taught by the father to his list'ning son, / Strange lays, whose power had charmed a Spenser's ear'. According to Henry Mackenzie's note of his interview with Home in the Highland Society's *Report*, Adam Ferguson, 'who understood Gaelic, had told Mr Home that there were, in the Highlands, some remains of ancient poetry in the Gaelic language'.[33] Ferguson, who was brought up in Atholl on the edge of the Highlands, mentioned one Gaelic poem which he had heard repeated and thought very beautiful.

The *literati* would all have seen the *Scots Magazine* of January 1756 in which was published a very free rhymed translation of a Gaelic ballad entitled 'Albin and the Daughter of Mey'. The poet/translator was Jerome Stone, a young man from Dunkeld on Tayside, at the southern gates of the Highlands, who had been collecting old ballads. More striking than his eighteenth-century verse was a preface, dated 15 November 1755, in which Stone claimed for Gaelic poetry all the new and fashionable literary virtues:

> Several of these performances are to be met with, which, for sublimity of sentiment, nervousness of expression, and high-spirited metaphors, are hardly to be equalled among the chief productions of the most cultivated nations. Others of them breathe such tenderness and simplicity, as must be greatly affecting to every mind that is in the least tinctured with the softer passions of pity and humanity.[34]

Who knows but that MacPherson also saw those words, and in them the possibility of 'genteel bread'? A year and a half later, Gray's 'The Bard: A Pindaric Ode', which had its setting in Wales, supplied the character of an old, blind Celtic poet.

So Home and MacPherson met, like ship and iceberg:

> Conversing with Mr Macpherson, Mr Home found that he was an exceeding good classical scholar; and was not a little pleased that he had met with one who was a native of the remote Highlands, and likely to give him some information concerning the ancient poetry of his country. Accordingly, when Mr Macpherson was questioned on that subject, he said that he had in his possession several pieces of antient poetry. When Mr Home desired to see them, Mr Macpherson asked if he understood the Gaelic? 'Not one word.' 'Then, how can I show you them?' 'Very easily,' said Mr Home; 'translate one of the poems which you think a good one, and I imagine that I shall be able to form some opinion of the genius and character of the Gaelic poetry.' Mr Macpherson declined the task, saying, that his translation would give a very imperfect idea of the orginal. Mr Home, with some difficulty persuaded him to try, and in a day or two he brought him the poem on the death of Oscar.[35]

The simple language and rhythmical prose of 'The Death of Oscur' entranced Home. Thus, in Laing's acid comment upon the Edinburgh *literati*, MacPherson was 'conducted, step by step, to other fabrications, by their importunity and zeal'.[36]

Events moved quickly. Over the winter of 1759–60, according to Ramsay of Ochtertyre, the 'Fragments' were circulating in the

Scots capital and 'exceedingly admired by philosophers and ladies'.[37] Hume, as might have been expected, was a little sceptical as to their authenticity, but was reassured by John Home.[38] Hume in his turn reassured Thomas Gray, who had been sent some of the pieces before publication by way of Sir David Dalymple and Horace Walpole; had written to MacPherson; and had been not at all charmed by the replies. Adam Smith pitched in to tell Gray that 'the Piper of the Argyleshire Militia repeated to him all those wch Mr Macpherson has translated, & many more of equal beauty'.[39] In June the 'Fragments' were published, with an unsigned introduction by Hugh Blair.

Blair, now at the peak of his career at the High Kirk and the College, was the doyen of literary taste in Edinburgh. He had ability but lacked humour, and was fussy about his appearance and standing. Burns skewered him: 'In my opinion Dr Blair is meerly an astonishing proof of what industry and application can do. Natural parts like his are frequently to be met with ... He has a heart, not of the finest water, but far from being an ordinary. – In short, he is truly a worthy and most respectable character.'[40]

Beneath a pose of frigidity, Blair could not conceal his excitement over MacPherson's work. 'The public', he began rashly, 'may depend on the following fragments as genuine remains of ancient Scottish poetry.'[41] In the manner of the philosophical historians, he went on to claim that the spirit of the poems and the ideas and manners belonged to the 'most early state of society'.[42] From there he bustled on to what was his chief interest, the hypothetical epic from which he had convinced himself the *Fragments* derived. He said there was reason to hope that 'one work of considerable length, and which deserves to be styled a heroic poem, might be recovered and translated'.[43]

Yet MacPherson proved very reluctant to go in search of the grail-epic, and remained in Balgowan. The reason was not long concealed. 'But, Sir,' he wrote to Blair from there on 16 June 1760, 'a journey thro' the Highlands and isles is attended with risque and Expence that are not proper for me to incur on my own Bottom.'[44] In other words, he wanted money. Though like others he later deprecated the eruption of money into feudal society, he was not prepared to rely on traditional Highland hospitality for his maintenance during this search. Blair arranged a dinner at which Lord Elibank, Robertson, Home and Ferguson, among others, pledged themselves to defray MacPherson's expenses. At least a hundred pounds was raised in Edinburgh. Hume, Boswell, the pious and hard-working lawyer Sir David Dalrymple and Lord Kames were among the subscribers, as was the poet Thomas Gray.

With money in his pocket, and accompanied by his clansman Lachlan MacPherson of Strathmashie, a good scholar and a popular man in the Highlands, James set off in August for the Western Isles, by way of Inverness.

The two men were entering a world being transformed. In the lands north of the Forth and Tay, where according to Alexander Webster's statistics of 1755 some 652,000 out of 1,265,380 Scots were living,[45] oral poetry was dying or dead. Edward Burt, the surveyor sent north by General Wade to build roads to pacify the clans in the 1720s, had reported that each Highland chief kept a bard who 'celebrates, in Irish verse, the Original of the Tribe, the famous Warlike actions of the successive Heads'.[46] By the 1760s, of these only the MacMhuirichs, hereditary bards of the old Lords of the Isles before their incorporation in the Kingdom of Scotland, survived, as pensioners of the Macdonalds of Clanranald at Benbecula in Uist.[47]

In his 'Dissertation on the Antiquity of the Poems of Ossian', prefaced to *Fingal*, MacPherson gave a 'philosophical' analysis of the change in the Highlands since the Forty-five. 'The introduction of trade and manufactures has destroyed that leisure which was formerly dedicated to hearing and repeating the poems of ancient times ... When property is established, the human mind confines its views to the pleasure it procures. It does not go back to antiquity, or look forward to succeeding ages. The cares of life increase, and the actions of other times no longer amuse.'[48] In his evidence to the Committee of the Highland Society, MacPherson's friend Andrew Gallie, minister at Brae Badenoch, reinforced the point: 'The pride of ancestry, the *fortia facta patrum* [heroic deeds of our forefathers: Virgil], are obsolete themes: the pressure of the times, the change of system, have brought forward other feelings and speculations.'[49]

The missionaries of the Society in Scotland for Propagating Christian Knowledge (SSPCK) were set not merely on eradicating Roman Catholicism in the Highlands, but on inculcating loyalty and English. No Gaelic was taught at SSPCK schools. As Adam Ferguson wrote at the end of the century, Gaelic was the language of the cottage not the parlour, 'its greatest elegancies ... to be learned from herdsmen or deer-stealers ... connected with disaffection and proscribed by government'.[50]

None the less, at the house of the minister Donald MacLeod at Glenelg, on the mainland opposite Skye, James and Lachlan MacPherson heard from two men of the glen the description of Cuchullin's horse and chariot that appears, very feebly according to MacLeod, in Book I of *Fingal*.[51] The transcription seems to have been done by Lachlan MacPherson, who told Blair that he 'took down from oral tradition, and transcribed from old manuscripts, by far the greatest part of those pieces he [James MacPherson] has published'.[52] James, by his own admission, was not good at reading Gaelic handwriting.

In Skye they called on Dr John MacPherson, a well-known Gaelic scholar and minister at Sleat, but the visit was not productive. But at Portree, the landlord James MacDonald summoned Alexander MacPherson, a blacksmith who was noted for his knowledge of Ossianic poems. In giving evidence to the Highland Society, Alexander's brother Malcolm later testified before Justices of the Peace that James MacPherson spent four days and nights in Portree taking down Alexander's recitations.[53] Malcolm also gave evidence of the poet's rapacious way with manuscripts. He deposed that Alexander had possessed a

Gaelic manuscript in quarto, and about an inch and quarter in thickness; that he procured the said book at Lochcarron, while an apprentice there: that he heard his said brother almost daily repeat the poems contained in the said manuscript, which wholly regarded the

Fions or Fingalians ... That before Mr Macpherson parted with the declarant's brother, the said Mr Macpherson observed that, as the declarant's brother could repeat the whole of the poems contained in the manuscript, he would oblige him if he would give him the said manuscript, for which he might expect his friendship and future reward: That his said brother informed the declarant he had accordingly given the said manuscript to Mr Macpherson, who carried it with him; since which time, the declarant never heard of it. Farther declares, That he heard his father often reprimand his brother for answering the frequent calls upon him to the house of Portree, to repeat the poems of Ossian to gentlemen...[54]

At some point Lachlan MacPherson returned to the mainland and his place was taken by Ewan MacPherson. This young man, also from Badenoch, a schoolmaster with a much better grasp of Gaelic orthography than James MacPherson, had been staying with the minister at Sleat. He wanted to be off home, but reluctantly agreed to accompany James to the MacLeod seat at Dunvegan. Once there, he 'was compulsively obliged'[55] by Colonel MacLeod of Talisker and others to embark for Uist. Still sour years later at the memory of his detention, Ewan painted an unflattering portrait of James's Gaelic scholarship as exhibited in Uist, deposing in 1800

That they landed at Lochmaddy, and proceeded across the Muir to Benbecula, the seat of the younger Clanranald: That on the way thither, they fell in with a man whom they afterwards ascertained to have been Mac Codrum the poet: That Mr Macpherson asked him the question 'A bheil dad agad air an Fhéinn?' by which he meant to enquire, whether or not he knew any of the poems of Ossian relative to the Fingalians; but that the terms in which the question was asked, strictly imported whether or not the Fingalians owed him anything.[56]

Mac Codrum replied, wittily, that if they did, alas, it was a bit late to hope for repayment.

Over three to four weeks, Ewan said, he recorded several recitations of Ossianic poems, 'which he gave to Mr MacPherson, who was seldom present when they were taken down'.[57] They visited the older Clanranald at Ormiglade, spent a week with the younger Clanranald at Benbecula, and then moved to the house of the minister, Angus MacNeill. There they met the bard Niel or Neil MacMhuirich, who evidently recited to them the poem of 'Dar-Thula', which MacPherson later published with *Fingal*, and handed over a manuscript of another poem, 'Berrathon' and, according to MacNeill, 'three or four more'.[58]

Niel's son Lachlan MacMhuirich, eighteenth in direct descent from the founder Murach, could justly claim to be the last of the bardic line. He later testified that his father had 'custody' of works of Ossian written on parchment, some loose, some stitched up into books. In addition, there was a book 'near as thick as a Bible, but that it was longer and broader', called the 'Red Book' or *Leabhar Dearg*, made of paper, which Niel 'had from his predecessors, and which, as his father informed him, contained a good deal of the history of the Highland Clans, together with part of the works of Ossian'.[59] He added that he 'remembered well that Clanronald made his father give up the red book to James Macpherson from Badenoch'.[60]

It seems after all that MacPherson not only found some manuscripts of moderate age but also played a role in the preservation of old Gaelic poetry. Lachlan MacMhuirich told the Justice of the Peace at Barra in 1800 that 'none of those books are to be found at this day, because when they [his family] were deprived of their lands, they lost their alacrity and zeal'. Some of the parchments were cut up for tailors' measures or patterns.[61] Andrew Gallie remembered Clanranald telling him that he had not known of the existence of the manuscripts 'till, to gratify Mr MacPherson, a search was made among his family papers. Clanronald added, that, since Mr MacPherson's visit, more volumes were recovered.'[62] Altogether James MacPherson collected nineteen Gaelic manuscripts, some of which he later deposited with his publishers in London for the sceptics to examine. None contained the epic.

By 27 October 1760 MacPherson was back in Badenoch, for he wrote from Ruthven that day to the minister at Amalrie, James

MacLagan, asking to see his collection of Gaelic poems. In his letter, he stated his belief that the old poems had been corrupted by oral transmission and needed to be 'restored'.[63] At some point he joined Andrew Gallie at Brae Badenoch, for the minister's young wife Christian remembered the two men arriving with 'two ponies laden with old manuscripts'. She barely saw her husband for six weeks.[64] Lachlan MacPherson of Strathmashie appears to have joined them.

Gallie's letters to the investigating committee, written forty years later, give a picture of *Fingal* that winter.

> At that time I could read the Gaelic characters, though with difficulty, and did often amuse myself with reading here and there in those poems, while Mr Macpherson was employed on his translation. At times we differed as to the meaning of the certain words in the original.[65]

MacPherson may genuinely have thought there was a complete Gaelic epic to be found on the West Coast and the Isles, of which the ballads were but the *disiecta membra*, the scattered limbs. He was not satisfied with the ballads he found, and thought the MacMhuirichs were like the copyists of a classical text, for ever garbling or corrupting the authentic poesy of antiquity. As for MacPherson's own language, Gallie wrote to a colleague:

> I remember Mr Macpherson reading the MSS. found in Clanronald's, execrating the bard who dictated to the amanuensis, saying, 'D—n the scoundrel, it is he himself that now speaks, and not Ossian.' This took place in my house, in two or three instances.[66]

He was impressed less by MacPherson's transcriptions than by his compositions:

> I recollect (it was often matter of conversation), that by worm-eating, and other injuries of time there were here and there whole words, yea lines, so obscured, as not to be read; and I, to whom this was then better known than to any else, one excepted, gave great credit to Mr Macpherson; concluding, that if he did not recover the very words

and ideas of Ossian, that the substitution did no discredit to that celebrated bard; and this, as I told you, I then considered one of Mr
Macpherson's chief excellencies.[67]

By 16 January 1761 MacPherson was in Edinburgh, staying in
the apartment below Dr Blair's in Blackfriars' Wynd. Adam
Ferguson called, and examined various fragments. He noted that
'the paper was much stained with smoke, and daubed with Scots
snuff'.[68] Ramsay of Ochtertyre was taken by the merchant Robert
Chalmers to visit the poet, now 'at the back of the Guard', in a
small room littered with books and manuscripts, some of which
bore 'marks of the rust of antiquity'. Ramsay found MacPherson
a 'plain-looking lad, dressed like a preacher. What he said was sensible, but his manner was starch and reserved.'[69]

MacPherson was still having trouble reading the manuscripts.
He told the Reverend James MacLagan: 'I find none expert in the
Irish orthography, so that an obscure poem is rendered doubly so,
by their uncouth way of spelling.' Even so, he added – not truth-
fully, it appears – 'I have been lucky enough to lay my hands on a
pretty complete poem, and truly epic, concerning Fingal.'[70]

In February he set off for London with Chalmers, his manuscript, and letters of introduction to various publishers. With
more generosity than wisdom, Hume had recommended him to
the bookseller William Strahan as 'a sensible, modest young
Fellow, a very good Scholar, and of unexceptionable Morals'.[71]
According to one scurrilous source,[72] MacPherson actually spent
his free time in London besieging a rich cheesemonger's widow,
but was not accepted. In the summer he returned to Scotland for
a second manuscript hunt in the west, this time with John Home
and concentrated in Mull. The tour appears to have been much
less productive, but by December 1761 Macpherson was no longer
in need of either Gaelic manuscripts or a well-provided widow.

Fingal, an Ancient Epic Poem ... by Ossian son of Fingal, one of
the most influential books ever composed in Scotland, was no

flimsy pamphlet with a diffident, unsigned preface. Issued by Thomas Becket in London that December but dated 1762, it was a sumptuous quarto decorated with a copperplate frontispiece showing the bard and his amanuensisis, Malvina, in an Arcadian landscape, the bard's harp strung from an oak tree. The text was no mere collection of fragments but a continuous narrative in six books, abounding in epic similes and crowded with literary-historical footnotes, and preceded by a learned introduction. An immediate sensation, it was reprinted early in 1762.

As a performance, *Fingal* is the opposite of mysterious. It resembles one of those old-fashioned clocks in which the movement is visible through glass sides: the epic mechanism can be seen turning round and round and round. This mechanism can also be seen in a lecture on *Fingal* Blair delivered at the College, which he subsequently expanded and published in January 1763 as *A Critical Dissertation on the Poems of Ossian, the son of Fingal.*

At no point is it possible to say for certain whether Blair in the *Critical Dissertation* was analysing MacPherson, or whether MacPherson in *Fingal* was fictionalising Blair. It is not possible to reconstruct the evenings in the Blackfriars' Wynd that gave rise to this co-operation. To use an optical metaphor from Hume's *Treatise*, Blair's enthusiasm and MacPherson's ambition bounce off one another, as if two mirrors had been placed face to face – back and forth and back and forth they go, to the point where "'tis difficult to distinguish the images and reflections, by reason of their faintness and confusion'.[73] Dr Blair wanted a sublime and sentimental epic, and that was what Dr Blair got.

For him, epic was 'of all poetical works, the most dignified, and, at the same time, the most difficult of execution'.[74] Luckily, the world possessed a manual on epic writing in the form of Aristotle's *Poetics*. In Blair's reading of Aristotle, an epic might be based on historical events but must be a fiction; its action must be 'one, compleat and great';[75] it should be enlivened with characters and manners; and be 'heightened by the marvellous', not excluding ghosts.[76] All the better if, in the language of Horace's commentary on Aristotle, the poet were to begin by plunging right in, *in medias res;*[77] and to end his poem on the proper note of calm repose.[78] To his astonishment, Blair found that *Fingal* met these criteria to the letter.

Fingal tells the story of the invasion of Ulster by Swaran and his Norsemen, and their defeat and humiliation of Cuchullin, general of the Irish tribes; of Cuchullin's call to Fingal, King of Morven (Scotland), for reinforcements; and of Fingal's triumph over the Scandinavians. There is a beginning and a middle and an end. The poem opens *in medias res* with Cuchullin leaning on his spear, watching the Norsemen land. It ends with Fingal sparing Swaran's life and sending him home.

Whereas only one or two of the *Fragments* were based on ballads that exist in other forms today, the modern scholar Derick Thomson has identified echoes of nine ballads in twelve passages in *Fingal*. Yet where the ballads are generally terse, dramatic and clear-cut, *Fingal* is copious, solemn and monotonous. An example of MacPherson's embellishments, of the type Andrew Gallie so admired, occurs in Book III in the story of Fingal's expedition to Norway and his escape from a treacherous attack. In the original ballad analysed by Derick Thomson, Finn (Fingal) is invited there by the King of Norway's sister. *Fingal* expands that reference into a tragic love story, decorated with tapestry scenes – a boar hunt, an ambush – and luxuriant description:

> The daughter of snow overheard, and left the hall of her secret sigh.
> She came in all her beauty, like the moon from the cloud of the east.
> Loveliness was around her as light. Her steps were like the music of songs.[79]

Blair saw in this description the tenderness of Tibullus combined with the majesty of Virgil. No doubt Dr Johnson had that sort of passage in mind when he said: 'Sir, a man might write such stuff for ever, if he would *abandon* his mind to it.'[80]

In his *Critical Dissertation*, Blair adopted the four-stage history of the development of society that was such a feature of Scottish thought of this period, in which hunting gives way first to pasturage, from which private property emerges, then to agriculture and commerce. From the absence from *Fingal* of such amenities as windows, agriculture, cities, of any professions other than navigation and iron-working, of military discipline and abstract ideas, Blair concluded that Ossian lived 'in the first of these periods of society'.[81] Warfare was as natural to the men of this ancient period as it was to the martial Edinburgh clergy agitating for a Scots militia. As Fingal says when he releases Swaran to return home at the close of the poem, 'Our families ... loved the strife of spears.'[82] (It was surely no accident that the Edinburgh personalities least interested in a Scots militia – Wallace, Hume, Lord President Robert Dundas – were also those most sceptical about Ossian.)

In treating of the remote past, Blair sounded the new tone of regret. In the spirit of his times, which had already brought to birth an idyll of primitive language and manners in Rousseau's prize essay for the Academy at Dijon in 1750, *Discours sur les sciences et les arts*, Blair complained that in humanity's progress much had been lost as well as gained. 'As the world advances, the understanding gains ground upon the imagination,' he had written in his *Critical Dissertation*. The genius and manners of men 'undergo a change more favourable to accuracy than to sprightliness and sublimity'.[83] Ossian had squared that particular circle, displaying the 'fire and enthusiasm of the most early times, combined with an amazing degree of regularity and art'.[84] As well he might.

'The two great characteristics of Ossian's poetry,' Blair wrote in the *Dissertation*, 'are tenderness and sublimity.'[85] Ossian was endowed with an 'exquisite sensibility of heart';[86] or again, 'his poetry, more perhaps than that of any other writers, deserves to be stiled *The Poetry of the Heart*.'[87] The extreme sensitivity of both the poet and his heroes and heroines, far from being anachronistic, as MacPherson's critics claimed, was in Dr Blair's view typical of a world uncorrupted by money and luxury. 'Barbarity ... though ... it exclude polished manners,' he wrote, 'is, however, not inconsistent with generous sentiments and tender affections.'[88] Years later, embittered by the improvements and clearances of the Highlands, Hugh MacDonald of South Uist told the investigating committee of the Highland Society that 'Those men are much mistaken who believe that neither kindness nor hospitality ... nor sympathy of soul, were conspicuous among the *Féinne* [Fingal's men]: that neither the knowledge nor the practice of virtue existed in their times; but these have lately been introduced into our country. In direct opposition to such conjectures, we can easily prove, that the noblest virtues have been ruined, or driven into exile, since the love of money has crept in amongst us ... Before this change, our chief cherished humanity.'[89]

If, for Adam Smith, sympathy had a social function, for Blair it was principally a source of pleasure. As he said from the pulpit of the High Kirk, 'When the heart is strongly moved by any of the kind affections, even when it pours itself forth in virtuous sorrow, a secret attractive charm mingles with the painful emotion; there is joy in the midst of grief.'[90] In this unctuous strain, Blair recommended that his listeners weep at the death of friends, and attend assiduously at what Ecclesiastes 7: 2–4 calls The House of Mourning to give 'relish' to the joys of life: 'such hours of virtuous sadness brighten the gleams of succeeding joy'.[91] The very evanescence of the world, he said, gave to the contemplative heart 'a certain kind of sorrowful pleasure'.[92] Blair was captivated by the morbid atmosphere of the poems of Ossian, and by the phrase 'The Joy of Grief' which occurs in two passages in Homer and all over the place in Ossian.

As for Sublimity or 'The Sublime', it was a wildly fashionable quality which had its origins in the antique – or, rather, the pseudo-antique. In the early sixteenth century there had been printed in Venice a manual on rhetoric, written in very difficult Greek, entitled on one of the manuscripts *Peri Hupsous*, 'On Height' or 'On the Elevated Style', but more generally known by the Latin translation, *De Sublimitate*. One of the manuscripts also carries the name of Dionysius Longinus, who may or may not have been its author. From internal evidence, the work appears to have been composed under the oppressions of the later Roman Empire.

This work caused a sensation when it was first printed, for its author or authors quoted in full a poem by Sappho which was known up to that time only in a Latin translation.[93] In the seventeenth century the work was translated into French by Boileau. What attracted the eighteenth century to this little book was not merely its astonishingly sure and confident literary taste – even now, Longinus is in a class of one – but also its plangent tone of regret for a golden age of virtue and art. For Longinus, sublimity had been achieved in the *Iliad*, from which point mankind had declined until all liberty and real learning and glory had been extinguished. And the cause of this decline? – *he gar philochrematia*,[94] love of money or, as Blair glossed it, 'covetousness and effeminacy'.[95] Longinus is the missing link in the eighteenth-century transformation of literary criticism from a set of antique rhetorical rules to a suite of aesthetic and emotional responses.

In 1757 Edmund Burke, the Irish writer and politician who so greatly admired Adam Smith's *Theory of Moral Sentiments*, expanded 'The Sublime' into a full-blown aesthetic theory. For Burke the sublime was the objective counterpart of terror, in the same way that for Francis Hutcheson beauty was the objective counterpart of pleasure. Burke located the sublime in obscurity, power, vastness, the infinite, difficulty, magnificence, colour, clamour, suddenness. 'An immense mountain covered with a shining green turf, is nothing in this respect [of grandeur] to one dark and gloomy; the cloudy sky is more grand than the blue; and night more sublime and solemn than day.'[96] He recommended 'sad and fuscous colours, as black, or brown, or deep purple and the like':[97] precisely the colours of the Highland hills. Taste had certainly changed since Edward Burt described the Highlands in the 1720s as a 'a dismal gloomy Brown, drawing upon a dirty Purple; and most of all disagreeable, when the Heath [heather] is in bloom'.[98]

'What are the scenes of nature,' Blair asked his students, 'that elevate the mind in the highest degree, and produce the sublime sensation? Not the gay landscape, the flowery field, or the flourishing city; but the hoary mountain, and the solitary lake; the aged forest, and the torrent falling over the rock.'[99] Or, from the *Critical Dissertation*: 'Amidst the rude scenes of nature, amidst rocks and torrents and whirlwinds and battles, dwells the sublime. It is the thunder and lightning of genius.'[100] A glance at *Fingal* reveals oaks, crags, moss, wind, torrents, moonlight, clouds, mist, sun-beams, blue waves and green hills. Blair even managed to supply the yawning absence of religion in Ossian. As much a religious aesthete as Hutcheson, he saw the sublime as mere evidence of the existence of a benevolent deity. Seeing and hearing might have been restricted to distinguishing external objects, but the Author of Nature, 'for promoting our entertainment', had benevolently allowed them to convey those 'refined and delicate sensations of Beauty and Grandeur, with which we are so much delighted'.[101]

The sublime in nature was not enough for the later eighteenth century. In *Julie, ou la nouvelle Héloïse* of 1761, one of the best-selling novels of its time, Rousseau mingled sexual passion, senti-ment, and the untamed landscape of his native Alps in a new configuration that has influenced literature to the present. In this view, acts of magnanimity or courage or devotion produce the same effect on the reader's emotions as such grand objects in nature as a waterfall or an abyss. In his lectures, Blair christened this connection the 'sentimental sublime'.[102]

Thomas Sheridan, last seen giving lectures on elocution in Edinburgh, told Boswell on 8 February 1763 that 'Mrs Sheridan and he had fixed [Ossian] as the standard of feeling, made it like a thermometer by which they could judge of the warmth of every-body's heart; and that they calculated beforehand in what degrees all their acquaintance would feel them [the Poems], which answered exactly.'[103] In that passage Ossian (like sentimentality in general) is revealed as a novelty for the sexes to share: the badge of membership of a secret society of the heart spanning the sexual divide.

In *Die Leiden des jungen Werthers* (*The Sorrows of Young Werther*), which caused a storm when it came out in 1774, Goethe was more alert than Blair and Sheridan to the black depression underlying Ossian's poems and to the risks, for young men and women in societies that liked to regulate commerce between the sexes, of the cult of sensibility. When Werther reads to Lotte his

translation of Ossian's 'Songs of Selma' from *Fingal* they recognise they are, catastrophically, in love. The poems stretch beyond
endurance the emotional tensions of Lotte's marriage and household, and there is no relief except in suicide. In place of the community of the sensitive are outlaws of the heart.

At first *Fingal* was a brilliant success in London. 'Thermometer'
Sheridan, unsurprisingly, told Boswell that he preferred Ossian 'to
all the poets in the world, and thought he excelled Homer in the
Sublime and Virgil in the Pathetic'.[104]

Other Celts were enraged. Irish and Welsh scholars were
incensed by both the Scottish and the sentimental character of the
poems, an anger compounded by *Temora*, published in 1763, in
which in a 'Dissertation' truculent even by his own elevated standard MacPherson attacked Irish 'pretensions'[105] not merely to
Ossian but to being the mother-nation of Scotland. 'In matters of
antiquity', he wrote, the Irish bards were 'monstrous in their
fables'.[106] In his Appendix to the collected *Works of Ossian* of
1765, Blair saw this crude abuse not as belligerence but as evidence of the poems' authenticity: for what 'person of common
understanding' would engage in 'controversy with the whole Irish
nation?'[107] Indeed.

Meanwhile, in the Highlands MacPherson was accused not of
passing off his own work as translations but of having failed to do
justice to the 'strength and sublimity' of the originals.[108] Captain
Alexander Morison, who had entertained MacPherson for one
night in Skye, thought he was 'no great poet, nor thoroughly con-
versant in Gaelic literature' and 'could as well create the Island of
Skye, as compose a Poem like one of Ossian's'.[109]

Scotland was fast going out of fashion. Londoners had been
resentful of the Prince of Wales's Scots favourites, particularly the
Earl of Bute, who as his tutor had acquired great influence. At
George III's coronation in 1761, Dr Johnson wrote to his friend
Giuseppe Baretti in Milan that the new King 'has been long in the
hands of the Scots, and has already favoured them more than the

English will contentedly endure'.[110] Bute rose rapidly through court appointments and became First Lord of the Treasury in May 1762.

Effectively Bute was Prime Minister, with control of the patronage available to the Crown, and he was thought to be favouring his fellow-countrymen. The presence of so many Scots in town – not just John Home and David Hume but such hell-raisers as MacPherson and Boswell – must have confirmed Londoners' worst suspicions.

Worst of all, Bute had displaced the Englishman William Pitt, later Earl of Chatham, who had led British arms to such spectacular victories since 1759, and inclined the King towards peace with France and Spain. Matters reached a head when on 3 November Bute's ministry signed the preliminaries of peace, which were seen as much too generous.[111] On Wednesday 8 December 1762, at a performance at Covent Garden of the comic opera *Love in a Village* (by Isaac Bickerstaffe), two Highland officers recently returned from Havana came in to the pit and were greeted from the upper gallery with cries of 'No Scots! No Scots! Out with them!' and pelted with apples.[112]

Boswell, who was there, lost his temper, jumped up on the benches and cried out to the galleries: 'Damn you, you rascals.' Later he went up to the two officers, who complained bitterly, 'And this is the thanks we get – to be hissed when we come home. If it was French, what could they do worse?' For all the brilliant service of the Highland regiments in the Americas, the men themselves were still the wild mountaineers who had broken a royal army at Prestonpans and whose presence at Derby had caused a run on the Bank of England. In a fit of gloom Boswell wrote: 'I hated the English; I wished from my soul that the Union was broke and that we might give them another battle of Bannockburn.'[113]

This was the heyday of *North Briton*. Edited by the poet Charles Churchill and the rakish MP John Wilkes, this scurrilous anti-Bute newspaper inflamed English xenophobia. In its issue of 11 November 1762 the *North Briton* lampooned both MacPherson and Home. Early the next year Churchill published *The Prophecy of Famine: A Scots Pastoral*, which portrayed Scotland as a nightmare country where 'half-starv'd spiders prey'd on half-starv'd flies'.[114] He insinuated that famine was sending Scots pouring down on the fat lands of the south. Leading the way were the former minister of Athelstaneford and his Highland protégé:

> *Thence*, HOME, disbanded from the sons of pray'r
> For Loving plays, tho' no *dull* DEAN was there.
> Thence issued forth, at great MACPHERSON'S call,
> That *old*, *new*, *Epic Pastoral*, FINGAL.[115]

As for Ossian, it was a mere forgery:

> OSSIAN, *sublimest*, *simplest* Bard of all
> Whome *English Infidels* MACPHERSON call.[116]

Hugh Blair, when he arrived in London at the beginning of April 1763, ran into a storm. Thomas Sheridan, of course, praised the *Critical Dissertation* extravagantly, and Blair became so mortified he seemed to Boswell like 'a Scotch fornicator rebuked on the stool of repentance'.[117] But he soon discovered that most English men of letters regarded Ossian as an imposture and the Scots as pests. Dr Johnson was particularly rude to him.[118]

The Whig campaign reached its climax with the conclusion of the Treaty of Paris and George III's speech to Parliament on 18 April. On 23 April there appeared *North Briton 45* – significant number – in which Wilkes alleged that the king's speech was a tissue of lies. He was arrested for seditious libel on 30 April, only to be released six days later on the ground of parliamentary privilege. In the uproar, Bute resigned.

The portrait of MacPherson sketched in Boswell's *Journal* of that year is of a man ambitious, bored, cynical, truculent, disagreeable and licentious. Boswell recounts how, one Saturday in May 1763, Blair asked MacPherson why he stayed in England, for surely he could not stand John Bull. 'Sir, I hate John Bull,'

MacPherson replied, 'but I love his daughters.'[119] Privately he told Boswell that he 'had no relish for anything in life except women, and even these he cared but little for'.[120] Where were the loving couples of *Fingal*, monogamous, companionable, faithful to death? Where was Comal, who slept 'with his loved Galvina at the noise of the sounding surge', their 'green tombs seen by the mariner, when he bounds in the waves of the north'?[121] Their author was a prolific generator of bastards. He and Boswell might have been the incarnation of Churchill's paranoia:

> Into our places, states, and beds they creep:
> They've sense to get, what we want sense to keep.[122]

David Hume had come to London in August and was staying in Leicester Fields. Alarmed by what he heard, in September he wrote a long and careful letter to Blair, warning him that London was calling the poems 'a palpable and most impudent forgery'.[123] The point to prove, as Hume saw it, was not that the poems were 'so antient as the age of Severus' – the great Roman Emperor who campaigned against the Caledonians in the early third century AD – 'but that they were not forged within these five years by James MacPherson'.[124] In a passage that is a model of the experimental approach to ancient literature   philosophy, said Hegel, comes too late to teach the world how it should be – Hume begged Blair to seek not argument, but testimonies from the Highlands that such poems were 'vulgarly recited in the Highlands'.[125] He added, with his habitual kindness: 'Your connections among your brethren of the clergy may here be of great use to you.'[126]

But act Blair must. MacPherson himself was at best no help, at worst a calamity. 'The absurd pride and caprice of MacPherson himself, who scorns, as he pretends, to satisfy any body that doubts his veracity, has tended much to confirm this general scepticism.[127] ... The child is, in a manner, become yours by adoption, as MacPherson has totally abandoned all care of it.'[128] When informed by Hume of his letter to Blair, MacPherson flew into a rage. By now, Hume had thoroughly shed his good opinion of the

poet: 'I have scarce ever known a man more perverse and unami-
able.'[129] MacPherson was, anyway, on the point of being sent
through Bute's influence to Florida, as secretary to Governor
George Johnstone at Pensacola, the first step in what proved a successful public career. Hume prayed that he might travel among the
'Chickisaws or Cherokees, in order to tame him and civilize him'.[130]

Hume betook himself to Paris in October, having been invited
to act as personal secretary to the pious and churchy new ambassador, Lord Hertford. Hume was as bitter about the prejudices of
London as he was about the intolerance of Edinburgh. His scheme
of anglicising the Scots, one of the projects of his life, was in ruins.
On 22 September 1764 he wrote from Paris to Gilbert Elliot of
Minto: 'Am I, or are you, an Englishman? Will they allow us to be
so? Do they not treat with Derision our Pretensions to that Name,
and with Hatred our just Pretensions to surpass & to govern
them?'[131] If he could not stay in Paris, he wrote, he would retire to
'Thoulouse, or Montauban, or some Provincial Town in the South
of France; where I shall spend, contented, the rest of my Life, with
more Money, under a finer Sky, & in better Company than I was
born to enjoy.'[132] (Had he chosen Montauban, and left his bones
there, he would have shared that pretty Roman town with the masterpiece of the Ossian cult. *La Songe d'Ossian*, painted by Ingres
for Napoleon's bedroom, portrays the bard bowed under the
weight of his memories while arrayed on each side, in shades of
unvarying spectral whitness, are the melancholy warriors and
lovely women of a vanished age.)

Hume, to his own astonishment and the disgust of the London
wits, was a brilliant social success in Paris. The fat philosopher's
popularity with women, including such famous hostesses as the
comtesse de Boufflers, Mme Geoffrin and the marquise du
Deffand, was to men such as Horace Walpole peculiarly offensive.
Hume continued loyally to champion Ossian in the salons, though
he warned Blair that among the admirers there were critics, not
just of its oddities and tedious uniformity,[133] but of its authenticity, and chivvied him to produce his testimonies. On 9 November
1763 he wrote from Fontainebleau that the duchesse d'Aiguillon

'was amusing herself with translating passages of Ossian; and I have assured her that the authenticity of those poems is to be proved soon beyond all contradiction'.[134] He was backed up by young Sir James MacDonald of Sleat in Skye, a brilliant young man who knew both the Gaelic and classical worlds.[135]

In 1765 a new edition of *Fingal, Temora*, the shorter poems and the *Critical Dissertation* was published as *The Works of Ossian*. It opened with a pompous flourish from MacPherson, and closed on a muted note from Blair. Before leaving for Florida MacPherson had written a new dedication to Bute, now out of office, which drew a foolish parallel between their conditions. 'There is a great debt of fame', he wrote, 'owing to the EARL of BUTE, which hereafter will be amply paid: there is also some share of reputation with-held from Ossian, which less prejudiced times may bestow.'[136] In his Appendix, Blair quoted testimonials to Ossian's authenticity from Highland 'gentlemen of fortune' and such clergymen as MacPherson in Sleat, Maclean at Kilmuir, MacLeod at Glenelg, and MacNeill in South Uist. There is something touching about those good men taking such pains for a rascal – or, rather, for the honour of old Scotland, which had eloped with him to London and was now parading, high-coloured and high-painted, on his arm through Vauxhall Gardens. Hume wrote from Compiègne on 20 July that he was 'much satisfy'd with the Appendix',[137] but in truth Blair had not done as he had asked: there was more argument than testimony.

MacPherson quickly quarrelled with Governor Johnstone and by 1766 was back in London. He seems to have learned no manners from the native Americans. He indulged his theories about the Celts in *An Introduction to the History of Great Britain* (1772) and was proposed by Strahan to complete Hume's *History*, about which Hume was distinctly unenthusiastic.[138] He was inveigled by Blair and others to publish a translation of the *Iliad* into Ossianic prose in 1773 – Hume thought it plain bad[139] – and by Kames to 'amend' his own Ossian in a revised edition.

In the section about Skye in his *Journey to the Western Islands of Scotland*, published in 1775, Dr Johnson laid out his attack on the poems, which he said had 'never existed in any other form than that which we have seen'.[140] He was right, but not for the right reasons. As a devotee of literature, Johnson had no belief in oral transmission, and kept coming back to the question of manuscripts: he thought (quite wrongly) that Gaelic 'never was a written language',[141] and that 'a nation that cannot write, or a language that was never written, has no manuscripts'.[142] *Ergo*, Ossian was an 'imposture'. Yet although there are no old manuscripts of the *Iliad*, nobody suggests that is an imposture.[143] Having denied Scotland any ancient written culture, Johnson then called all the Scots liars: 'A Scotchman must be a very sturdy moralist, who does not love Scotland better than truth.'[144] The tenor of MacPherson's reply can be guessed from Johnson's counter-reply of 20 January, the sort of letter many dream of writing, but few have the opportunity or the guts to write:

> MR JAMES MACPHERSON – I received your foolish and impudent letter. Any violence offered me I shall do my best to repel; and what I cannot do for myself, the law shall do for me. I hope I shall never be deterred from detecting what I think a cheat, by the menaces of a ruffian.
>
> What would you have me retract? I thought your book an imposture; I think it an imposture still. For this opinion I have given my reason to the publick, which I here dare you to refute. Your rage I defy. Your abilities, since your Homer, are not so formidable; and what I hear of your morals, inclines me to pay regard not to what you shall say, but to what you shall prove. You may print this if you will.
>
> SAM. JOHNSON[145]

The next year, Edmund Gibbon's *History of the Decline and Fall of the Roman Empire*, treating of the Emperor Severus's campaign in Scotland, handled Ossian as, at best, a piece of anachronistic Edinburgh whimsy.[146] Hume took it on himself to apologise to Gibbon on behalf of all of his Edinburgh friends:

> I see you entertain a great Doubt with regard to the Authenticity of the Poems of Ossian. You are certainly right in so doing. It is, indeed, strange, that any men of Sense could have imagin'd it possible, that

above twenty thousand Verses, along with numberless historical Facts, could have been preserved by oral Tradition during fifty Generations, by the rudest, perhaps, of all European Nations; the most necessitious, the most turbulent, and the most unsettled.

But then, he added sadly, 'Men run with great Avidity to give their Evidence in favour of what flatters their Passions, and their national Prejudices.'[147] He had put it less diplomatically on 6 March 1775 when Boswell called to pay his half-yearly rent for James's Court: 'He said if fifty barea—d Highlanders should say that *Fingal* was an ancient poem, he would not believe them.'[148] The Highlanders were too busy avoiding starvation and the rope to compose a poem in six books. Hume wrote an essay of his own on Ossian, which was critical of Blair, but did not publish it.[149]

MacPherson did not care. He was launched on a prosperous career as a political journalist for Lord North's government and, from about 1780, as London agent for the Nawab of Arcot. In the circle about Lord Advocate Henry Dundas of Arniston, he helped engineer the Scots dominion in the East India Company. In 1780 he became MP for Camelford in Cornwall but, having made his pile as the Nawab's agent, he returned to Inverness-shire, bought an estate in Badenoch and in the 1790s had Robert Adam build him a house at Balavil. He was buried in Westminster Abbey.

Blair published his sermons, with a dedication to the Queen, no less. They ran through more than a dozen editions and in 1780 he received a royal pension of £200 a year.[150] Yet a faint ridicule attends his memory. In the rush to Ossian, all Edinburgh's vaunted philosophical weapons – the experimental method, 'ocular inspection' (Wallace), common sense – were left on the battlefield. Edinburgh critics were determined there should be no repetition. Beginning with Gilbert Stuart, a brilliant and acerbic mind ruined by drink and feuding, they became aggressive, dry-eyed and scurrilous. As for the English, they never trusted Scottish criticism again. MacPherson's gain was Burns's loss.

Yet with his chivalrous warfare and passionate monogamy,

MacPherson's Ossian built a bridge between a brutal and starveling past and a respectable, citified present. He gave to Edinburgh, as later to other societies without much notion of their own history – Germany, Napoleonic France, divided Italy, even Latin America – a phantom identity. Within two or three generations, *Fingal* had been displaced by more or less authentic creations of national imagination: the *Nibelungenlied* of Germany and the *Märchen* collected by the Brothers Grimm, the Anglo-Saxon poem *Beowulf*, the Finnish *Kalevala*, the Irish and Icelandic sagas.

History can be hard to suffer. Ossian permitted Scotland to mourn its lost independence through the medium of the supernatural, the sentimental and the unhistorical. The world is grown old and decrepit. Fingal is gone, and all the heroes, and only Ossian remains in a shattered world from which all glory and good name and vigour and pleasure and fellowship and beauty have long since departed. There is no God, no purpose to existence. Ghosts crowd in, helpless and intolerable. 'Roll on, ye dark-brown years, for ye bring no joy on your course. Let the tomb open to Ossian, for his strength has failed. The sons of the song are gone to rest: my voice remains, like a blast…'[151]

Torrents of Wind

Daniel Defoe, a London journalist who had been in his time both a hosiery merchant and a government secret agent, looked at the ancient capital of Scotland with a modern commercial eye. In his *Tour thro' the Whole Island of Great Britain*, which he began to publish in 1724, he wrote that such a steep and inconvenient and cramped site served just a military purpose. In a more secure world, the city would long ago have spread out on the plains to north and south.

'On the North Side,' he wrote, 'were not the North Side of the Hill, which the City Stands on, so exceeding steep, as hardly ... to be clamber'd up on foot, much less to be made passable for Carriages ... the City might have been extended upon the Plain below; and fine beautiful Streets would, no Doubt, have been built there; nay, I question much whether, in Time, the high Streets would not have been forsaken, and the City, as we might say, run all out of its Gates to the North.'[1]

That, approximately, and including the beautiful last phrase, is what happened in Edinburgh in the second half of the eighteenth century.

Edinburgh New Town is intriguing not merely as a suite of

handsome buildings, but as the material expression of ideas of civilian life created out of the ruins of Jacobite rebellion and Presbyterian theocracy. All those George Streets and Princes Streets and Charlotte Squares, though they seem so classic and serene, are in reality frantic gestures of loyalty from a city with only the shallowest attachment to the House of Hanover and to Great Britain. They embody a new social existence that is suave, class-conscious, sensitive, law-abiding, hygienic and uxorious: in short, modern.

There was no rush to live in the New Town. The poor lingered in the old neighbourhoods, as did most of the town's celebrities, with the exception of David Hume and the town historian Hugo Arnot. New residential squares in the milder south of Edinburgh – particularly George's – were the fashionable addresses until the turn of the century. It was not until after Waterloo that the centre of Edinburgh's gravity shifted north of Princes Street. In time, the ravines holding in the Old Town were all spanned by stone bridges – the North, the South, Waterloo, King George IV – and Edinburgh followed its own private ways north, south, west and east, enveloping the old villages of Dean and Corstorphine and Portobello and Duddingston. The philosophical view from the High Street – Adam Smith's 'lawns and woods, and arms of the sea, and distant mountains'[2] – filled up with slates and chimneys.

As early as 1681, when James Duke of York was living at Holyrood, there was talk of extending the city's Royalty – that is, the exclusive rights and privileges granted the city by the Crown – to the north.[3] In 1688 the town obtained a charter from him, as the new King James VII/II, for an extended Royalty, but his flight that year ensured that it would not be ratified by Parliament. In the matter of capital cities, as in so much else, Andrew Fletcher of Saltoun was radical. He said the government of Scotland should abandon Edinburgh. Just as London had brought glory and riches to England, he wrote in his second *Discourse* of 1698, 'so the bad situation of Edinburgh, has been one great occasion of the poverty and uncleanliness in which the greater part of the people of Scotland live'.[4]

In 1716 the Town Council acquired a piece of ground on the north bank of the North Loch known as Barefoot's or Bearford's Parks. In 1723 the magistrates persuaded Parliament to allow it to levy its twopenny ale tax on the suburbs as well as the town, and to use the proceeds for, among other capital purposes, 'narrowing the noxious lake on the north side of the said City, commonly called the North Loch, into a canal of running water', and laying a road to the land owned by the town on the north bank.[5] During his Continental exile John Earl of Mar, who had been rebel commander in the Fifteen, indulged his home-sickness thus:

> All ways of improving Edinburgh should be thought on: as in particular, making a large bridge of three arches, over the grounds betwixt the North Loch and Physic Garden, from the High Street at Halkerstone's Wynd to the Multursey Hill, where many fine streets might be built, as the inhabitants increased.[6]

But the other landowners to the north were hostile to an extension of the City revenue system into their ground, and the usual shortage of investment capital, the Porteous Affair and then the Forty-five put paid to these ideas.

Edinburgh was not entirely stagnant. In the Lawnmarket can still be seen Mylne's or Milne's Court, built in the 1680s and 1690s by Robert Mylne, of a family of hereditary master-masons to the Stuart court. Nearby is most of the eight-storey James's Court, of the 1720s, where both Hume and Boswell later lived. From the 1740s onwards there were small developments outside the Royalty in the south, such as Argyle's Square, Brown's Square and Adam's Square, which boasted separate dwelling-houses on the London model, advertised (according to Walter Scott, writing 75 years later in *Guy Mannering*) as houses 'within themselves'.[7] All of these schemes were designed to give a little more light, air, propriety and social exclusivity to the newly self-conscious inhabitants of the town.

What Edinburgh missed were the meeting-places of a society that had other purposes than to pray, drink itself senseless or litigate. Merchants did their exiguous business by the Cross beneath the scudding clouds, the trades met in chapels, while during the crisis of September 1745 there was nowhere to accommodate the town meeting or the Volunteers except the High Kirk. As for entertainments, after 1762 the musical society had an elegant new hall for its weekly concerts in Niddry's Wynd, designed by one of the Mylnes on the pattern, it was said, of the Parma opera house, but the new Assembly Rooms off the High Street would not have passed muster in Beau Nash's Bath. 'The door is so disposed,' wrote an asthmatic Hugo Arnot, 'that a stream of air rushes through it into the room; and as the footmen are allowed to stand with their flambeaux in the entry, before the entertainment is half over, the room is filled with smoak almost to suffocation.'[8]

The Forty-five showed that the city's narrow site and cramped walls no longer served even to defend it. Yet the New Town, and modern Edinburgh itself, might not have arisen but for an accident in September 1751, when the side wall of a six-storey land, 'in which several reputable familes lived, gave way all of a sudden'.[9] The Town Council commissioned a survey of the old houses, 'and such as were insufficient were pulled down; so that several of the principal parts of the town were laid in ruins'.[10] The next summer, the Convention of Royal Burghs, meeting on 8 July in Parliament Close, resolved to use this 'favourable opportunity' to build a hall for their meetings, a covered exchange where merchants could meet to transact their business and a repository for the public records of Scotland, and to carry out other improvements. To gain support for the scheme, the Convention resolved to issue a pamphlet 'explaining and recommending the design'.[11]

This pamphlet, entitled in the modest and wordy manner of the period *Proposals for carrying on certain Public Works in the city of EDINBURGH*, was as important for Scotland as any work from the philosophers, and for more practical reasons. It is attributed to Sir Gilbert Elliot, who was just embarking on his brilliant political career as Edinburgh's agent at Westminster, but in its Continental horizons, commercial sagacity and historical philosophising it might have been written by any of the circle of Hume and Elibank, Smith and Kames.

The pamphlet began:

> Among the several causes to which the prosperity of a nation may be ascribed, the situation, conveniency, and beauty of its capital are surely not the least considerable. A capital where these circumstances happen fortunately to concur, should naturally become the centre of trade and commerce, of learning and the arts, of politeness, and of refinement of every kind.[12]

The author proceeded to compare London and Edinburgh, very much to the latter's disadvantage in point of health, fresh air, convenience and accommodation. He complained of the lack of public buildings in Edinburgh, the ungenteel food markets in the High Street and the shambles, tanners and leather-workers on the banks of the North Loch. At a time when cities were valued chiefly for their aristocratical residents, 'to these and such other reasons it must be imputed, that so few people of rank reside in this city; that it is rarely visited by strangers; and that so many local prejudices, and narrow notions, inconsistent with polished manners and growing wealth, are still so obstinately retained.'[13]

Next, the author considered the modern history of Scotland: the disorder and backwardness of the seventeenth century, the false dawns of the Unions of the Crowns and Parliaments, the stagnation of the first decades of the eighteenth century, the Forty-five. 'But since the year 1746,' he wrote, 'when the rebellion was suppressed, a most surprising revolution has happened in the affairs of the country. The whole system of our trade, husbandry and manufactures ... began to advance with such a rapid and general progression, as almost exceeds the bounds of probability.'[14] As examples, he quoted a near 70 per cent increase in the value of linen cloth sold in Scotland since 1742,[15] a more than doubling in the tonnage of shipping at Leith, new public companies for whale- and herring-fishing, for sugar, glass, linen, rope and sail-cloth, iron and other manufactures, and (since 1745) a four-fold increase in the whisky distilled in Edinburgh.

To create a capital city worthy of this new commercial country, it was now proposed to build a merchants' exchange 'upon the ruins on the north-side of the high street'; to 'erect upon the ruins in parliament-close, a large building' to provide a robing room for the justices, a borough room, a council chamber, offices for the court clerks, magazines for the public records scattered about town, and a new Advocates' Library; 'to obtain an act of parliament for extending the royalty; to enlarge and beautify the town, by opening new streets to the north and south, removing the markets and shambles, and turning the *North-Loch* into a canal, with walks and terrasses on each side.' Significantly for a town that hitherto had had to finance its own building, the pampleteer held that 'the expense of these public works should be defrayed by a national contribution'.[16] Of the proposed improvements, he said, 'the extending the royalty, and enlargement of the town, make no doubt the most important article'.[17]

To sum up:

A nation cannot at this day be considerable, unless it be opulent. Wealth is only to be obtained by trade and commerce, and these are only carried on to advantage in populous cities. There also we find the chief objects of pleasure and ambition, and there consequently all those will flock whose circumstances can afford it.[18]

By the operation of a sort of moral perspective, the modest pleasures of an improved Edinburgh would appear to the Quality as enticing as the flesh-pots of faraway London.[19] Any fear that they might abandon the Old Town was anticipated by the author's worldliness; professional people would continue to live near the Exchange and the Courts of Justice, he thought, as they did in Turin and Berlin:

In these cities, what is called the *new town*, consists of spacious streets and large buildings, which are thinly inhabited, and that too by strangers chiefly, and persons of considerable rank; while the *old town*, though not near so commodious, is more crouded than before these late additions were made.[20]

Montesquieu had recognised that concentrating people in capital cities increased their commercial appetites.[21] The author of the *Proposals* argued that their consumption would become competitive. 'As the consumption is greater so is it quicker and more discernible. Hence follows a more rapid circulation of money and other commodities, the great spring which gives motion to general industry and improvement. The example set by the capital, the nation will soon follow.'[22] The result? Happiness.

> The certain consequence is, general wealth and prosperity: the number of useful people will increase; the rents of land rise; the public revenue improve; and, in the room of sloth and poverty, will succeed industry and opulence.[23]

That this castle in the air was built, almost to the last aerial stone, is the strangest of all the strange things about modern Edinburgh. Elliot's arguments are quite obviously back to front. As the antiquarian lawyer Sir David Dalrymple (later Lord Hailes) pointed out in a scatological parody of the *Proposals*, London's elegant quarters were not the cause of her prosperity, but the effects of it.[24] More to the point, the revenues of the Good Town were deteriorating fast.

The execution of the scheme was due not to Elliot, who made his career in London; not to Dalrymple or the Duke of Argyll or Lord Milton or Lord President Dundas or the philosophers, and not, at least in the 1750s, to the Adam brothers. It was due to a man last seen gathered up on his old cob in a rout of fleeing redcoats at Prestonpans:

> PEACE to thy shade, thou wale [best] o'men,
> DRUMMOND![25]

George Drummond, six times Lord Provost of Edinburgh, was born probably on 27 June 1687, probably at Newton Castle, Blairgowrie. Employed as his secretary at the age of eighteen by old Sir John Clerk of Penicuik, one of the Union commissioners, to rate the country for its amalgamation with England, he made his way up through the new-fangled Excise, and by 1715 was one of the Commissioners of Customs at a salary of £1,000 sterling a year. In other words, Edinburgh's great benefactor was neither nobleman nor merchant nor craftsman but one of the new breed of office-holders required by the British administration, wholly dependent on the Crown and its agents in Scotland. At eighteen if not before, George Drummond had committed himself to Hanover and the Union, and he never deserted them. 'By the Union with England,' wrote Adam Smith, 'the middling and inferior ranks of people in Scotland gained a compleat deliverance from the power of an aristocracy which had always before oppressed them.'[26] Drummond's life is partial evidence for that trenchant claim.

By October 1717 he was City Treasurer and on 5 October 1725, to the disgust of the Jacobites, was elected Lord Provost with expenses of £300 a year. In this first period of municipal control Drummond was occupied in establishing the medical school at the College, helping found the Royal Bank of Scotland, and managing the General Assembly of the Church. In 1727 he withdrew from the Town Council to concentrate on building the Royal Infirmary and to serve (with Forbes of Culloden and Sir John Clerk, among others) on the Board of Trustees for the Encouragement of the Fisheries, Arts and Manufactures of Scotland, set up to invest excise revenue in the country's infant industries.

In the course of the 1730s George Drummond went into decline. The manuscript diary he kept between June 1736, as he entered his fiftieth year, and November 1738 throws a smoky light on his condition. The diary is now in the Edinburgh University Library; and to read it there, high over George Square and its gaggles of polyglot students, is truly to experience the Scots eighteenth century. The five hundred and twenty-three folio pages reveal a life of grinding financial difficulty, depression, terrifying religious doubt, heavy drinking, and aristocratic bullying.

Widowered for the second time in 1732, with ten living children, Drummond about 1734 came under the influence of a woman he refers to as 'R.B.' A good half of the diary consists of her spiritual journal, carefully copied out by Drummond as his 'favourite employment'. Even by the sociable standards of the Edinburgh Whigs R.B. was on intimate terms with the Almighty, interceding for Drummond in what seem quite worldly affairs. Never in history was so much prayer and mortification devoted to £500 a year.

Almost every day Drummond met some 'shocking dun ... I dare not tell a ly to any one creditor, nor shift my hands to save my credite'.[27] He found no satisfaction in his prayers: he was 'weighed', 'in bonds', 'dead', 'not distinct', 'not breathed on'. Ilay and Milton were putting pressure on him, evidently to abandon his support for the High-Flyers. On 19 July 1736 Drummond reluctantly gave evidence at Porteous's trial for murder, and that night got badly drunk. On 8 September he heard – at dinner, he was careful to say – of the lynching of Porteous the night before and was 'much concerned at the consequences which I apprehend may follow upon it'.[28] The King was absent in Hanover, and Queen Caroline was incandescent. With the arrival of her messenger from London, the administration decided it must consult Drummond. On 16 September Drummond set down his political impressions with admirable precision:

> The administration in this countrey is supported by fear, not love, and the tools who are employed are hated and contemnd, by almost all people high and low. The steps taken to undermine serious Religion in this church, in Edr. especially, is attributed to My Lord Ilay. He has thrown the towns affairs into the hands of men void of religion, and little respected in the place. I am lookt on as an enemy to their measures, and have carefully avoided medling with them, its surely dangerous for me to speak my sentiments. May I have light from the Lord to be faithfull and prudent, for I dwell among lions, and among them that are set on fire whose teeth are spears and arrows, and their tongues a sharp sword [Psalm 57] I spent the evening turning over this matter in my thoughts, in all its different lights, & laying myself before the Lord for light. My heart was engaged in family worship, but my mind was weak, closed the day with God.[29]

Ilay and Milton were also threatening to cut his salary as Commissioner of Customs to £500. R.B. insisted God would not permit that, and on Christmas Day Drummond wrote: 'My soul trusts to what the Lord has said to R.B. about our salarys.'[30] When the following autumn, on a visit to Inveresk, fearing to be 'torn to pieces by my creditors',[31] he heard he was to be turned out of the Customs, he suffered a crisis of religious faith. He expostulated with God 'about the clashing of his providence and promise in this affair'.[32] 'Why will you,' Drummond raged on 13 October 1737, the day he was turned out, 'or how can you have any dependence on any promise in the bible!'[33] On 8 December a letter from Ilay, the first for many years, raised his hopes. On 26 January 1738 he heard not only that he had been appointed a Commissioner of Excise, but that his salary was to be back-dated to October. God's Providence and His promise to R.B. were miraculously reconciled. On 6 February the *London Gazette* carrying the notice of his commission arrived in Edinburgh, and he went proudly to the High Street to receive congratulations.[34] Even so, his cousin sent troubling reports from London on 19 March: 'Ilay says, if I don't behave right now, he'll do with me as he has done with [a dismissed colleague], turn me out and put in my brother, poor man!'[35]

Drummond's consolations at this time were the preaching of Alexander Webster at the Tolbooth Kirk, and the building of the Infirmary. On 10 October 1738 he was at Duddingston, cajoling the Duke of Argyll's tenants to help transport stone, with some success. 'They signed for 600 carriage draughts,' he wrote. Afterwards he 'supt in a crowd, where there was some excess in drinking'.[36] On the 13th he took stock of himself, with his characteristic mixture of arrogance and unctuous humility: 'Forwarding the building of the Royal Infirmary is my only amusement. The Lord gives remarkable success to all our applications. At first it was somewhat uphill work, but now it is the favourite undertaking, among all ranks of people … I am distinguished and called the father of it, &c with which, alas, I have too much pride and vanity, not to be pleased; yea, I am afraid I am puft up, woes me. I can neither be humble under success, nor bear up under discouragements, O what a poor worthless creature am I!'[37] On 2 November he spent the day among the 'free Masons – setting the work of the Infirmary – to which their countenance [support] is of great use'.[38]

By the close of the diary, Drummond had come up off his knees. On 16 October he called after dinner on Mrs Fenton, the widow of a colleague on the Council, who mentioned 'a widow, with an estate large enough to relieve me of my distresses'.[39] Mrs Fenton feared the lady might not have 'grace', but was rebuked by a voice: 'What do you know if this woman's money is not given to her, to be a Blessing to him, and, if he is not to be a blessing to her by being a means of her conversion.'[40] On 23 November Drummond was introduced to the lady and found 'nothing disagreeable either in her manner or person'.[41] With the advent of the widow there was no more about R.B.

In 1744 Drummond's eldest daughter married Dr John Jardine, a forceful church politician, thus completing the old Provost's change of ecclesiastical sides.[42] In the crisis of the Forty-five Drummond proved both loyal and popular, and was re-elected Lord Provost by a vote of the burgesses on 26 November 1746. In effect, the Forty-five had discredited the old system of managing Scotland under Ilay (who had by now succeeded his brother as Duke of Argyll), and thus made Drummond's prominence possible.

In 1755, in the midst of his fourth term as Lord Provost, Drummond was married for the fourth time to a Quakeress with £20,000. Evidently he had by then shed all his sectarian scruples. With her fortune he bought a property in the village of Broughton to the north of Edinburgh, known as Drummond Lodge and commemorated in the present Drummond Place. From 1758 to 1760 he was Lord Provost for a fifth time, and again, for the last time, from 1762 to 1764. He died in November 1766, and lies buried in the Canongate Kirkyard.

If George Drummond was no hero, he was a man for his age. In Drummond, the Scots come down to earth. If his fault, as

Alexander Carlyle noted, was an excessive 'deference to his superiors',[43] he probably knew what he was doing: he is remembered while the dukes and lords are all forgotten. His virtue lay not merely in his intense pride in Edinburgh, but in his gradually unfolding vision of where its future lay: a great university, a world-famous medical school with its own teaching hospital, a cultivated professional population.[44] That, essentially, is Edinburgh today.

Thomas Somerville, a minister in the Borders and a protégé of Sir Gilbert Elliot, remembered in the summer of 1763 standing at the window of Jardine's apartments on the top floor of a building at the north corner of the Exchange, and looking out beyond the Loch to Barefoot's or Bearford's Parks, where 'there was not a single house to be seen'.[45] The Lord Provost stood beside him. 'Look at these fields,' said Drummond; 'you, Mr Somerville, are a young man, and may probably live, though I will not, to see all these fields covered with houses, forming a splendid and magnificent city. To the accomplishment of this, nothing more is necessary than draining the North Loch, and providing a proper access from the old town. I have never lost sight of this object since the year 1725, when I was first elected provost.'[46]

The first of Drummond's improvement schemes in his second period of municipal eminence, the Royal Exchange on the High Street, was an unlikely project. Seventy years earlier Edinburgh's merchants had been built a covered piazza in Parliament Close, on the pattern pioneered in the medieval Low Countries. That burned down in the great fire of 1700 and they were built another; but the 'marchantis of renoun' (Dunbar) obstinately adhered to the Cross and the open air.

In the course of Drummond's third term as Lord Provost (1750–52), according to the Town Council Minutes, there was agitation 'by a great Number of the Princll Inhabitants Merchants and Burgesses of this City setting forth that the want of a fforum or Convenient place of exchange in this Metropolis had long been regreted.'[47] When 'severall tenements near the Cross became ruinous it occurred to many that so lucky an Opportunity for a well situate exchange ought not to be lost'. Part of the site was occupied by a disreputable relic of old Edinburgh, the grim and gousty Mary King's Close, abandoned by all its inhabitants but Satan in 1645 and ravaged by fire in 1750. Taking the building line back one hundred to one hundred and fifty feet from the High Street pavement, the Council heard, would give the town the opportunity to create a square with 'a stately covered walk att the back part' for the merchants and 'ane handsome range of shops on each side'. But with so many proprietors on the site, only an ambitious plan would persuade them 'to unite their interest'. Otherwise, 'the opportunity of executeing a work so much to the Honour of the Town and Interest of the Inhabitants may be intirely lost'.[48]

At a Council meeting on 1 July 1752 Drummond reported that he had 'caused Mr John Adam architect make out a plan'[49] for the exchange and covered walk and also, in the ruined area to the south of Parliament Close, for a large hall for the annual Conventions of Royal Burrows or Burghs, Council Chambers and a Provost's residence. The problem was money. According to the Minutes,

> The City's revenue is not sufficient for carrying on these necessary good works, and of making ane easy & convenient access to the high street from the south & north – which in the view of extending the Royalty of the City is absolutely necessary to be done nor was there any ffund for following out the plan of making the lake called the North Loch a beauty & ornament to the City in place of the hatefull nuisance it now is.[50]

The twopenny ale and beer duty, levied first in 1653, was extended to the suburbs in 1723 to permit the town to borrow up to £25,000 for public works. But ale was no longer the sole drink of Edinburgh and the take from the tax had fallen from £8,000 a year to less than £5,000. 'Such', wrote Arnot, who presented these sums, 'to the propagation of idleness, vice, and disease, has been, among the poor, the increased consumption of tea and *whisky*.'[51]

As for the 'proper' revenues or Common Good – duty on wines, shore dues at Leith, market fees and rents and feu-duties from tenanted property – those were much the same as they had been in 1690. (Even in 1768, with income from the sale of lots in the 'extended royalty' to the north, the town's annual revenue was little more than £28,000.[52])

Drummond retreated forwards. According to the Town Council Minutes, he 'therefore had talked with some persons of quality judges & others upon these subjects' and resolved that in addition to what was proposed the town would undertake to build a library for the Faculty of Advocates, a robing room for the Lords of Session and offices for the clerks, and a register house. With all the professional constituencies in Edinburgh thus involved, 'there was no room to doubt that money enough might be raisd by voluntary subscriptions to carry on the whole'.[53] Equally, the scheme was now grand enough to attract national attention.

Drummond was proved right. By the end of 1752 more than £6,000 had been subscribed to the enlarged scheme. The chief contributors were the Crown, which offered £1,934 from confiscated Jacobite estates, and the Royal Burghs, which gave £1,500 (but never received their meeting hall). Drummond put himself down for £50. The next summer, the City Council obtained an Act of Parliament 'For Erecting several Public Buildings in the City of Edinburgh; and to impower the Trustees therein to be mentioned to purchase Lands for that Purpose and also for Widening and Enlarging the Streets of the said City, and certain Avenues leading thereto.'

In August it was decided to begin at once with the Exchange and complete it first. A plan by the eldest of the Adam brothers, John, was accepted and five contractors or 'undertakers' – a mason, three wrights and an 'architect' – were commissioned. On 13 September 1753, with all manner of municipal and Masonic pomp, the foundation stone of the Exchange was laid.

Not much else seems to have happened until the following June, when a building contract was signed. The agreement specified a structure 111 ft 6 ins in length running west to east, consisting of a Customs House with a broad piazza across the inner side of the courtyard to form the Exchange, and two wings or jambs projecting 131 ft to the south, comprising shops, dwelling-houses and coffee-houses to create an open courtyard. The undertakers agreed to clear the site and construct the buildings for £19,707 16s. 4d., not including some £11,749 7s. 8d. already paid out by the Magistrates to assemble the site. (The money was borrowed from the two town banks and secured on the ale duty.) In its turn, the Council engaged to advance the undertakers £18,000 in working capital over the course of the project. The shops and other properties were to be sold as completed, the whole to be finished by 15 May 1762.

It seems to have been a disastrous bargain for the undertakers. The town advanced only £4,100 in working capital, and that during the years 1756 and 1757, and the undertakers had to borrow the balance at rates of interest higher than the town would have had to pay. The stubborn merchants refused to use the piazza. Even after Drummond had the Cross dismantled in 1756, they continued to congregate in the street where it had stood. The banker Sir William Forbes of Pitsligo recorded them foregathering in the open air as late as 1772, where they were joined by 'many who had no concern in the mercantile world, such as physicians and lawyers, who frequented the Cross nearly with as much regularity as the others for the sake of gossiping and amusement merely'.[54] (Boswell was in the habit of going to the Cross almost every day before dinner, to see and be seen.) The result was that many of the shops and coffee-houses, which had been valued at a premium for their proximity to the new Exchange, were sold below their valuation. The undertakers owed the Town Council £2,006 in 1766, and probably still do.

The Magistrates, on the other hand, did very well. An account of 1765 showed a surplus of £211 14s. 5¹/₂d., which the Trustees hoped to assign to the New Bridge. The Adam plan was not completed, an old building in Writers' Court being incorporated into the western side of the square, but the Corinthian pillars and pediment, the deep loggia on the inner side of the courtyard and the screen wall along the High Street were well enough. The Council had reserved the right to keep the Custom House, and later let it for £360 per annum to the Government, when it accommodated, among others, Adam Smith as Commissioner.[55] It should be no surprise that the Edinburgh City Council is now the tenant of this handsome development with its magnificent exterior and baffling internal geography. The site is so steep that going down to the city archives in the basement is like descending into hell by way of Mary King's Close.

As the Exchange neared completion, Drummond moved on to more ambitious schemes. In 1758 the town had purchased another piece of ground, Allan's Parks (under what is now Charlotte Square), but a second attempt the next year to gain an Act to extend the Royalty foundered on the opposition of country landowners.[56] For the next few years the question of a northern residential suburbs was raised under various cyphers, such as 'scheme for communication with the fields in the North', or a new road to Leith.

Though the country was at war, the rest of the town did not stand still. About 1760, Dr Thomas Young developed a private street from the Canongate to the foot of Calton's Hill, called New Street or Young's Street, where lived two Lords of Session – Lord Kames and Lord Hailes (formerly Sir David Dalrymple). Allan Ramsay junior, flushed with his success as a portrait painter, in 1763 bought a property beside the Guse-pie on Castle Hill and had Robert Adam build two houses 'in the London fashion for the accommodation of two genteel families'.[57]

In 1761 the 26-acre policies of Ross House, on a comparatively sunny slope looking south over The Meadows, were offered by John Adam to the Town Council for £1,200. Inexplicably the Magistrates rejected his offer and it was taken up by one James Brown, who began to sub-divide or 'feu' the ground to tenants. The Council changed its mind and tried to buy Brown off for £2,000; but, for reasons which will become clear, he was not prepared to sell even for ten times that amount.[58]

The north side of what was to become George's or George Square, named for Brown's beloved brother and still a fine space for all the ravages of time and Edinburgh University, was begun in 1766. The mason was Michael Nasmyth, father of the painter of a famous portrait of Burns, who mixed blue basalt in the mortar in a chequerboard pattern that seemed to Hugo Arnot like the 'stuff that sailors' shirts are made of, and having, upon the whole, a very bad effect'.[59] The naturalist and traveller Thomas Pennant, who visited Edinburgh in July 1769, reported that 'a small portion is at present built, consisting of small but commodious houses, in the *English* fashion. Such is the spirit of improvement, that within these three years sixty thousand pounds have been expended in houses of the modern taste, and twenty thousand in the old.'[60] Brown wanted a genteel neighbourhood and stipulated that the 'feuars' – that is, his tenants under the feudal system of Scottish land rights – should not indulge in trade or retailing. The square was to be cleansed and lighted at a cost of a shilling in the pound of rental. By 1779 the west and east sides were finished, and in 1785 a range was built across the south side to complete the square.

Brown saw a return on his investment of something like 100 per cent per annum, which shows just how risky the enterprise was: nothing like that is recorded for the New Town.[61] He created a new fashionable quarter with a suburban character and its own entertainments, such as a riding school, designed in 1763 for the new Nicolson Street by Robert Adam and known as the Royal Academy for Teaching Exercises, and Assembly Rooms for dancing in Buccleuch Place, at the south-east corner of George's Square.

Walter Scott, who grew up at number 25 on the west side of George's Square, remembered that the southern districts of Argyle Square, Brown's Square and George's Square 'formed, about twenty or thirty years since, a little world of their own, and had their own Assembly-rooms, and society of an excellent quality, in some degree apart from the rest of Edinburgh'.[62] In his *Redgauntlet*, which is set in the 1760s, the old Jacobite Herries jeers at Fairford senior in Brown's Square for his 'new fashionable dwelling'.[63] Scott's own father's Jacobite clients, who included Murray of Broughton, might well have said the same about number 25 George's Square. The new districts generated their own manners, and the cosy Old Town supper party in time gave way to the formal George's Square dinner.

There was a new urgency to life. From 10 October 1763 the post went to London every weekday, instead of three times a week.[64] Dinner was delayed, according to Scott, so that lawyers and merchants could 'answer their London correspondents'.[65] The President of the Session still set the tone in Edinburgh, and from 1760 the President was Robert Dundas the younger. Scion of one of Edinburgh's great legal and political dynasties – his father, grandfather and great-grandfather had been Lords of Session and his father Lord President – Dundas was no philosopher but a businesslike man. He encouraged his fellow Senators to be brief or not speak at all, and was particularly impatient with Kames's metaphysical subtlety. 'Lord Kames,' he once said, 'I do not understand a word of what you have been saying all the while; it is too deep for me.'[66] Poor Kames: the old lawyer-philosopher-agriculturist was a dying breed in this busy new Edinburgh, though in his case it was the law and not philosophy that lost out (to reduce the time he spent on Session papers, he had his clerk read him only the facts).[67] To clear the arrears built up under his wordy predecessor, Lord Craigie, Dundas sometimes kept the court sitting until four in the afternoon. The result, according to Ramsay of Ochtertyre, was to push the hour of dinner even later.[68] Some men, of course, were incorrigible, and in his memoirs Lord Cockburn reported that even at the turn of the century some of the Edinburgh justices preferred to drink their dinner on the bench.[69]

The Peace of Paris that ended the French war was celebrated in Edinburgh on Tuesday, 29 March 1763. Twenty-one guns were fired on the Castle esplanade and the Magistrates and the principal citizens marched in procession to the Assembly Hall amid pealing bells to drink their Majesties' healths.[70] The Council now had no excuse for delay. On 2 July there appeared an advertisement

for tenders for a new way to Leith. It was the largest public work
in Europe until the canal mania of the late 1760s:

> As it is greatly desired, for the public utility, that a road of communi-
> cation be made betwixt the high-street of Edinburgh, and the adja-
> cent grounds belonging to the city and the other neighbouring fields,
> as well as to the port of Leith, by building a stone bridge over the east
> end of the North Loch, at least forty feet wide betwixt the parapets
> of the said bridge, and upon an equal declivity of one foot in sixteen
> from the high street at the Cap-and-Feather close, in a straight line to
> the opposite side leading to Multrees-hills. As the proposal for carry-
> ing on the above work was some time ago made to the town-council,
> and they having chearfully agreed to the same, this advertisement is
> publicly given to all who are willing to undertake the said work, to
> give in plans, elevations and estimates...[71]

A subscription was opened, stone was found and foundation piles
– both actual and philosophical – were driven in. In a letter to the
newspapers of 27 July an anonymous author presented one of the
most widespread of the town theories of the eighteenth century:
that a capital aspiring to be a centre of 'arts and politeness'[72]
needed a leisure class. Edinburgh was a burgess town and its
inhabitants were

> people of business, who have little leisure for any improvements, but
> what are connected with their proper occupation. A city of this kind
> is not qualified to be a capital. To make it a centre of arts and of
> politeness, a large proportion of its inhabitants ought to be men of
> rank and figure, who have leisure and money to cultivate the fine arts,
> and every sort of embellishment. Such a mixture in a town-society
> never fails to produce good effects: taste and politeness are com-
> municated by the men of leisure to the men of business, and solidity
> of judgement by the latter to the former, while their different talents
> are united in making discoveries, and in promoting whatever is pub-
> licly useful or beneficial.[73]

Thus luxury, far from demoralising the uncouth burgesses, will be
channelled by the gentlemen and ladies into correct taste. 'All that

seems necessary at present ... is to add to the town a fine extensive
field to the north, by an easy passage from the high street over the
lower end of the North Loch', and all for less than £5,000.[74]

On Friday 21 October the town's six hundred Freemasons met at
the Parliament House and at 3 p.m., accompanied by two companies of the military and the City Guard, paraded in reverse order
of Lodge seniority towards the site of the bridge. Last of all came
the Grand Lodge, preceded by a musical band of the fraternity 'who
all the way sung several fine airs, accompanied by French horns
&c'[75] parading behind George Drummond, standing in for the
Grand Master, Lord Elgin. Down the High Street they marched, by
the Netherbow and Leith Wynd to the site, where Drummond laid
the foundation stone 'amidst the acclamations of the brethren, and
of a prodigious number of spectators' ranged on scaffolds round
the scene. Beneath the stone he buried three commemorative
medals, one with a view of the bridge, a second with a bust of the
King and the third with a Latin inscription that translates thus:

> *By the favour of Almighty God*
> George Drummond, Esq.,
> Lord Provost of this city,
> Laid
> This Foundation-Stone.
> Of the Bridge leading towards Leith,
> The sea-port of the city of Edinburgh,
> In the twelfth year of his Provostship,
> Upon the 21st day of October,
> In the year of our Lord 1763,
> And of the aera of Masonry 5763,
> (The Right Honourable and Most Worshipful
> Charles Earl of Elgin
> Being Grand Master Mason of Scotland),
> And of the reign of George III.
> King of G. Britain, France, and Ireland,
> The third year
> *Which may the supreme God prosper.*

After an anthem the brethren processed back, and passed the evening at the Assembly Rooms 'with that social cheerfulness for which the society [Freemasonry] is so eminently distinguished'.[76]

There matters once again rested, amid a welter of different schemes, while the economic argument for extension became ever stronger. At the Council meeting of 7 November 1764 a letter from Baillie John Brown was read into the Minutes to the effect that there was now 'a Complaint or rather outcry for the scarcity of Houses'.[77] On 16 January 1765 'the Committee on the Scheme of the Communication with the Fields in the North' duly posted a notice that invited 'Architects and others' to submit plans for the bridge, for a reward of thirty guineas.[78] The committee, along with 'several other noblemen and gentlemen of taste and know-ledge in architecture',[79] met on 13 February and awarded the prize to the plan with the shallowest drop from south to north: number 7, submitted by David Henderson. But under the usual Town Council fog, on 10 July the committee changed its mind and selected the runner-up, number 5, submitted by Deacon William Mylne (who just happened to be a Member of Council), with alterations to the piers and arches by John Adam. The ostensible reason for the change was that Henderson could not provide security for completion on such a colossal work.[80]

John Adam drew up the building contract, which was signed in August 1765. It specified a bridge 1,134 feet long with no less than 40 feet between the parapets, over three great arches of 72 feet width. The contract price was £10,140, payable £500 on signature and the rest according to progress, with client inspections every Saturday and completion by Martinmas, 11 November 1769. To meet the expense, Lord Provost James Stuart pledged the town's holdings of some £6,000 in 'bank stock'. He hoped another £2,500 would be available from subscription. As for the balance, savings in expenditure 'gave the fairest prospect of having suffi-cient funds for finishing the whole without any new contractions'.

Work began in October, and proceeded fairly steadily. By December 1768 the third arch was complete, and the bridge was open to foot-passengers in the New Year. But on 3 August 1769 part of the side-wall of the south end of the bridge at Halkerston's Wynd collapsed, killing five people. Mylne had apparently piled up earth on the southern approach in order to reduce the gradient from the High Street, and the bridge proved too heavy for such unstable foundations. In the midst of recriminations the project fell hopelessly behind its November completion deadline. The young Robert Fergusson bitterly criticised Drummond's successors:

> The spacious *Brig* neglected lies,
> Tho' plagu'd wi' pamphlets, dunn'd wi' cries;
> ...For POLITICS are a' their mark,
> *Bribes latent and corruption dark.*[81]

Repairs were carried out by Mylne, and by June of 1771 no fewer than 158 pairs of hands were working on the bridge. The cost of the remedial work and of modifications and improvements was borne by the town. Arnot reported in 1779 that £17,354 had been spent, or nearly three-quarters as much again as had been contracted.[82] The bridge was finally completed in 1772 and was open to traffic at least by the time of a Town Council report the following February.

Though today it is sheltered from the north and east by a large city, the North Bridge still demands some fortitude of the pedestrian. When it was first built, according to Robert Chambers, it was so exposed that a man visiting his mistress in the New Town compared his ordeal to that of Leander, who each night swam the Hellespont to visit Hero, till one night he perished in the waves.[83] The caricaturist Kay loved to stand and watch the Quality making their way through the wind, as in his sketch of Dr James Graham and Miss Dunbar on their way from Gilbert Stuart's funeral, she holding on to her hat. It was said that the Irish giant Charles Byrne, who was 8 ft 2 ins tall, once lit his pipe from the bridge-lamps.[84]

By the spring of 1766 Drummond's successor Gilbert Laurie could be confident not only of the extension of the Royalty but also of the commercial demand in Edinburgh for a polite residential suburbs. There was to be no repetition of the embarrassment over George's Square. In a town so starved of investment capital, new banking houses were springing up and the British Linen Company, chartered in 1746 by the Argyll–Milton–Drummond interest, was abandoning the linen trade for the more lucrative business of banking. In his memoirs Sir William Forbes reported Scots as speculating in the sugar islands acquired at the peace of 1763, in agricultural improvements in the Lowlands, and in the new residential districts of Edinburgh.[85] Among large landowners in the Lowlands there were plans for a new bank to supply the want of capital. Douglas, Heron & Co., usually known as the Ayr or Air Bank, was founded in 1769 with capital of £160,000 and immediately began a policy of aggressive lending on land security and discounting of commercial bills. Meanwhile the Carron Iron Works, established in 1759–60 near Falkirk to exploit English blast-furnace and coke-smelting processes, was on its way to employing ten thousand men.[86]

In March 1766 the Magistrates announced that the ground for the new development had been surveyed, and in April called for plans. On 21 May 1766 it was noted in the Minutes that six plans had been submitted.[87] A seventh came in after the deadline. The Minutes of 2 August record that the plan submitted by Mr James Craig had been judged the best by Lord Provost Laurie and Mr John Adam.[88] Craig, 21-year-old son of a town merchant and nephew of the poet James Thomson, was then unknown. A Council committee that included Lord Kames and John Adam considered several amendments, some of them suggested by Craig himself, before his plan was finally adopted on 29 July 1767. By then Parliament had passed the Act extending the Royalty.

For one modern historian, Craig's New Town was 'the cold, clear and beautiful expression of the rational confidence of the eighteenth-century middle class'.[89] For another, it was the heavenly city of the Edinburgh philosophers.[90] Craig's plan, as it was engraved that summer, does indeed carry a quotation from the panegyric of Industry and Commerce in his uncle James Thomson's *Autumn* of 1730:

> August, around, what PUBLIC WORKS I see!
> Lo', stately Streets! Lo', Squares that court the breeze!
> See! long Canals and deepened Rivers join
> Each part with each and with the circling Main,
> The whole entwined Isle.[91]

Yet what is also striking about the plan is its extreme modesty and parsimony. A.J. Youngson, in his standard history of the New Town, *The Making of Classical Edinburgh, 1750–1840*, was hard-pressed to find much trace of the then newly-fashionable linked *places* being built at Nancy and Richelieu in France, or of the crescents and circuses of John Wood's Bath.

In the plan in the Huntly Museum in Edinburgh which was signed by Provost Laurie on 29 July 1767 – 'This is the Plan to which the Report of the Committee of Council of this date relates' – three unnamed streets run east to west, the middle connecting two formal squares, the outer two running on. Between them are two narrow 'Back Streets' (now Thistle and Rose Streets) for servants and artisans. The east–west streets are intersected by cross-streets at precise intervals, giving a gridiron plan. The main street along the axis of the New Town was at some stage in 1766–7 named George Street, in honour of the king, and the two squares were named after the patron saints of the united kingdoms, St Andrew's Square at the east and St George's Square at the west. The outer streets were named Queen Street and Prince's Street. Unionism was turned up a pitch in 1785, when St George's Square was renamed Charlotte Square in more particular reference to George III's queen.

St Andrew's Square was to be no Place Vendôme, George Street no Bath Royal Crescent. The Continental practice of planning a unified frontage for a terrace of houses was not adopted until Robert Adam produced his elevations for Charlotte Square twenty-five years later. The Town Council was willing only to level and causey – pave or cobble – the streets, pipe in water and lay sewers along the main streets – 'common Shores' – and divide the frontages into plots of ground, or stances, to be feued under the broadest conditions as regards the width of the pavements, the number of storeys, and so on. In fact, it was soon found that nobody in Edinburgh had sufficient experience to lay a modern sewer, and in October Craig was paid thirty guineas in expenses to travel to London to study the latest works.[92] By this point, incidentally, the Town Council Minutes were referring to the extended Royalty as a 'new Town'.[93]

In August the first stances were feued in St Andrew's Square, and over the next two years further plots were taken up in South St Andrew's Street, South St David's Street and the east end of Prince's Street. The Town Council received just £972 14s. 7d. from the sale of building plots in the first full year to November 1768.[94] By July of 1769, as Pennant reported, houses in St Andrew's Square cost from £1,800 to £2,000 each, with one or two from £4,000 to £5,000.[95] These would have been large sums for speculative builders, and many plots were feued by the potential owners themselves, such as David Hume and Sir Laurence Dundas, a distant kinsman of Arniston who had made a fortune supplying Cumberland's redcoats and who, as befitted the new MP for the town, took over the prime site in the middle of the east side of St Andrew's Square.

The feuing procedure transferred the financial risk from the Council, but at the cost of a rather plain effect. The magnificent house designed by Sir William Chambers for Laurence Dundas (who had grand political ambitions), and now occupied by the Royal Bank of Scotland, was very much the exception. The east end of Prince's Street, exposed to the traffic to Leith and the butchers and fleshers now established under the North Bridge, was not exactly genteel. George Street was, according to a visitor of 1788, 'so wide in proportion to the height of the buildings, that in the declining line of perspective they appear like Barracks'.[96] The merit of the first New Town arose out of pure parsimony. On Craig's plan the two outer streets are marked with houses only on their inner sides, while the outer sides are open space. The result

is spectacular views to both north and south. Prince's Street (or
Princes Street, as it now is) looks across to the Castle and High
Street, while Queen Street faces the Firth and the Fife hills. In fact
the Town Council at first permitted building south of Prince's
Street, on a steep site just west of the North Bridge, and almost
squandered that advantage. Nearby was the new Theatre Royal,
put up at the east end of Prince's Street in 1768, at the spot in the
park belonging to the Orphan's Hospital where George Whitefield
had once preached to thousands in the open air. Times had
changed.

Craig's commission was for a rather dour residential district.
His plan showed no public buildings except for a church at the far
side of each square. The *Proposals* of 1752 had mentioned the
building of a registry for the public records of Scotland, which
were then mostly kept in the basement of the Parliament House.
Boswell took Johnson there and was pleased to see the doctor
'rolling about in this old magazine of antiquities' until they quarrelled about the Union of the Parliaments.[97] The government
offered £12,000 from forfeited estates for a new repository. A site
was found at the north-west corner of Heriot's Hospital, but that
was deemed to be too far from the Courts. With the bridge complete and the Magistrates anxious to promote the feuing of its
north end,[98] the Council made available a site at a cluster of
houses known as Mutrie's or Multree's Hill.

Robert Adam and his brother James were appointed architects, and in 1774 the foundation stone was laid. But funds
were short throughout the 1770s, and Robert Adam was in
London. In 1778 work stopped altogether, and the building
gradually became derelict. The government finally voted a further £15,000 in 1784, and four years later the building was
occupied. For all its tribulations it made a magnificent focus to
the vista over the bridge, and a more than sufficient invitation
to cross it.

In 1775 James Craig designed a new hall for the Royal College
of Physicians, to stand on the south side of George Street, midway
along the first block of buildings. In 1783 no less than £6,300 was

subscribed – how rich the town was becoming – for a new suite of Assembly Rooms in the equivalent position in the second block, to designs by John Henderson.[99] The ninety-two-foot ballroom with its seven crystal chandeliers can still be admired.

On 31 January 1781 Old Provost James Hunter Blair reported to the Council that 'since the extension of the Royalty, a great number of familys of distinction reside in the new Town, where the number of inhabitants are yearly encreasing; that the inhabitants of the extended Royalty not only find the greatest difficulty of being accomodated with seats in the present Churches, but the situation of these Churches with regard to them is so very inconvenient that they are desirous of a new church being built there.'[100] The site for the St Andrew's Square church having been snatched by Sir Laurence Dundas, the Town Council chose a plot on the north side of George Street, immediately opposite the Physicians' Hall, and borrowed £3,000 from Heriot's Hospital (secured on the ale duty) to build a 'decent handsome church'. Designed by David Kay and Captain Andrew Fraser of the Engineers, it was completed in 1783, and a tower and spire were added a couple of years later. The church's modest position makes its classical oval interior and its ceiling plasterwork all the more rewarding.

The migration to the 'draughty parallelograms', as Robert Louis Stevenson described them, was slow enough. Professionals were anxious about the cost of living and entertaining in the modern or 'English' style, and feared they might be too far from the Courts and Cross. The collapse of the side-walls of the south end of the bridge in the summer of 1769 was the opposite of reassuring.

That same year David Hume returned from Paris and London to find that he had, in every sense, outgrown James's Court. The set of rooms was, he said, 'too small to display my great Talent for Cookery, the science to which I intend to addict the remaining Years of my life'.[101] He thought of taking one of Allan Ramsay's houses on Castle Hill or moving to George's Square before boldly resolving to feu a small plot at the corner of St Andrew's Square and the street that became known, by intention or irony, as St David's Street. With the bridge out of service he was obliged to pick his way to his chosen site across the drained North Loch, and at one point fell into the bog. He was rescued, it was said, by an old fish-wife, but only after she had forced 'Hume the Atheist' to recite the Lord's Prayer.[102]

In the summer of 1772 several Scottish banking houses, including the fast-growing Douglas, Heron and Co., failed. The 'Ayr Bank', which had opened for business on 6 November 1769 with the Duke of Queensberry and Adam Smith's pupil Buccleuch as directors, had financed its partners' land speculation by means of short-term borrowing at ruinous rates in London.[103] When the London Scottish banking house of Neale, James, Fordyce and Downe collapsed on 10 June 1772 with liabilities of £243,000, there was a flight out of Scottish banking paper and the Ayr Bank, whose balance-sheet had swelled to no less than £1.25 million, stopped redeeming its bank-notes on 25 June. The only recourse was the sale of annuities at usurious rates – 'junk bonds' – in London, but that bought respite only until the spring.[104] Without limited liability, the partners were called on to find more than £600,000. 'A couple of our Ayr shire Noblemen, and the major part of our Knights and squires are all insolvent. A miserable job [swindle] of a Douglas, Heron, & Co.'s Bank ...' Burns observed ten years later.[105] In Edinburgh itself only the Royal Bank, the Bank of Scotland, the British Linen Company and three private banking houses survived. The rising politician Henry Dundas, brother of the Lord President, lost his wife's dowry, amounting to as much as £100,000. The crisis stopped the New Town in its tracks and the outbreak of war with the American colonists arrested development for the remainder of the 1770s.

'Indian' Peter Williamson, whose varied career had included being kidnapped, shipwrecked, sold into indentures in Pennsylvania, captured and tortured by Indians – and then escaping and writing a blood-curdling account of his life, *French and Indian Cruelty* – published an indispensible trade and social *Directory* of Edinburgh in the 1770s. In this, among the musty

Old Town parade of spatterdash(gaiter)-makers, limners and makers of 'grave cloaths', the New Town residents are few and far between. Of the *literati* there are only Hugo Arnot and David Hume, who entertained Benjamin Franklin 'with the greatest Kindness and Hospitality' for a month in St David's Street in the autumn of 1771; of the full members of the Town Council, Baillie Thomas Elder (Prince's Street) and an old dean of guild, Charles Wright (St Andrew's Street). There are some merchants and half a dozen advocates and Writers to the Signet, and also a few architects, wrights and masons who are, presumably, actually developing the site. Williamson, who had started an hourly penny post for the town and suburbs, had but one 'post office' (a shop-keeper to take in letters) north of the bridge.[106] By now, according to Arnot, the population of the town and suburbs had risen to 82,000.[107]

The new St James's Square, designed by Craig with unified frontages between St Andrew's Square and the road to Leith, began feuing in the middle of the 1770s and attracted, as might have been expected, some sentimental Jacobites (who entertained Burns in 1787). A new style of lodging-house, which Arnot called a *hotel*, was open in Prince's Street by 1779, but its rates were unheard-of for the city: for three bedrooms, dining-room and 'closet', no less than three guineas (£3 3s.) a week.[108]

Meanwhile, the feuars of St Andrew's Square and its surrounding streets, including David Hume and Sir William Forbes, in 1771 sought an injunction against the cottages and workshops sprouting to the south of Prince's Street. The lawsuit wound its way through Court to Session to the House of Lords, where Lord Chief Justice Mansfield pronounced against the Magistrates but sent the case back to the Court of Session. In the end, it went to arbitration. In his Decreet Arbitral in 1776 David Rae, later Lord Eskgrove, ruled that the houses under construction might be completed, but the rest of the ground westward to Hanover Street was to remain 'pleasure ground'.

What with the city taxes in the Extended Royalty, the gales, this expensive litigation and a 'crazy bridge', Arnot reported, the

gentry preferred the new squares to the south of the Cowgate. In
Williamson's *Directory* of 1775–6 the Dowager Countess of
Leven is listed as living in St Andrew's Street, as are a handful of
country gentlemen, but the vast majority of noble families are
still living cheek-by-jowl with hairdressers in the closes or in the
new squares, and particularly in George's. These were joined to
the High Street by the new South Bridge over the Cowgate in
1788.[109]

In his *Memorials of his Time*, Henry Cockburn said that even
after the building of the new Assembly Rooms in George Street
(1787), 'the whole fashionable dancing, as indeed the fashionable
everything, clung to George Square'.[110] Over the years George's
Square housed (apart from Scott at No. 25) Admiral Duncan, who
defeated the Dutch fleet off Camperdown in 1797 (No. 5); the
great heiress, the Countess of Sutherland (No. 14); Lord Melville
(No. 5 and then No. 38) and his nephew Lord Advocate Robert
Dundas (No. 57); the Draconian justice Lord Braxfield, who characteristically tried to extend the cellar at No. 13 and flooded it;
Burns's Clarinda's cousin the judge Lord Craig at No. 23; in the
top storey of No.39 Dr Adam, the rector of the High School; and
at No. 55 the Man of Feeling, Henry Mackenzie.[111] Jane Welsh,
who married the writer Thomas Carlyle, often stayed at No. 22.
Walter Scott, writing in the next century, reported that one
respectable gentleman lived and died south of the Cowgate without ever seeing the New Town.[112]

By 1784 St Andrew's Square was built, but the sale of lots that
year brought in only £1, 217 11s. 8d. while the Magistrates spent
£1,287 11s. 6d. on civil works. Meeting on 3 March the committee on the New Town found that 'abstracting from the great
expence the City was put to in purchasing Grounds and building
the Bridge, the purchase money hitherto received has but barely
answered the expence laid out upon levelling and paving the
Streets common sewers &c and therefore that it would be necessary to deliberate upon this matter before any more ground is
feued.'[113] In total, the town had spent nearly £16,000 on civil
works in the Extended Royalty, not including the cost of £6,000

for assembling the land and the £3,000 borrowed to build St Andrew's Church.

Fortunately, the end of the American War in 1783 saw the beginning of a period of rapid expansion in agriculture and the textiles trade. There was a surge of interest in the New Town on the part of country families and, in the time-honoured manner, their lawyers. The town's revenue from sales jumped to £4,605 13s. 8d. in the twelve months to November 1787.[114] The New Town began to make its way westwards. By the autumn of 1791, the dung of humans and horses collected from the Extended Royalty was worth a full £100 a year to the Council as manure.[115] Hanover Street was largely built by 1790, Frederick Street a couple of years later, and building had started in Castle Street. When Scott and his bride moved there from George's Square in 1798, the New Town had triumphed. The aristocratic society prescribed in the *Proposals* of 1752 had taken root in the New Town, and stayed there until the railway age brought London within easy reach.

In 1791 Robert Adam, the most famous architect in the country, was asked to provide elevations for a palace-like, unified frontage in Charlotte Square. The outbreak of war with revolutionary France and the consequent rise in the rate of interest brought work to a halt with just two-thirds of the north side built, as it did Adam's great designs for the University across the ravine. Not until the peace of 1815, long after Adam's death in 1792, was the University completed on a reduced scale, Charlotte Square built, and the Moray, Raeburn and Heriot estates developed under its enduring influence.

For much of this period, from before Drummond's death in 1766 until 1796, the Council's purse-strings were in the practised hands of Mr Hugh Buchan of Mylne's Court, the City Chamberlain. Buchan managed the town's finances in a manner that was understood only by himself, possibly not even by him. His impenetrable book-keeping did not hinder the New Town, University and South Bridge from being built, and enabled the Town Council to scorn calls for reform until, one day in 1833, suddenly and quite comprehensively, the City of Edinburgh went bust. But that is another story.

—◦—

Edinburgh is a paradox: a Classical town rescued from the frigid by a Gothic town rescued from the grotesque. As A.J. Youngson put it in the language of architecture, the true singularity of Edinburgh is the 'physical separation and the visible conjunction of the Old Town and the New'.[116]

There is no city like Edinburgh in all the world. 'It is what Paris ought to be,' Stevenson mused. 'It has the scenic quality that would best set off a life of unthinking, open-air diversion ... If the climate were but towardly ...'[117] He thought Edinburgh so romantic that it was let down by its inhabitants, and should be abandoned to the sun and wind and birds, and gypsies camping in the thoroughfares. 'These citizens, with their cabs and tramways, their trains and posters, are altogether out of key.'[118]

In fact, Princes Street has been full of hotels, steamship offices and drapers since the 1830s. It is precisely the people of Edinburgh and their 'little air of possession' of their sensational domicile that is the great charm of the city. Cockburn and others mourned that the developments after 1800 robbed the New Town of its country character. Yet even today, on a summer evening, looking down from Telford's Dean Bridge into the trees waving below, you feel the country racing through.

The rise of the New Town had its counterpart in the decline of the Old. The bookseller William Creech recorded this process with precision: 'In 1783 ... The Lord Justice-Clerk Tinwald's house was possessed by a French Teacher – Lord President Craigie's house by a Rouping-wife or Sales-woman of old furniture – and Lord Drummore's house left by a [Sedan] Chairman for want of accommodation [that is, he found it too small].'[119] A passage in Chambers's Traditions of Edinburgh composed in the 1820s sounds Dickensian, but only because it was precisely such human archaeology, the recovery of city types obliterated by social cataclysm, that Dickens made his own:

Did he ascend a stair and enter a floor, now subdivided perhaps into four or five distinct dwellings, he might readily perceive, in the massive wainscot of the lobby, a proof that the refinements of life had once been there ... Passing into one of the best rooms of the old house, he would find not only a continuation of such wainscoting, but perhaps a tolerable landscape by Norie[120] on a panel above the fireplace, or a ceiling decorated by De la Cour, a French artist, who flourished in Edinburgh about 1740. Even yet he would discover a very few relics of gentry maintaining their ground in the Old Town, as if faintly to show what it had once been. These were generally old people, who did not think it worth while to make any change till the great one ... Any one ascending a now miserable-looking stair in Blackfriars Wynd would have seen a door-plate inscribed with the name MISS OLIPHANT, a member of the Gask family.[121]

Drummond and his friends presided over the destruction of the Royal Porch at Holyrood in 1753. The Mercat Cross in the High Street followed in 1756. The Netherbow Port, having survived the retribution of Queen Caroline after the Porteous Affair, was taken down to make way for carriage traffic in 1764. No doubt there were protests, and *The Scots Magazine* carried a series of satires by 'Claudero', James Wilson, on the destruction of those buildings. 'Your porch falls a sacrifice to luxury,' he wrote in his 'Sermon' on the condemnation of the Netherbow. 'Let that be the butt of your just resentment.'[122] He added: 'Our gates must be extended wide for accommodating the gilded chariots, which, from the luxury of the age, are become numerous.'[123]

'Claudero' was an eccentric who dedicated his scurrilous verses to 'Indian' Peter Williamson 'of the Mohauk Nation', copied college lecture notes for a precarious living, was bullied by his wife, and once spent six weeks locked up in the Tolbooth. Behind his Jacobitical nostalgia lay a very modern sense of outrage:

What then? My Echo loud did cry.
Must Scots antiquity now die?
Yes, cry'd AULD REIKIE, die you must
For ―― [?Cumberland] at you has a disgust.[124]

The Potterow Port and West Port were taken down to accommodate turnpike traffic in 1786, and much of the city wall to the south in 1787. The Tolbooth and Luckenbooths were destroyed in 1817.

On the night of Monday, 15 November 1824, fire broke out in an engraver's shop in the Old Assembly Close, and by eleven o'clock had spread to three adjoining tenements and was making its way through the crowded masses of buildings to the Cowgate. At 11.30 a.m. on the Tuesday the steeple of the Tron Church was aflame, and burned with such fury as to melt the two-ton church bell. That night a new fire broke out in the apartment of 'a female of abandoned character' by Parliament Square and, with a gale blowing from the west, by 5 a.m. on Wednesday the whole east side of the square had either fallen or was in blackened ruins. It was feared that 'the Old Town is doomed to be burnt down in detail'.[125]

In the 1830s the most old-fashioned street in Edinburgh, the West Bow, was partly obliterated by Victoria Street. The west side of James's Court, with its mementoes of Hume, Blair, Boswell and Dr Johnson, burned down before 1868. Between then and the Great War, approximately two-thirds of the Old Town buildings were destroyed by fire or by intention, including Milton House, Hope House and Gordon House.

'It is easy', said Stevenson, 'to be a conservator of the discomforts of others.'[126] What may be regretted is the loss in the capital of Scotland not of masonry but of a certain unity of social feeling those modest buildings embodied. Separated from the genteel quarters by Princes Street Gardens, the Old Town sank into poverty and then, more recently, gave itself over to subsidised drama, tartan, and goth gear. As Stevenson put it, people take 'to the High Street, like a wounded animal to the woods'.[127] The Canongate is once again, for all the new Parliament building, Allan Ramsay's 'puir eldritch hole'. The New Town has Harvey Nichols.

Robert Chambers tried to be optimistic about the city's divided destiny. 'It was only too accordant with that tendency of our present form of civilisation to separate the high from the low, the intelligent from the ignorant – that dissociation, in short, which would in itself run nigh to be a condemnation of all progress, if we were not allowed to suppose that better forms of civilisation are realisable.'[128] This modern Edinburgh, where people are united not by blood, familiarity or topography but by their common commercial appetites, is the subject of the next chapter.

8

The Savage and the Shopkeeper

During the Jacobite occupation of Edinburgh in 1745, *The Scots Magazine* reported, Highlanders would hold up pedestrians in the High Street for a penny.[1] When the rebel Lord Balmerino went out for execution on the green above the Tower of London the following January, he apologised for under-tipping the headman: 'Friend, I never had much money; this is all I now have; I wish it were more for your sake.'[2] Cameron of Lochiel, Adam Smith wrote more in astonishment than regret, 'whose rent [income] never exceeded five hundred pounds, carried, in 1745, eight hundred of his own people into the rebellion.'[3]

For Edinburgh, the Forty-five accelerated a process which had been in train for a long time and is now all but complete. Values of ancient or obscure social origin were displaced by prices. Honour, salvation, feudal allegiance, divine right, custom had lost conviction before the certainties of money. Land and family were matched as sources of distinction by money made from commerce and the Funds (national debt).

The people of the new districts of Edinburgh appeared to be as passionately concerned with their wishes as with their needs; indeed, could no longer distinguish them as separate orders of satisfaction, for both were discharged by money. A new Scotland based not on animal requirements but on fancy and fashion was coming to birth. Pious Whigs, feudal Jacobites, backwoods agriculturists and civic republicans were fascinated and troubled by what was happening.

This new Scotland was analysed and understood in eighteenth-century Edinburgh under such headings as 'luxury', 'opulence' and – a new term devised by Sir James Steuart – 'political oeconomy'. For some time the discussion was encumbered by moral, religious and political nostalgia: for moneyless societies such as those of ancient Sparta or the Ossianic isles, for holy poverty, for the feudal order, for the parsimony and plainness of old Scotland, even for the exiled Stuarts.

David Hume and Robert Wallace in Edinburgh, Francis Hutcheson in Glasgow, Adam Smith in Kirkcaldy and Sir James Steuart on the Continent sought to examine the material interests of society free of those moral and censorious entanglements. Yet a regret for a plainer, more direct and more organised society tinges even that great apology for luxury, *The Wealth of Nations*. It was this ambivalence in the Edinburgh school of political economy that attracted Karl Marx and his successors, to the great disturbance of the peace of the twentieth century.

Because the eighteenth century looked to the classical world for its universals and to ancient authors for reliable information, the discussion was conducted at first on quite remote literary ground. The parsimony of the ancient Romans was set against the extravagance of the present, or the military communism of Sparta against the empire-building democracy of Athens. Yet these highfalutin' debates were also code for the bitter and long-standing rivalry of Scotland and England.

By reason of its poverty, aristocracy and military traditions, Scotland felt a decided affinity for Sparta and a society which, according to the ancient authors Plutarch and Xenophon, outlawed for free-born males all property, money, superfluity, foreign trade, and bathing. Under laws promulgated by the half-legendary king Lycurgus, Spartan men lived together in military messes and saw their wives only in secret, children were taken from their parents, menial work was done by slaves, and theft was encouraged. It helped that Spartan or Lacedaemonian communism had achieved unparalleled feats of arms against the Persian invasions of the early fifth century BC, broken the commercial and naval might of Athens at the turn of the fourth, and survived until defeat by the Thebans and the extinction of all the independent Greek city-states by Alexander the Great in the 330s. Of the Scottish Lacedaemonians, Andrew Fletcher of Saltoun in 1698 actually recommended slavery for destitutes and military training for all Scottish males.[4]

It was Hume who attempted to introduce reality into this romance of ancient republican virtue. In his *Political Discourses*, published in the same year (1752) as the *Proposals* for the City's improvement, he painted a sunny picture of commercial modernity. In his 'Of the Populousness of Ancient Nations' he took issue with Robert Wallace, who had attempted to argue, in a paper read to the Philosophical Society in the 1740s, that the parsimony of antique times had increased the prevalence of marriage, and the ancient world was therefore more populous (and, by extension, more prosperous) than the modern.[5] Hume reminded his readers that ancient societies such as Sparta were founded on a slavery that was not merely cruel but inhibited the growth of population. Ancient 'liberty' was fatally prone to factionalism and demagoguery. Hume was the first to contend, however tentatively, that the modern world was more populous than the ancient.

In 'Of Commerce', Hume attacked the new wave of Spartan nostalgia set off – O *Sparte*! – by Jean-Jacques Rousseau in his prize essay for the Dijon Academy in 1750, usually termed *Discours sur les sciences et les arts*. Societies which directed the surplus income from agriculture to commerce and manufacture were not merely more natural and happy than societies which deployed it in military adventures, they were also more powerful.[6] In 'Of Luxury', which in a new edition of 1758 he renamed 'Of Refinement in the Arts' to remove the last tincture of censoriousness from the discussion, *le bon David* made a tour of 'the modern city', one that sounds very much like Edinburgh:

> The more these refined arts advance, the more sociable men become; nor is it possible, that, when enriched with science, and possessed of a fund of conversation, they should be contented to remain in solitude, or live with their fellow-citizens in that distant manner, which is peculiar to ignorant and barbarous nations. They flock into cities; love to receive and communicate knowledge; to show their wit or their breeding; their taste in conversation or living, in clothes or furniture. Curiosity allures the wise; vanity the foolish; and pleasure both. Particular clubs and societies are everywhere formed: Both sexes meet in an easy and sociable manner; and the tempers of men, as well as their behaviour, refine apace. So that, beside the improvements, which they receive from knowledge and the liberal arts, it is impossible but they must feel an encrease of humanity...[7]

Wallace was won round. At the time of the staging of *Douglas*, he desired to inform the Edinburgh Presbytery that luxury was 'necessary for promoting an honest and laudable industry & the support of the poor'.[8] He opened his *Characteristics of the Present Political State of Great Britain* of 1758 with a quotation from Ovid: 'Let others delight in ancient things, I congratulate myself that I was born in modern times.'[9] In this essay, which appeared when the war with France was going badly, Wallace took issue with the doom-sayers who claimed that Britain was too luxurious to confront the French monarchy. He distinguished between private luxury, which was 'always bad', and national luxury, which would have as its counterpart increased national wealth.[10] Commerce, public credit and the moneyed interest could all be sources of national strength. As for Spartan communism, it could be established in Great Britain or France only 'by means of an extraordinary concurrence of circumstances, at the time of a grand revolution',[11] when 'a spirit of patriotism and a love of equality may be accidentally raised, and run high'.[12] It was an uncanny premonition of the events of 1789 in France.

Sir James Steuart, the only Edinburgh philosopher active in the

Jacobite cause in Forty-five, was another man torn in his attitudes and allegiances. In his great *Inquiry into the Principles of Political Oeconomy*, on which he laboured for eighteen years in stagnant Continental backwaters, Steuart gave a name and a method to a new technique of examining modern commercial society while never wholly suppressing a hankering after Spartan communism as 'the most perfect plan of political oeconomy'.[13] Attacked as absolutist, foreign and Jacobitical on its publication in 1767, the *Inquiry* found few readers in Scotland and England[14] and its influence has been confined to Continental philosophers, notably Hegel and Marx.

Sir James Steuart of Goodtrees and Coltness was the unluckiest of philosophers. Born on 10 October 1713 into a family at the very heart of legal and political Edinburgh,[15] he was regarded as one of the most brilliant young men of his generation. After attending the College he passed advocate in 1735. The first of his several misjudgements was to make the then-fashionable Grand Tour, in the course of which, at Avignon, he came under the influence of Jacobite exiles. In the winter of 1739–40 he made his pilgrimage to the Stuart court in Rome. His second error was to fall foul of the powerful Dundases of Arniston, first in love and then in politics.[16] Steuart married Lord Elcho's sister, Lady Frances Wemyss, in October 1743 and the house at Goodtrees, three miles to the east of Edinburgh, was soon the 'rendez-vous of all the supporters of the house of Stuart'.[17] Strongly opposed to Prince Charles Edward coming to Scotland unless at the head of a French army,[18] Steuart joined the Rebellion only after Prestonpans, and with only half a heart.

Steuart arranged to have himself arrested at the Cross by the Prince's forces and taken to Holyrood, which fooled nobody in Edinburgh or in London.[19] Behind the scenes, he almost certainly wrote the conciliatory Proclamation of 10 October 1745[20] which apologised for the 'miscarriages' of previous Stuart reigns, promised a repeal of the Union, a limited monarchy and parliamentary government for Scotland, and guaranteed the holders of government debt.[21] Charles Edward refused to sign it.[22] At the end of

October Steuart set sail from Stonehaven for France, again in the guise of a prisoner but in fact as Charles Edward's special envoy to plead for an immediate French invasion of the east coast of England. Plagued by gout, he energetically pressed the Prince's cause at Versailles, but soon came to doubt the sincerity of French intentions. With the collapse of the Rebellion, Steuart was excepted by name from the Act of Oblivion of 1747,[23] and on 13 October 1748 was indicted for high treason by a Grand Jury at the Court of Justiciary.[24] The indictment was resented in Edinburgh, and put down to Dundas malice; but Ramsay of Ochtertyre exonerated the Arnistons: 'In truth, it originated from the Ministers of State, who were much incensed against him.'[25]

The Steuarts lived at first in Paris, then at Angoulême in the south-west. Goodtrees was sold. With the prospect of war between Britain and France, and not wishing to compromise himself further, Steuart moved to Brussels, then to Spa for treatment of his gout, then on to Frankfurt, before coming to rest in 1757 at Tübingen. In that quiet university town he lived happily with his wife and young son and began writing the *Inquiry*. In 1763, after a mysterious period in French custody which did him no harm in the eyes of his supporters in London, Steuart was at last permitted to return quietly to Scotland, where he put his estate at Coltness in Lanarkshire in order and saw the *Inquiry* through the press.[26] The London Scottish bookseller Andrew Millar paid £500 for the title, which appeared in two quarto volumes at a guinea apiece early in 1767. Reviewers commended Steuart's scope, originality and industry but hinted darkly at an incorrigible Jacobitism. The volumes had, at first, a modest sale. Steuart was formally pardoned in 1771, yet he kept aloof from Edinburgh intellectual society.[27] He interested himself in the administration of India, wrote on finance, and devoted his last years, in the old Scots way, to metaphysics and agriculture. He died in 1780.

In Sir James Steuart's political economy, human history is not a tissue of contingencies but a regular progress from 'great simplicity to complicated refinement'[28] governed by certain permanent human impulses. The first is sexual desire, that causes populations to rise in line with the supply of food at any given time. (This argument was developed by Thomas Malthus in England later in the century.) The second is self-interest or self-love.

Steuart examined three fundamental forms of society – pastoral, agrarian and commercial – which progress from one to the next through the operation of these two principles. Rising population demolishes the pre-agrarian economy as men seek to secure food for their children by appropriating part of the land and enslaving others to work it for them. Any surplus they produce from the soil they learn to barter.[29] Trade and industry develop, 'luxury' appears, and slavery becomes outmoded. The master–slave relationship gives way to a web of mutual interdependence in which relations of money become ever more the norm.

> When once this imaginary wealth (money) becomes well introduced into a country, luxury will very naturally follow; and when money becomes the object of our wants, mankind become industrious, in turning their labour towards every object which may engage the rich to part with it.[30]

As industry flourishes, so populations rise. 'I am no patron either of vice, profusion, or the dissipation of private fortunes,' Steuart noted at the foot of the page, 'although I may now and then reason very coolly upon the political consequences of such diseases in a state, when I consider only the influence they have as to feeding and multiplying a people.'[31]

In a brilliant passage Steuart distinguished luxury in the modern commercial economy from the luxury of the Roman Empire:

> Ancient luxury was quite *arbitrary*; consequently could be laid under no limitations, but produced the worst effects, which *naturally* and *mechanically* could proceed from it. Modern luxury is *systematical*: it cannot make one step, but at the expence of an adequate equivalent, acquired by those who stand most in need of the protection and

assistance of their fellow citizens; and without producing a vibration [alteration] in the balance of their wealth.[32]

The connecting-rod of the modern system was money. As 'the adequate equivalent for every service',[33] money emancipates the labouring classes, not merely from slavery but from feudal service and counter-service. Money frees humanity from everything except – and here is a point the Marxians later picked up – itself. 'Men were then forced to labour because they were slaves to others; men are now forced to labour because they are slaves to their own wants.'[34]

Unlike Hume and Smith, who regarded money as simply a utensil of trade,[35] Steuart had the landed man's vivid sense of money and credit as dynamic forces setting people and property in violent motion, 'their lands, their houses, their manufactures, nay their personal service, even their hours'.[36] Further, the money economy, by its very complexity, limited political power:

> When once a state begins to subsist by the consequences of industry, there is less danger to be apprehended from the power of the sovereign. The mechanism of his administration becomes more complex, and, as was observed in the introduction to the first book, he finds himself so bound up by the laws of his political oeconomy, that every transgression of them runs him into new difficulties.[37]

Or again:

> Modern oeconomy, therefore, is the most effectual bridle ever was invented against the folly of despotism.[38]

The problem for Steuart was that this modern economy is not perfect. Public authority – what he called first 'The Statesman' and then 'the legislature and supreme power'[39] – must intervene to protect the fragility of modern systems. 'Were industry and frugality', he wrote in the second book of the *Inquiry*, 'found to prevail equally in every part of these great political bodies, or were luxury and superfluous consumption every where carried to the same height, trade might, without any hurt, be thrown entirely open.'[40] But that was not the case. Meanwhile, limited and democratic government, through its interventions in the commercial and domestic economy, its income and sales taxes, its bounties and subsidies and rates of interest, controlled the lives and property of the citizenry far more completely than the 'most despotic and arbitrary authority can do'.[41]

For Steuart, the value of the Spartan system was its longevity, consistency and simplicity:

> It is of governments as of machines, the more they are simple, the more they are solid and lasting; the more they are artfully composed, the more they become useful; but the more apt they are to be out of order. The Lacedemonian form may be compared to the wedge, the most solid and compact of all the mechanical powers. Those of modern states to watches, which are continually going wrong; sometimes the spring is found too weak, at other times too strong for the machine: and when the wheels are not made according to a determinate proportion, by the able hands of a Graham, or a Julien le Roy, they do not tally well with one another; then the machine stops, and if it be forced, some part gives way; and the workman's hand becomes necessary to set it right.[42]

Steuart had been affected by his long sojourn in Continental princedoms. In Book II of the *Inquiry* he gave space to such government measures – protection, import controls, bounties, subsidies and employment policies – as his British merchant readers were beginning to resent, and Adam Smith in *The Wealth of Nations* was to reject. 'We believe,' one reviewer of the *Inquiry* sniffed, 'that the superiority which England has at present over all the world, in point of commerce, is owing to her excluding Statesmen from the executive part of all commercial concerns.'[43] There was 'too much of a foreign cast in this author's ideas'.[44] Another contemporary reviewer was free with Jacobitical aspersions: Steuart's politics, 'the effect of being educated in arbitrary principles' on the Continent, were 'by no means consistent with the present state of England, or the genius of Englishmen'.[45]

It was on the Continent that Steuart exercised most influence.

Hegel devoted several months of the spring of 1799 to writing a commentary on the *Inquiry*.[46] From it he drew reinforcement not only for his very Scots view of history, in which Mind or Spirit – *Geist* in German – came both to know and to reveal itself, but also for the idea that the modern economy based on money was horribly distinct from the antique *polis*, or city-state. In lectures given at Jena he portrayed money as a 'monstrous system of community and reciprocal dependence' which operated as a sort of zombie – elemental, spastic and blind – and must be chained up like a wild beast.[47] Marx, in the notes written in Paris in the early 1840s which are now known as the *Economic and Philosophic Manuscripts*, portrayed the world of money as a phantasmagoria in which human beings can no longer distinguish what is real from what is not.

Adam Ferguson went even further in the Spartan cause. In a career that was at various times military, pastoral, scientific and metaphysical, from his Gaelic sermon on the Rebellion in 1745 to his *Essay on the History of Civil Society* of 1767 and on to *The History of the Progress and Termination of the Roman Republic* in 1782, he pursued an antique theme that had been revived by the Florentine philosopher Niccolò Machiavelli in the Renaissance, by Puritan thinkers in seventeenth-century England and by Andrew Fletcher of Saltoun: that a nation cannot measure its prosperity by mere population or commerce or industry. Something else, called *virtú* or 'virtue', was required to secure a rich nation and its citizens from servility.[48]

Unlike Hume, Smith or Steuart, Ferguson had known a living Highlands where communal bonds were not mediated by money. His model societies – even his Sparta – have the tang of peat and heather about them.[49] His career as a military chaplain with the Black Watch led him to insist on military valour as the cornerstone of civic virtue, and to play the leading role – among a notably bellicose group of Edinburgh parsons – in the agitation during the second half of the eighteenth century to allow the Scots to muster their own militia or volunteer army. For Ferguson, conflict was indispensable to human society. But he was no Andrew Fletcher reborn, no truculent Scots patriot with a taste for brawls and Italianate learning: the passage of two generations must be accommodated. Ferguson was not a sceptic of Union with England, but a passionate adherent.

Adam Ferguson was born the youngest son of the minister at Logierait on Tayside, on the main eastern road into the Highlands, on 20 June 1723. He attended college at St Andrews and studied for the ministry at Edinburgh, where he came to know Blair, Carlyle, Home, Wedderburn and Robertson and join in their debates. A protégé of the loyalist Dowager Duchess of Atholl (who had the presentation of the living at Logierait), at the age of twenty-two he was appointed deputy chaplain of the recently embodied 42nd Regiment, The Black Watch, commanded by the Duchess's younger son Lord John Murray. More than three-quarters of a century later Sir Walter Scott recounted a story that Ferguson was in the front of the column at Fontenoy on 11 May 1745, armed with a broadsword, before being ordered to the rear.[50] The Gaelic sermon he preached to the regiment at Camberwell in London on 18 December 1745, at the height of the Rebellion, made such an impression that the Dowager Duchess asked him to translate it and had it printed. In it he sounded the characteristic note of his mature work when he told the soldiers that 'publick Calamities are the Effect of publick Corruption'.[51] He rose to be the regiment's principal chaplain and served in Ireland and on the Continent until 1754–5 when the regiment was ordered to North America.

Disappointed, it seems, of a living in the presentation of the Duke of Atholl, Ferguson returned to Edinburgh in search of civilian employment. Like David Hume before him, what he wanted was the College chair of Ethical and Pneumatical Philosophy, currently occupied by James Balfour of Pilrig. What he received, in January 1757 as a successor to Hume, was the Keepership of the Advocates' Library at £40 per annum. That income was quite inadequate, and he was soon applying through John Home to the Earl of Bute to serve as tutor to his son. In 1759, in a complex manoeuvre of the type so beloved of Lord Provost Drummond,

Ferguson was appointed to the vacant chair of Natural Philosophy.[52] According to Carlyle, Ferguson mastered physics 'in three months' of a single Edinburgh summer, 'so as to be able to teach it'.[53]

In this period Ferguson was involved in the projects of the *literati*, rehearsing and defending *Douglas*,[54] promoting Ossian, presiding over the meetings of the Select Society. From Carlyle's pen portrait and from his own letters Ferguson appears as a now-extinct type of mingled clergyman and military officer: peppery, 'indignant against assumed superiority',[55] possessed of a secret vein of humour, haughty, occasionally obscure. His chief interest was, as might be expected, the campaign for the Scots militia, literary Edinburgh's dominant preoccupation until the lawsuit known as the Douglas Cause of 1767.

The ministry in London was happy to run the Press through Edinburgh and Leith for sailors and to despatch Scottish regiments to North America, but dared not trust the Scots with their own local defence against invasion or insurrection. With the outbreak of war with France a Militia Act for conscription by ballot passed through Parliament in 1757 but applied only to England and Wales. The discrimination touched several raw nerves, both philosophical and national, in Ferguson's Edinburgh circle.

For Fletcher of Saltoun, a Scottish militia had been as much a moral as a political good. 'The Lacedemonians', he wrote in his *Discourse of Government with relation to Militias* of 1697–8, 'continued eight hundred years free, and in great honour, because they had a good militia.'[56] The fight at Prestonpans and the meek surrender of the Scots capital to a handful of Camerons seemed to confirm prevailing theories of the corrupting effect of commerce and luxury. From the time of his sermon at Camberwell Ferguson remained fascinated by the moral fragility of successful societies. In his *Reflections Previous to the Establishment of a Militia*, printed in London in 1756, he wrote that 'a few Banditti from the Mountains, trained by their Situation to a warlike Disposition, might over-run the Country, and, in a Critical Time, give Law to this Nation'.[57] By 1759 he was circulating an essay under the title 'Treatise on Refinement'; David Hume had been shown a work of that name at least twice by April 1759.[58]

In the summer of that year, a French flotilla under the corsair François Thurot broke through the British blockade of French ports, sailed round the north of Scotland and cruised off the west coast. Hawke's brilliant twilight action at Quiberon Bay on 20 November put paid for the moment to the danger of French invasion, but not before the proponents of the militia in Edinburgh had gathered to agitate for the extension of the Act to Scotland. In the chair was Lord Milton, the Duke of Argyll's under-minister (and Fletcher of Saltoun's nephew). On 4 March 1760 the two Scots MPs with closest links to the *literati*, Sir Gilbert Elliot and James Oswald of Dunnikier, introduced an appropriate Bill in the House of Commons. In his *The Question relating to a Scots Militia*, Alexander Carlyle attempted to stiffen them to their task. He argued that Scotland needed a militia relatively more than England, and concluded grimly: 'I do maintain, that if the militia-bill, now brought into parliament, does not pass, it had been good for Scotland that there had been no union.'[59] The Bill was opposed not just by English MPs but even by Robert Dundas the younger, Lord Advocate and member for Midlothian, and was roundly defeated at its second reading in April 1760.

Ferguson replied with a satire (which has also been attributed to Hume) in which John Bull (England) and his sister Margaret (Scotland) are backwoods gentry who quarrel with their neighbour, Lewis Baboon. The satire, entitled *The History of the Proceedings in the Case of Margaret, Commonly called Peg, only lawful sister to John Bull, Esq* but known as 'Sister Peg', is full of the downstairs domestic detail missing from the novels of the period. Early in 1762, in the time-honoured manner of Edinburgh men, the agitators formed a club. Scholars have followed John Hill Burton in deriving its name 'Poker Club' from the members' wish to 'stir up' the militia question, but the Edinburgh clubs of the period delighted in impenetrable and frivolous names, and this is far too straightforward. In 'Sister Peg' Margaret's garret lodgers, loyal to John Bull's predecessor, break into the dining-room, overset the china, drink the cream and generally run amok, slapping one of John Bull's gamekeepers; Peg throws a poker at them. 'Poker' thus revives, in convivial and coded form, the days of the College Company and campaigning in the Forty-five.[60] The club was an extension of the Select Society, minus the riff-raff that had gradually accumulated in the larger body.[61]

In 1764 Balfour of Pilrig was persuaded to transfer to another chair, and Ferguson was at last appointed Professor of Moral Philosophy. Two years later he married Katharine Burnett, niece of the chemist-physician Dr Joseph Black. Copies of the finished 'Essay on Civil Society' were by now circulating among his friends. Blair and Robertson were enthusiastic, but Hume, who had been less than whole-hearted in support of the 'Treatise on Refinement' in 1759, took strongly against the 'Form and Matter' of the 'Civil Society' essay and argued that its publication would injure the 'class': that is, the Philosophy School at the College.[62] Sir Gilbert Elliot disliked the work. Alexander Carlyle thought it merely a 'college exercise', yet recognised that there was something very appealing about it: 'a turn of thought and a species of eloquence peculiar to Ferguson'.[63] Whatever their feelings, nobody dared dissuade its curmudgeonly author from publication. Boswell, in a letter to his friend William Temple, a parson in Devon, reported on 8 March 1767 that 'there is a pretty book just now published, *An Essay on the History of Civil Society*, by the Moral Philosophy Professor here.'[64]

By then the *Essay* had reached London, where it quickly found readers. Blair, in particular, felt vindicated. The book ran through seven editions in Ferguson's lifetime, but then fell prey (like Steuart's *Inquiry*) to the prestige of *The Wealth of Nations*. Like Steuart's *Inquiry*, it left a deep impression on the Continent, not least in Germany, where it gave to German philosophy the phrase 'civil society', *bürgerliche Gesellschaft*. Marx took from Ferguson not only his pessimism but also his paradoxical view of history as both wholly willed and wholly involuntary. Indeed, the *Essay* forms the essential bridge between Machiavelli and Marx: between an aristocratic dream of civic participation and the Leftist nightmare of an atomised state and 'alienated' personality.

At first sight, the *Essay* is one of those Edinburgh compendiums in which humanity proceeds in a natural and unforced manner from an original rudeness to polish, and history converges on the Edinburgh High Street. Yet very soon a new tone is heard. Ferguson rejects, as confounding 'the provinces of imagination and reason', the sort of hyper-logical anthropology so beloved of Edinburgh.[65] Progress is neither linear nor inevitable, and later is not necessarily better. On the other hand, there is no golden age or Rousseauan state of nature from which humanity has fallen: a man is as 'natural' in Niddry's Wynd as in the forest.[66] Ferguson's human being – ingenious, cautious, obstinate, restless – as outlined in Part One, 'Of the General Characteristics of Human of Nature', is a most vivid creation:

> He would be always improving on his subject, and he carries this intention where-ever he moves, through the streets of the populous city, or the wilds of the forest. While he appears equally fitted to every condition, he is upon this account unable to settle in any. At once obstinate and fickle, he complains of innovations, and is never sated with novelty. He is perpetually busied in reformations, and is continually wedded to his errors. If he dwell in a cave, he would improve it into a cottage; if he has already built, he would still build to a greater extent.

His achievements do not die with him; 'the species has a progress as well as the individual; they build in every subsequent age on foundations formerly laid'.[67] As he reaches backward and forward in time, so he extends in space. Human beings are sociable and can only properly be examined 'in groupes, as they have always subsisted'. The organisation and institutions of these groups, their growth and decay over generations, the bonds that hold them together and the multiplicity of their pursuits comprise what Ferguson calls 'civil society'.

The next two parts – 'Of the History of Rude Nations', 'Of the History of Policy and Arts' – tell the story of human history through 'savage' and 'barbarian' stages. History proceeds, as it did for Andrew Fletcher, in a sort of fog, conjured up by Ferguson in a language as much Latin as English, in which even the rhythms of Tacitus are preserved. 'Every step and every movement of the multitude, even in what are termed enlightened ages, are made with equal blindness to the future; and nations stumble upon establishments, which are indeed the result of human action but not the execution of any human design ... No constitution is formed by consent, no government is copied from a plan ... They proceed from one form of government to another, by easy transitions, and frequently under old names adopt a new constitution.'[68] Simple explanations will not do. As for those who hold self-interest as the simple engine of history, they are like the foreigner at a performance of Shakespeare's *Othello* who thought the Moor was enraged by the loss of a valuable handkerchief.[69]

The process of social evolution brings liberty and the rule of law, but also evils that threaten it. Much is lost as well as gained. Ferguson warns his readers not to consider 'commerce and wealth ... the sum of national felicity, or the principal object of any state'.[70] A good state is not something of the present, or of the future, but can be found among the Spartans, the early Romans and the modern Britons. His readings of the Jesuit missionaries in North America had convinced him that there was a sort of American Sparta out in the woods.[71] Meanwhile, societies that pride themselves on virtue often rest on foundations of the most profound injustices. The free-born Spartan had as his counterpart the helot slave, and 'we are apt to forget, like themselves, that slaves have a title to be treated like men.'[72]

In the last three parts – 'Of Consequences that result from the Advance of the Civil and Commercial Arts', 'Of the Decline of Nations', 'Of Corruption and Political Slavery' – Ferguson at last unveils his secret purpose. 'We are now to enquire, why nations cease to be eminent; and societies, which have drawn the attention of mankind by great examples of magnanimity, conduct and national success, should sink from the height of their honours.'[73]

He applauds the modern 'separation of the arts and professions' – the division of labour – as it adds to the prosperity and efficiency of nations, though at the risk of breaking down the personality and weakening the state. Manufacturing reduces the human being to a simple moving hand or foot,[74] men become narrow-minded and specialised, they lose their notion of public good and 'we make a nation of helots and have no free citizens'.[75] Yet it is in matters of defence that Ferguson brings together his double preoccupation with the destruction of the moral personality and the dissolution of the state:

> By having separated the arts of the clothier and the tanner, we are the better supplied with shoes and with cloth. But to separate the arts which form the citizen and the statesman, the arts of policy and war, is an attempt to dismember the human character, and to destroy the very arts we mean to improve. By this separation, we in effect deprive a free people of what is necessary to their safety; or we prepare a defence against invasions from abroad, which gives a prospect of usurpation, and threatens the establishment of military government at home.[76]

Even in technically free and prosperous societies slavery is an unremitting peril, as luxury and money inculcate values of their own making and lull us into false security. Wages and liberty are not synonyms. 'Sensibility', 'delicacy', 'politeness' are mere fashionable virtues, and real merit becomes impossible to distinguish. In another vivid passage in which the eighteenth century is paraded in all its new-found ostentation, the very consciousness of 'being' in modern society vanishes behind 'having', as 'the idea of perfection' is transferred 'from the character to the equipage'. The moral personality becomes 'a mere pageant, adorned at a great expence, by the labours of many workmen'.[77] In the hopeless confusion of values, slavery seems not so bad a thing at all.

Hume seems to have found such moralising tedious and outmoded. In *Humphry Clinker*, published in 1771, Tobias Smollett lampooned this strand of truculent and high-minded Edinburgh republicanism. The novel is a sort of comic shadow-play of Edinburgh philosophy in which the old soldier Lismahago extols the Lacedaemonian poverty of the Scots and argues against the Union.[78]

The nature of commerce was such, that it could not be fixed or per-
petuated, but, having flowed to a certain height, would immediately
begin to ebb, and so continue till the channels should be left almost
dry ... Mean while the sudden affluence occasioned by trade, forced
open all the sluices of luxury and overflowed the land with every
species of profligacy and corruption; a total pravity of manners
would ensue ... practice of buying boroughs ... the crown would
always have influence ... supported by a standing army ... overthrow
all the bulwarks of the constitution ...[79]

Only Smollett could have seen the propriety of a character that is
half Adam Ferguson, and half 'Indian' Peter.

In 1766 Ferguson published a handbook or syllabus of moral
philosophy for his students which he expanded in 1769 into
Institutes of Moral Philosophy. In 1774 he breezily gave up teaching to act as travelling tutor to the young Earl of Chesterfield, at
£400 per annum with a promise of an annual pension of £200.
He told Adam Smith: 'I can justify my conduct to the world, who
rate men commonly as they do horses, by the price that is put
upon them.'[80] The Town Council dismissed him as professor but
was forced, two years later, to reinstate him.

All about him were new species of republican virtue, but
Ferguson failed to recognise them. He opposed the American colonists on grounds both selfish and pessimistic:

Is Great Britain then to be sacrificed to America; the whole to a part,
and a state which has attained high measures of national felicity, for
one that is yet only in expectation, and which, by attempting some
extravagant plans of Continental Republic, is probably laying the
seeds of anarchy, of civil wars, and at last of a military government?[81]

He travelled to America in 1778 as secretary to a commission to
negotiate with the rebels, but George Washington refused him a
passport through the rebel lines. Back in Edinburgh, he published
*The History of the Progress and Termination of the Roman
Republic* (1783). Amid outbreaks of military indiscipline among
the Highland regiments during the American War, including

mutinies on the crags of Arthur's Seat and at Leith, a second Militia Bill failed. The Poker Club expired in 1784. In 1785 Ferguson resigned as professor in favour of Dugald Stewart but took over the emoluments of the chair of Mathematics while John Playfair acted as professor. In 1792 he refurbished his college lectures as *Principles of Moral and Political Science*.

Ferguson was fascinated by the 'democratical' character of the French revolutionary army and predicted, correctly, that there would be years of conflict.[82] When the Militia Bill was finally enacted for Scotland in 1797, it proved the worst instance of answered prayers. It fell heavily on the labouring classes and provoked bloody riots, most notably in Tranent, near Prestonpans. As Ferguson wrote to Carlyle, it was 'a sort of Press act'.[83] By now Ferguson was the Scottish Cato, an obsolete moralist of antique character, living in a house at Sciennes, south of The Meadows, which from its distance from town as well as his ethnographic interests was known as 'Kamtschatka'. Writing Ferguson's medical notes in 1797, his friend and physician Joseph Black reported that his patient wore 'an uncommon quantity of clothing'.[84] His diet was restricted to broth, boiled cabbage and roots. (Black's son Adam remembered the two philosophers 'rioting over a boiled turnip'.[85]) Ferguson sat to Raeburn and also, as it were, to his neighbour in The Meadows, young Henry Cockburn:

> His hair was silky and white; his eyes animated and light blue; his cheeks springled with broken red, like autumnal apples, but fresh and healthy; his lips thin, and the under one curled. His raiment, therefore, consisted of half boots lined with fur, cloth breeches, a long cloth waistcoat with capacious pockets, a single-breasted coat, a cloth greatcoat also lined with fur, and a felt hat commonly tied by a ribbon below the chin.[86]

He outlived all his generation, and heard the news of the victory of Waterloo before dying at the age of ninety-three on 22 February 1816.

Adam Smith took modern commercial luxury – or opulence, as he preferred to call it – as a fact. What interested him was how it came about: how, as he put it in a draft of *The Wealth of Nations*, 'a common day labourer in Britain or in Holland' enjoyed 'luxury ... much superior to that of many an Indian prince, the absolute master of the lives and liberties of a thousand naked savages'.[87] The answer, he decided, was specialisation, or what he called the division of labour.

Smith was greatly influenced by the commercial preoccupations of his friend Hume, and of his teacher Hutcheson, who in his *System of Moral Philosophy* wrote eloquently on value as expressed in price and interest. Smith had promised, at the close of *The Theory of Moral Sentiments* in 1759, to write another book dealing with society in its organisation: the 'general principles of law and government', and the different forms they have taken in 'what concerns police [public order], revenue and arms'.[88] After some delay spent handling objections to the *Theory* from Hume and Sir Gilbert Elliot, he was devoting substantial parts of his lectures to these matters by 1762–3. From students' notes of the lectures it is possible to watch Smith gradually freeing his enquiries from their ethical and jurisprudential harnesses.

In 1763 Smith resigned his chair at Glasgow to act as tutor to the young Duke of Buccleuch at £500 a year. The twentieth-century scholar W.R. Scott – author of the phrase 'Scottish Enlightenment' – found among the papers at Dalkeith Palace, seat of the dukes of Buccleuch, a manuscript which he entitled 'An Early Draft of Part of *The Wealth of Nations*' and published in 1937. Smith's modern editors date the draft to shortly before April 1763.[89] In it, even before his exposure to the French economic philosophers during his travels on the Continent with the young duke between 1764 and 1766, Smith displayed his wonder at the sheer complexity of the commercial age. Plato in *The Republic* had well grasped the consequences of the division of labour, but Smith's mastery of the mechanical style is worthy of comparison even with the greatest of all philosophical writers.

Luxury for Smith, as for Hume, was not a moral or a sensual

category but a process – and a damned complex process at that –
that resulted in an object that would have seemed magical to an
earlier age:

> The woolen coat which covers the day labourer, as coarse and rough
> as it may appear to be, could not be produced without the joint
> labour of a multitude of artists. The shepherd, the grazier, the clip-
> per, the sorter of the wool, the picker, the comber, the dyer, the scrib-
> bler, the spinner, the weaver, the fuller, the dresser, must all join their
> different arts in order to make this very homely production. Not to
> mention the merchants and carriers, who transport the materials
> from one of those artists to another, who often lives in a very distant
> country; how many other artists are employed in producing the
> tools even of the very meanest of these. I shall say nothing of so very
> complex a machine as the loom of the weaver or as the mill of the
> fuller; much less of the immense commerce and navigation, the ship
> building, the sail-making, the rope-making, necessary to bring
> together the different drugs made use of by the dyer, which often
> come from the remotest corners of the world; but consider only
> what a variety of labour is necessary to produce that very simple
> machine, the sheers of the clipper. The miner, the builder of the fur-
> nace for smelting the ore, the burner of the charcoal to be made use
> of in that operation, the feller of the timber of which that charcoal
> is made, the brickmaker, the bricklayer, the smelter, the mill wright,
> the forger, the smith, must all club their different industries in order
> to produce them ...[90]

Smith looked on the dawning machine age of the Carron Iron
Works with the same Stoical satisfaction as he looked upon the
astronomical universe.

Yet the 'Early Draft' betrays its ethical origins more clearly than
the finished text written in the quiet and solitude of Kirkcaldy.
What is all but submerged in the copious logic of the published
Wealth of Nations, its relentless concentration on objects and
relations in isolation and its Edinburgh anthropologies, are certain grim consequences of the division of labour which sound not
Smithian but Marxian. Amid the wonder of the 'Early Draft' there

is a note of regret as the intellect succumbs to the siege engines of commerce:

> In opulent and commercial societies, besides, to think or to reason comes to be, like every other employment, a particular business, which is carried on by a very few people, who furnish the public with all the throught and reason possessed by the vast multitudes that labour. Let any ordinary person make a fair review of all the knowledge which he possesses concerning any subject that does not fall within the limits of his particular occupation, and he will find that almost every thing he knows has been acquired at second hand, from books ... A very small part of it only, he will find, has been the produce of his own observations or reflections. All the rest has been purchased, in the same manner as his shoes or his stockings, from those whose business it is to make up and prepare for the market that particular species of goods. It is in this manner that he has acquired all his general ideas concerning the great subjects of religion, morals, and government, concerning his own happiness or that of his country.[91]

With the information that religion, right and politics are mere commodities, the continuous part of the 'Early Draft' gives way to a list of contents. A year or two later, according to a student's report of his lectures at Glasgow in 1766, Smith used the Forty-five to illustrate the 'disadvantages of the commercial spirit'. In a commercial country such as England or lowland Scotland, 'the minds of men are contracted and rendered incapable of elevation, education is despised or at least neglected, and heroic spirit is almost utterly extinguished.' He added: 'To remedy these defects would be an object worthy of serious attention.'[92]

In *The Wealth of Nations*, published in 1776 after a delay caused by the banking crisis of the early 1770s,[93] Smith made no mention of Ferguson's *Essay* or of Steuart's *Inquiry*.[94] Living with his mother and maiden cousin at Kirkcaldy, he had shaken off much of the pessimism of the 'Early Draft.'[95] Half a generation had passed and the study of wealth and population had shed the remainder of its moral encumbrances. The world of *The Wealth of Nations* is a wonderful machine whose parts are unconscious of their mutual connection, or of any larger purpose to their activity. Smith's state of nature is not pastoral or agrarian but commercial. Man has, it seems, a 'propensity to truck, barter, and exchange one thing for another'.[96] On that bold and questionable premise, Smith erects the contraption which he calls the 'obvious and simple system of natural liberty'.[97] The individuals, trades and classes that comprise commercial society pursue their selfish goals, which the machine converts into social benefit and commercial prosperity. Public spirit, or virtue, does nothing to assist the workings of the machine, and might impede it.

Having expelled Steuart's 'Statesman' from the world of commerce, Smith could now employ him to ameliorate the stultifying effects of the division of labour, and to defend the realm. Having been Fletcherian in his lectures,[98] and a member of the Poker Club, Smith was all but won over to the standing army. 'The security which it gives to the sovereign,' he wrote, 'renders unnecessary that troublesome jealousy which, in some modern republics, seems to watch over the minutest actions, and to be at all times ready to disturb the peace of every citizen.'[99]

With some prescience, he suggested that an American militia, once it had served long enough to achieve military discipline, might be a match for a British standing army.[100] He was immeasurably better equipped than Ferguson to understand the commercial potential of the United States of America. 'Such has hitherto been the rapid progress of that country in wealth, population and improvement,' he wrote in Book IV of *The Wealth of Nations*, 'that in the course of little more than a century, perhaps, the produce of American might exceed that of British taxation. The seat of the empire would then naturally remove itself to that part of the empire which contributed most to the general defence and support of the whole.'[101] That same year, Gibbon published in England the first volume of his *History of the Decline and Fall of the Roman Empire*. In this work, which marked the end of the one-way traffic in ideas across the Tweed, luxury was a self-evident benefit to society: 'the only means that can correct the unequal distribution of property'.[102]

Dugald Stewart, who admired both Smith and Ferguson and had good words for Steuart, tried to reconcile them. He believed that the printing press would function as a sort of factory of the mind, transferring the benefits Smith had divined in the division of industrial labour to the world of the mind. The passage foreshadows, if 'Internet' be substituted for 'Press', the boundless enthusiasm of modern Utopians:

> The progress of knowledge must be wonderfully aided by the effect of the press in multiplying the number of scientific inquirers, and in facilitating a *free commerce of ideas* all over the civilized world; effects, not proportioned merely to the increased number of cultivated minds, thus engaged in the search of truth, but to the powers of the increased number, combined with all those arising from the division and distribution of intellectual labour.[103]

That is more persuasive than the specialised idiocy of Ferguson's *Essay*. In modern society, labour is mobile, education generic, caste and guild monopolies weak or moribund. Faith in God has been replaced by faith in complex, fragile, expert and invisible systems, from air-traffic control to money itself. The instability of economic life means that even repetitive occupations are rarely permitted to become boring. The separation of citizen and soldier has not, in fact, ushered in Praetorian thuggery: indeed, the soldier is one more expert in an expert society. Society worships the rich, but delights to see them humbled.

Robespierre and Saint-Just revealed the terrorist within the virtuous man. Having watched the Soviet Union and the brutal and incompetent regimes of the Khmer Rouge and the Taliban come to grief, the West is suspicious of Lacedaemonian experiments. It has come to believe that virtue as an aim of state is fatal not merely to prosperity but to virtue itself.

***

There is one more voice to be heard, that of a notorious Englishman who passed through Edinburgh in 1773 on a secret philosophical mission to the Highlands.

Dr Samuel Johnson, whose visit to Edinburgh has left such an impression on posterity, had little to say about the capital of Scotland. The town was 'too well known to admit description',[104] while the 'men of learning ... and women of elegance'[105] who took such trouble to entertain him required, it seems, no commemoration or praise. Johnson disliked Adam Smith and loathed David Hume. He would have been pleased to have met the learned Jacobite Thomas Ruddiman, but as he had remarked sadly when Boswell showed him some awful Latin exercises, Ruddiman was dead.[106] What praise he had for Scotland would more often be called abuse: 'The conversation of the Scots grows every day less unpleasing to the English.'[107]

Johnson's *A Journey to the Western Islands of Scotland*, published in 1775, is a profound and melancholy meditation on the eruption of money into an antique sequence of values. In contrast, Boswell's much later *Journal of a Tour to the Hebrides* begins with the phrase 'When I was at Ferney, in 1764, I mentioned our design to Voltaire ...' and proceeds through a clatter of dropped names. It is in Boswell's *Tour of the Hebrides* that lemonade enters history. The accounts, like the men, complement each other.

For at least ten years, almost ever since Boswell had met Johnson in Davies's bookshop under the shadow of Ossian and *North Briton* No. 45 in the spring of 1763, the Hebrides had been in the air as a sort of goal or commemoration of their friendship. Boswell, whose Scots patriotism was warm and conflicted, wanted to show Johnson his homeland and his patrimony, including a father whom he could never please; just as much he wanted to show Scotland Dr Johnson, as evidence of his own progress beyond it. He had brought his first hero, the Corsican partisan General Pasquale de Paoli, to Edinburgh two years earlier and the visit had been a success, even if the general had not been made a freeman of the city. In the course of the spring of 1773 Dr Robertson, the Aberdeen poet and critic of Hume Dr James Beattie, and Lord Elibank were all persuaded to write enthusiastic letters of invitation which Boswell then read out to his friend. Johnson's interest in language, Toryism, divine right and the Stuart succession drove him towards the Highlands. His violent hostility to Scotland was both prejudice, and a curb for keeping the puppy Boswell at heel.

Early in the morning of Thursday 12 August Boswell finished an unsuccessful plea for two tinkers charged with murder. The Court of Session had risen the day before. On Saturday 14 August he received a message late in the evening at his spacious new apartment in James's Court that 'Mr Johnson sends his compliments to Mr Boswell, being just arrived at Boyd's.'[108] Boyd's was The White Horse Inn at the very bottom of the Canongate, where the post-chaise from London set down its passengers. Like all Edinburgh inns of the time, it was primarily a stables. Arnot, writing in 1779, described them as 'mean buildings: their apartments dirty and dismal; and if the waiters happen to be out of the way, a stranger will, perhaps, be shocked with the novelty of being shown into a room by a dirty, sun-burned wench, without shoes or stockings'.[109] By the time Boswell had hurried down the High Street from James's Court Johnson was incandescent. He had hurled a glass of lemonade out of the window when a waiter used his own dirty fingers to put in a lump of sugar. As they walked up arm-in-arm to James's Court, it being past ten they were assaulted by the smell of excrement from the gutters. 'I smell you in the dark,'[110] Johnson growled into Boswell's ear. Margaret Boswell's kindness in giving up her bedroom and providing tea at all hours did something to mollify him. They sat up until near two in the morning.

On Sunday Johnson attended a Church of England service, cast some pious aspersions on the infidel David Hume, and dined with the banker Sir William Forbes, among others. Dr Robertson, who had dined 'between sermons', joined them over their wine. After breakfast with Robertson on Monday the 16th, Johnson, Boswell and what was by now 'a pretty numerous circle' made a tour of old Edinburgh. In the undercroft of Parliament House, where the ancient records of Scotland were stored while Robert Adam's masterpiece across the bridge collected pigeon droppings, Boswell began to indulge 'old Scottish sentiments' and to regret the Union, and Scotland's lost independence. James Kerr, the Keeper of the Records, complained that the Scottish notables had been bribed with English money to ratify the treaty. 'Sir, that is no defence,' Johnson said, 'that makes you worse.' Boswell burst out that the English were glad enough to have the Scots fight their battles in the Seven Years' War. JOHNSON: 'We should have had you for the same price, though there had been no Union, as we might have had Swiss or other troops. No, no, I shall agree to a separation. You have only to go *home*.'[111] His companions, who were already at home, possibly wished the same about their rude guest. As they passed on, into and out of St Giles, down the Back Stairs to the Cowgate, up to the Infirmary and the College and round to Holyrood, Johnson missed no opportunity to tease his companions (who had been joined by Ferguson) about the poverty and dirt of old Edinburgh. (The town turned the other cheek: four Edinburgh physicians treated him by correspondence in his last illness, while Sir Alexander Dick sent rhubarb from his own garden.) Afterwards, they dined on grouse and made fun of Lord Monboddo. Johnson, who disapproved of Monboddo's cult of the savage life, his conjectural anthropology and his self-conscious Hellenism, cheered up.

Monboddo's reputation has never recovered. While interviewing witnesses in Paris in 1765 in connection with the sensational inheritance case known as the Douglas Cause, James Burnett (as he was before his elevation to the bench) had met the celebrated 'wild woman' known as Mademoiselle Le Blanc. Captured near Châlons in September 1731 at the age of nine or ten, she was evidently a Huron from Canada who had been kidnapped and sold into slavery in the Caribbean. Shipwrecked off the coast of France, she had made her way across the country by trapping game and drinking its blood. She spoke only in guttural sounds. Burnett had his clerk translate an account of her story by La Condamine and published it, with his own learned preface, in 1768 as *An Account of a Savage Girl, Caught Wild in the Woods of Champagne*.[112]

In the preface, Monboddo proposed that man was not naturally a social animal, and that there was a stage of development, anterior even to that of Mademoiselle Le Blanc, where man did not use language at all; and that those superior faculties of mind, which distinguish our nature from that of any other animal on this earth, are not *congenial* with it ... but *adventitious* and *acquired*, being only at first *latent powers* in our nature, which have been evolved and brought into exertion by degrees ... That the *rational* man has grown out of the mere *animal*, and that *reason* and *animal sensation*, however distinct we imagine them, run into one another by such insensible degrees, that it is as difficult, or perhaps more difficult to draw the line betwixt these two, then betwixt the *animal* and *vegetable*.[113]

At intervals between his work in the Court of Session and his 'learned suppers' in St John's Street, off the Canongate, Monboddo worked on this advanced theory. In November 1772 Boswell had taken Daniel Carl Solander, the Swedish botanist who sailed with Captain Cook on the *Endeavour* voyage of 1768–71, to breakfast at St John's Street and Monboddo had asked eagerly if the Australians had tails.[114] Early the next year, in the first volume of *Of the Origin and Progress of Language*, he argued that the 'Ouran Outangs that are found in the kingdom of Angola in Africa, and in several parts of Asia' were human beings who had not reached the stage of language: 'it appears certain, that they are of our species'.[115] That was too much for the town wits. Monboddo's tail became a standing joke. Boswell's friend Sir Adolphus Oughton called him a judge *a posteriori*.[116]

On Wednesday 18 August, accompanied by Boswell's servant Joseph Ritter, Boswell and Johnson set out. The two men, one sixty-three and one thirty-two, were to be in each other's company for twelve weeks. They travelled by post-chaise for twelve days, up the east coast by way of St Andrews to Aberdeen. On Sunday the 21st, avid for a literary 'scene' or fight, Boswell sent Ritter ahead from Montrose with a message proposing to call on Monboddo at his estate.

Monboddo met them at the gate, dressed as usual as a farmer, carrying a stalk of his own corn. The two philosophers almost came to blows. Monboddo pointed at the house:

'In such houses,' said he, 'our ancestors lived, who were better men than we.' 'No, no, my lord,' said Dr Johnson. 'We are as strong as

they, and a great deal wiser.' This was an assault upon one of Lord
Monboddo's capital dogmas, and I was afraid there would have been
a violent altercation in the very close, before we got into the house.[117]

But Monboddo was not to be provoked, even when Johnson tried
to give his young son some Jacobite Latin. They disputed a little
whether the savage or the London shopkeeper had the best existence.[118]

At Inverness Boswell and Johnson hired horses and rode down,
by the same route as 'Ossian' MacPherson, to Glenelg, where they
had a terrific row when Boswell struck on by himself, and then
took boat to Skye. Through September and October they criss-
crossed the Hebrides by horse or sail- and row-boat, to Raasay,
Coll, Mull, Ulva, Inchkenneth and St Columba's island, Iona. It
rained almost without interruption, and the accommodation left
much to be desired. On 22 October they regained the mainland at
Oban and passed on to a stay of six days at Auchinleck, where
actual and elective father at one point came into collision. The dispute was over the execution of Charles I – a touchstone for Whig
and Tory – which Boswell, this time, scrupled to report.[119] They
returned to Edinburgh on 9 November, and though from the next
day Boswell had to spend his mornings in court, Johnson stayed
on for ten days more. From ten till one, he had at James's Court 'a
constant levée of various persons' while the long-suffering
Margaret Boswell, who loathed Johnson's irregular hours and
uncouth habit of dropping candle-wax on her carpets, devoted her
mornings to 'the endless task of pouring out tea for my friend and
his visitors'.[120] They supped and dined and supped and dined,
'harassed by invitations'. But as Johnson put it, 'How much worse
would it have been, if we had been neglected?'[121]

What was this journey for? What had they come to see?

They arrived in the Hebrides by way of the Isle of Skye. Boswell
had arranged that they would stay first with Sir Alexander
MacDonald of Sleat, one of two principal lairds on the island. He
clearly hoped to show Johnson the old patrimonial style. Landing
at Armadale on 2 September, they were sorely disappointed. Sir

Alexander and his lady, who were on their way to Edinburgh, entertained the travellers not at home but at a tenant's house. Dinner was poor, no claret was served, and again – the horror! – no sugar-tongs. Bred at Eton, and thoroughly versed in modern agriculture, Sir Alexander was quite content to raise his rents and clear off long-standing tenants to the Americas. His talk of 'emigration and rack rents' bored and disgusted the two visitors. On retiring, Johnson compared the place to a 'lodging-house in London' but resolved to 'weather it out till Monday'.[122]

Boswell had two personal grudges against Sir Alexander. First, he seemed a mere shadow of his older brother, Sir James, who had passed like a shooting-star over Eton, Oxford, London and Paris, only to die aged just twenty-four of a consumption at Frascati, near Rome. Boswell had truly loved him, calling him 'the Marcellus of Scotland'[123] after the heir to the Roman emperor Augustus, who also died young. That Boswell might have tolerated, but Sir Alexander had also married a very pretty girl who had been in Boswell's matrimonial sights: Elizabeth Diana Bosville, daughter of the Yorkshire squire whom Boswell considered his clan chief.

On the Friday, chafing indoors because of the rain, Boswell cornered Sir Alexander and picked a quarrel. He complained of the mean figure the couple cut, and 'upon my lady's neither having a maid, nor being dressed better than one'.[124] The quarrel rumbled on:

> *Saturday, 4th September*: My endeavours to rouse the English-bred chieftain, in whose house we were, to the feudal and patriarchal feelings, proving ineffectual, Dr Johnson this morning tried to bring him to our way of thinking. JOHNSON: 'Were I in your place, sir, in seven years I would make this an independent island. I would roast oxen whole, and hang out a flag as a signal to the Macdonalds to come and get beef and whisky.' Sir Alexander was still starting difficulties. JOHNSON: 'Sir, I would have a magazine of arms.' SIR ALEXANDER: 'They would rust.'[125]

Johnson retorted: 'Your ancestors did not use to let them rust.'[126] (Johnson seemed to have forgotten that since the penal legislation after the Forty-five, a Highlander who went about armed risked transportation.) To Boswell, Johnson whispered: 'He has no more ideas of a chief than an attorney who has twenty houses in a street and considers how much he can make of them.' As for my lady, 'This woman would sink a ninety-gun ship. She is so dull – so heavy.'[127] Boswell excised the insults from the version he published, but even so Lord MacDonald (as he became) was enraged and Boswell was lucky to escape having to fight him.

In Johnson's account of his experiences, such scenes as these vanish in a general meditation on the breakdown of a feudal system. Deprived of their hereditable jurisdictions by Parliament after the Forty-five, the chiefs 'degenerate from patriarchal rulers to rapacious landlords.[128] ... When the power of birth and station ceases, no hope remains but from the prevalence of money.'[129] In his old-fashioned dialectical manner, he converts Sir Alexander into a set of queries: 'It may likewise deserve to be inquired, whether a great nation ought to be totally commercial? Whether amidst the uncertainty of human affairs, too much attention to one mode of happiness may not endanger others? Whether the pride of riches must not sometimes have recourse to the protection of courage? And whether, if it be necessary to preserve in some part of the empire the military spirit, it can submit more commodiously in any place, than in remote and unprofitable provinces.'[130]

Matters could only improve. The two travellers spent four days with John MacLeod and his ten daughters and three sons on the island of Raasay, where there was dancing and whisky, and Johnson was in fine spirits. On the Thursday, he confided to Boswell: 'This is truely the patriarchal life: this is what we came to find.'[131] On the Sunday following, on the northern shore of the island, they were received at the house of Kingsburgh by Allan MacDonald, whose father had sheltered the fugitive Prince Charles Edward after Culloden, despite the bounty of £30,000 on his head. To Boswell, Kingsburgh (as he was known) was completely 'the figure of a gallant highlander', a large man with jet-black hair tied at the back, side-ringlets, a tartan plaid and blue bonnet, a blueish filibeg or kilt and tartan hose.[132] He was on bad terms with

Sir Alexander, deeply in debt, and planning to emigrate to America. None the less, he received the travellers with great courtesy, and led them in to a roaring fire and a dram of Holland gin.

In the evening the party was joined by Kingsburgh's wife. Boswell described her as 'a little woman, of mild and genteel appearance, mighty soft and well-bred'.[133] This was none other than the Jacobite heroine Flora MacDonald, who as a young woman had brought Prince Charles Edward safely from Benbecula to Skye and subsequently been imprisoned in the Tower of London. The travellers were in romantic ecstasies. Later they were shown to an upper room, where Johnson slept in the very bed, with its tartan curtains, in which the Prince had lain the night of 29 June 1746. In the morning Boswell found a slip of paper on which Johnson had written in Latin – his first language, as it were, though his English was not bad – *Quantum cedat virtutibus aurum.*[134]

At this point, for Boswell, all the themes intersect, and the tensions are resolved. *How gold gives way before virtue!* In the endless drizzle, fidelity triumphs over luxury, Kingsburgh over Sleat, Flora over Miss Bosville, Jacobite over Hanoverian, the savage over the shopkeeper.

Johnson was less sure. 'We came thither too late', he concluded in the *Journey to the Western Islands*, 'to see what we expected, a people of peculiar appearance, and a system of antiquated life.'[135] The feudal lords had embraced money. 'For a pair of diamond buckles perhaps,' Adam Smith was writing in Kirkcaldy, 'they exchanged the maintenance ... of a thousand men for a year, and with it the whole weight and authority which it could give them.'[136] Kingsburgh and Flora emigrated to North Carolina in 1774 only to return, true Tories that they were, with the defeat of the British armies. John Walker, Professor of Natural History at the College in Edinburgh, warned that sheep farming and increased rents were depopulating the Highlands and Western Islands.[137] In the next century, the new style of estate management pioneered by Sir Alexander in the Highlands provoked Marx to his habitual Caliban sneer. 'But what "clearing of estates" really and properly signifies', he wrote in *Capital*, 'we learn only in the promised land of modern romance, the Highlands of Scotland.'[138] Johnson had searched hard and, in the phrase used by the Englishman Captain Edward Topham, who was in Edinburgh when the *Journey to the Western Islands* came out, 'found nothing more barbarous than himself'.[139]

Johnson ended his Scottish narrative with the sole 'subject of philosophical curiosity'[140] in Edinburgh that he felt merited attention. That was not David Hume or Dr Robertson or Dr Ferguson or Dr Blair or Lord Monboddo, but a college for the deaf and dumb operated by Thomas Braidwood on St Leonard's Hill in the fields beyond The Pleasance. At the close of the first volume of his *Of the Origin and Progress of Language*, published earlier that year, Monboddo had devoted a section to the school in support of his theory that language was an acquisition, not a natural attribute of humanity. Johnson himself was hard of hearing.

As so often with Johnson, what begins as a half-crazed insult is softened by a sort of cack-handed gallantry and ends in a state just this side of tears. In the process, Edinburgh turns to pure gold. One of the pupils was carrying her slate, and Johnson wrote down for her a problem of long multiplication:

> I pointed at the place where the sum total should stand, and she noted it with such expedition as seemed to shew that she had it only to write.
>
> It was pleasing to see one of the most desperate of human calamities capable of so much help: whatever enlarges hope, will exalt courage; after having seen the deaf taught arithmetick, who would be afraid to cultivate the Hebrides?

Or, returning to mathematics: 'Novelty and ignorance must always be reciprocal.'[141]

9

The Art of Dancing

The Edinburgh commemorated in its eighteenth-century annals and philosophy was a bachelor society. The philosophers wrote about what men had done or ought to have done and either ignored women or approached them at an angle, as self-appointed teachers or provincial gallants. To read Adam Smith is to do without women, except the wives of aldermen. There is no clue in *The Wealth of Nations* that women will help power the industrial revolution in Scotland. Oliver Goldsmith may have had Edinburgh in mind when he had his good Vicar of Wakefield resolve to marry and have children, rather than write about population.

Yet the eighteenth century was *the* women's century in Scotland. The more settled tone to civilian life after 1745, the decline in superstition, improvements in public health and greater prosperity, as well as a new domesticity, favoured the less dominant sections of society, notably women and children. Edinburgh recognised that. 'In proportion as real politeness and elegance of manners advance,' wrote the Edinburgh surgeon William Alexander in his *History of Women* of 1779, 'the interests and advantages of the fair sex not only advance also, but become firmly and permanently established.'[1] Indeed politeness itself, as the heading for a range of

eighteenth-century social purposes, was recognised to be, in however obscure and intimate a fashion, a creation of femininity. The
cult of sex naturally increased the estimation of women.[2]

In the course of the eighteenth century Edinburgh was dis-
armed, domesticated and refined. Wide tracts of civilian life and
mental activity opened to women, from spinning linen yarn to
painting portraits and writing novels. The household – hitherto
the province of the 'gude man' praying in his closet – was more
firmly allocated as a female territory. The house in St Andrew's or
St George's Square filled up with a companionable clutter of tea-
tables, china, silver, portraits. Even now, the debris of this domestic explosion litters whole streets of antiques shops in London,
Glasgow and Edinburgh.

As the household and its décor gained prestige, so did its actual
and potential mistresses. Outside the house, social life was recon-
figured as an activity shared by the sexes. Men, even judges, drank
a bit less. In the new 'entertainments' imported from England and
the Continent – the theatre, the musical concert, the dancing
assembly – women were indispensable. The pianoforte, a novelty
in Edinburgh at the time of Burns's visit in the winter of 1786,
and the penny post (set up in 1774), offered new social opportunities to women. The Sentimental movement, which accorded the
franchise of feeling to all creatures – but especially to well-bred
women and girls – and prized emotion over spelling, was an
avenue for women into intellectual life.

Of course, Edinburgh was not Paris. The High Street had no
salonnières, and even Hume was at first so shy of meeting the
comtesse de Boufflers that he fled to Harrogate to avoid her. Later
on the Edinburgh periodicals *The Mirror* and *The Lounger*, written by men living in George's Square and its surroundings,
betrayed misgivings about women's liberties, their levity, head-
dresses, extravagance, ankles, and free thought. Men such as
William Creech and John Gregory, Professor of the Practice of
Medicine at the College, saw women losing respect in proportion
as they shed their reserve, so that their 'drawing-rooms' were
'deserted' and men shooed them from the dinner-table.[3]

In the matter of women's lives, eighteenth-century Edinburgh has two claims to interest. The first is the Forty-five rebellion, in which the Jacobite women played brilliant and striking roles. In the recollections of the rebellion the sexes have become oddly intertwined, switch professions and even clothes, strike chivalrous or Quixotic poses. Hospitality and fidelity, the virtues of the Scots woman, triumph over the brutal and indifferent English man. Jacobitism had a life among Scots women – not only Mr 'Evidence' Murray's 'Ladyships and green girls' – long after it had died in Scots males. This feminised Jacobitism, which gave to Scotland its great laments such as the 'Flowers of the Forest', survived as a vestige or posture among certain Scottish women well into the twentieth century.

The second occurs in the realm of philosophy. The philosophers thought about women, they really did: but they kept getting in a muddle. They wanted to find in women a counterpart for masculine virtue, but their theoretical models, derived from Aristotle's discussions of 'oeconomy' or household management, were a little antiquated for a money society. As always, they looked for a state of nature, but their natural woman reads very much like a creature of the kitchen and parlour.

The philosophers thought that women were radically different from men. They sensed women were *for* something, but when they tried to discover what that something was, they became entangled in the intricacies of motherhood and sex. Repelled by Lord Chesterfield's description of women as 'only children of a larger growth',[4] they yet could not accept that women should enter public life on equal terms with men.[5] They wanted to relieve women of drudgery, but not of service.

Scotland came up with two approaches, neither at all satisfactory. The first, derived from *The Spectator* and favoured by Hume in his character of man-of-the-world, gave women a function in society beyond child-bearing and -nurture: a sort of *mission civilisatrice* in taming men in society. The company of women, Hume wrote, refined not only the behaviour of men, but 'their tempers'.[6] His predecessor as Lord Annandale's tutor in England, Colonel

James Forrester, known as 'The Polite Philosopher' after his excruciatingly arch compendium of manners of that title, defined politeness as 'acquaintance of the *Ladies*': for only women's company 'can bestow that *Easiness* of Address *whereby* the *fine Gentleman* is distinguish'd from the *Scholar*, and the man of *Business*'.[7] William Alexander said the company of women made men ambitious to please, sober and temperate, less quarrelsome and more polished. Men unused to female company were clownish and tongue-tied. William Smellie, writing to a friend in 1763 that he was to be married, was candid: 'I need scarcely mention how hard it is for a young man living singly in a room to be virtuous ... Every other evening he is obliged to crawl to bed with his body steaming with liquour, or his mind dissipated by nonsensical conversation.'[8] By the 1780s, this sort of argument was so widespread that Burns deployed it in his wooing letters.[9] In society in general, the mingling of the sexes promoted industry and the arts.

The second solution, promoted in sermons given by the Aberdeen clergyman James Fordyce in London, was for a sort of sexual division of emotion, described with the same mixture of conjectural anthropology and commercial reasoning as Adam Smith deployed on the division of labour in *The Wealth of Nations*. Women's bodies were too 'delicate' for the practice of war, commerce and politics, their minds for any hard study, abstract philosophy and the abstruse sciences. Women's empire must be the heart. The medical men, such as the three Doctor Monros, successive Professors of Anatomy at the College, and the obstetrician William Hunter, threw in some conjectural physiology in support. Henry Mackenzie, the Man of Feeling and theorist of Edinburgh sentimentality in the 1770s, doubted if there were really, as Addison had claimed in *The Spectator*, 'a sex in the soul', but accepted that 'custom and education have established one, in our [men's] idea'.[10] Men had fenced off for women a sort of emotional paddock: 'a little world of sentiment made for women to move in, where they excel our sex.'[11]

And the purpose of all this? To make men better, in the sense of both more comfortable and more virtuous. The poet James Thomson in *Autumn* told the 'British Fair ... Well-order'd Home Man's best Delight to make ... This be the female Dignity and Praise.' Lord Kames's educational objects for his young 'favourite' Kattie Gordon – or those at least that he was able to express to her and her mama – were that she should be 'a good wife, a tender mother, and a faithful friend'.[12] Dr Fordyce told his women listeners that their 'Business is chiefly to read Men'.[13] In James Hutton's moral philosophy at the end of the century, educated or 'informed' women, through their influence on children and the hearts of young men, rescued mercenary society from corruption.[14]

Yet all this heart and soul made no great difference to the allocation of power between the sexes. As the Douglas Cause reminded Edinburgh, most women were excluded from the world of property and infantilised by the law. Chastity was a double standard, always enforced for females, for males often suspended. The most chivalrous of the philosophers, Hutcheson and Hume, recognised that politeness was a veneer laid over a relation of force. 'Barbarous nations', Hume wrote in 1742, 'display their superiority, by reducing their females to the most abject slavery ... But the male sex, among a polite people, discover their authority in a more generous, though not a less evident manner; by civility, by respect, by complaisance and, in a word, by gallantry.'[15] For one of Mackenzie's heroes, the problem was how to love the inferiority of women and not despise it.[16]

As so often in the history of women's emancipation, what seemed to be a new field of activity brought with it new restrictions. Where men were in short supply, women were not too 'delicate' to labour in the bleach-fields, collieries and cotton and flax mills of the new industrial Scotland of the 1790s. For the better-off, sensibility had its pitfalls. The greater readiness of the sexes to read together, to weep together, to dress and speak informally, to sit down together at the piano, conflicted with notions of decorum and carried risks of compromise. Such were the themes of Henry Mackenzie's stories in *The Mirror* and of his Gothic-didactic novel *Julia de Roubigné*.

The superior moral sense became a new burden for women to carry, a sort of super-chastity. In the sermons of James Fordyce and the writings of John Gregory and James Hutton, the Victorian domestic angel – Esther of Dickens's *Bleak House* – began to materialise. Meanwhile, the cult of feeling threatened to convert better-off women into perpetual adolescents, fools not for folly but for love, like Dr Robertson's queens, Mary and Elizabeth: sex, which blinds Mary to the crimes of Darnley and Bothwell, also wreaks its havoc on Elizabeth, as she mourns Essex, sitting on a cushion for days and nights on end, staring at the floor, with her finger in her mouth.[17] (No wonder Robertson's *History* was a best-seller.)

The physicians wondered whether women were irrational by virtue of their baffling reproductive biology. William Hunter pleaded for sympathy in cases of the deaths of illegitimate children where the mothers' 'distress of body and mind ... deprived them of judgment, and rational conduct'.[18] In the great Edinburgh portraits of Sir Henry Raeburn, the restrained and affectionate Clerks of Penicuik, or Mrs Dundas defeating her husband at chess, are displaced by Mrs Scott Moncrieff in clouds of fantasy and sex. Woman as friend and companion gives way, with the French Revolution, to woman as unexploded ordnance.

~

To the suave Edinburgh of the late eighteenth century, the old town was as backward in the condition of its women as in all else. Arnot looked back in disgust at the old Scots belligerence towards women, as embodied in the witch hunts and cutty stools – stools of repentance – of the seventeenth century. 'Women, in the mean ranks of life,' he wrote, 'were in the most deplorable condition imaginable. The young, if they lost their chastity, were harrassed and terrified into crimes which brought them to the gallows; and the old, under the absurd imputation of witchcraft, were tormented by the rabble, till, by the confession of an imaginary crime, an end was put to their sufferings.'[19] He reported that the

Court of Justiciary in 1678 condemned ten women to be stran-
gled and burned, including one for 'carnal copulation' with the
devil,[20] while there were twenty-one convictions for child-murder
in Edinburgh between 1700 and 1706, including four on a single
day.[21] While these persecutions were abating – the Scottish
Witchcraft Act was repealed in 1736 – women could still be
treated with appalling violence. In 1732 Alexander Carlyle's
neighbour at Preston James Erskine, former Lord Justice-Clerk,
pillar of the General Assembly, drunkard and hypocrite Lord
Grange, had his disturbed wife Rachel abducted by Highlanders
from the neighbourhood of Niddry's Wynd and marooned on the
remote island of St Kilda for seven years. She died on Skye in
1745.[22]

The problem about women, at least until the revolutions at the
end of the century, was to do with not their rights but their duties.
While spirited men such as Hume and Kames or English visitors
such as Burt, Goldsmith and Topham delighted in the forthright
and easy manners of Edinburgh women, and even found them
superior to the restraint of English ladies, a more polished generation was appalled.[23] There was no Court to set the tone. The residence of the Duke of York (later James II/VII) and Princess Anne
at Holyrood in the 1680s had been too short to 'effect any revolution in taste or manners', according to Ramsay of Ochtertyre.[24]
Elizabeth Mure of Caldwell, who died in 1795 at the age of 81,
wrote an account of gentlewomen's manners before the Forty-five:

No attention was given to what we call accomplishments. Reading and
writing well or even spelling was never thought of. Musick, drawing,
or French were seldom taught to girls. They were allow'd to rune about
and amuse themselves in the way they choiced even to the age of
women, at which they were generally sent to Edinr for a winter or two,
to lairn to dress themselves and to dance and to see a little of the
world. The world was only to be seen at Church, at marriages, Burials,
and Baptisams. These were the only public places were the Ladys went
in full dress, and as they walkd the street they were seen by every body;
but it was the fashion when in undress allwise to be masked.[25]

Charity schools for girls had been opened by both the Merchant Company (the Merchant Maiden Hospital) in about 1695 and the Incorporation of Trades (the Trades Maiden) in 1704. From about that time, the gentry took to sending their daughters to Edinburgh for some modest education. An account book survives from as early as 1700 for thirteen-year-old Margaret Rose, of the old family of Kilravock of Nairn, who learned dancing, singing, playing the virginals, writing and fine sewing at a school run by Elizabeth Stratoun.[26] She already knew how to read. About the education of Lady Anne Hamilton, Mrs Anna Montgomerie, housekeeper, wrote to the girl's aunt, Lady Tweeddale, in 1716: 'Tho she is not so old, yet she is very big and womanly and it is pitty that nothing should be taught her that other ladys of her quality are teacht, such as wryting, dansing and Frainch, which she is very capable of and begins to be thoughtfull about.'[27] By the 1750s, the most respectable boarding school was run in Forrester's Wynd by Mrs Effie Sinclair, who advertised thus in the *Courant*:

> Mris Eupham Sinclair having given up the millinery business at Whitsunday last, still continues to board young ladies, and has for that purpose a good, well air'd House, in Swinton's Land, Middle of Forrester's Wynd, where she can conveniently lodge six or eight boarders, which is the utmost number she designs to take. They may be taught any Sort of Needlework, white or coloured, mending, joining, or dressing laces, clear stearching etc.[28]

The pupils, who included Walter Scott's mother Margaret Rutherford, were then 'finished off' at The Hon. Mrs Patrick Ogilvie's. Even at the age of almost eighty, Scott told Robert Chambers, his mother sat bolt upright without touching the back of her chair.[29]

For all these accomplishments, gentlewomen left school speaking broad Scots, then considered a fault. They were ignorant of grammar, mathematics, the classics, and the rudiments of spelling. As Scott later wrote of the women's letters of the period, 'strange orthography and singular diction form the strongest contrast to the good sense which their correspondence usually intimates'.[30] The prize goes to Anne, Lady Mackintosh who began a letter to the Jacobite Duke of Atholl on 16 October 1745, 'My Lord Douke: The Beraer of this is a veray Pretay fellow...'[31]

Women were barred from attending the College, but not from all lectures. When in early 1745 Colin MacLaurin gave a 'College of Experiments' as a benefit for a colleague's daughter who had been left almost destitute, 'her friends insisted on admitting Ladies'. MacLaurin at first resisted, but on the tactful advice of Lord President Forbes submitted, in immemorial tones: 'I at length connived at it, but take no Notice of their being there.'[32]

Social life was settling down. Duels were no longer an Edinburgh routine.[33] Women, neglected in an age of heavy drinking, were now sought out. Alexander Abercromby of Tullibody told Ramsay of Ochtertyre that between 1730 and 1740 he and his married friends agreed to sup at one another's houses, as being 'more agreeable and less expensive than the tavern'.[34] True, after the Assembly or an oratorio of Handel at St Cecilia's in Niddry's Wynd the men might go on to Fortune's and stupefy themselves in 'saving the ladies' – that is, drinking toasts to their favourite women. But that practice, too, was dying out.

Tea, supposed to have been introduced at Holyrood by the Duke of York, had been stamped as the social province of women. After dinner, women served tea to men and women visitors, either in their bedrooms or, in the new residential squares, in what were known as drawing-rooms.[35] Cheap, elegant and modern, the tea-party had conversation, compliments and display for its purpose, rather than the satisfying of animal appetites: what Boswell called 'a true Edinburgh tea-drinking, where lady comes in after lady, and all the little news are told.'[36] Some men opposed the practice as new-fangled, English, extravagant and unwomanly, including Lord President Forbes himself, but by the time Robert Fergusson was writing at the turn of the 1770s, such superannuated attitudes were for yokels:

> But hyn awa' to Edinbrough scoured she
> To get a making o' her fav'rite tea;
> And cause I left her not the weary clink [money]
> She sell't the very trunchers [plates] frae my bink [dresser].[37]

Ramsay of Ochtertyre thought that 'The drinking of tea is an important era in female manners as well as housekeeping'[38] which 'contributed not a little to soften and polish manners'.[39]

As to relations between the sexes, Edinburgh had been celebrated, in Hugo Arnot's words, for the authorities' 'gloomy and morose contempt of the social pleasures'.[40] Edward Burt, writing in the 1720s, reported:

> There are some rugged Hills about the Skirts of that City, which, by their Hollows and Windings, may serve as Skreens from incurious Eyes; but there are Sets of Fellows, Enemies to Love, and Lovers of Profit, who make it a Part of their Business, when they see two Persons of different Sexes walk out to take the Air, to dog them about from Place to Place, and observe their Motions, while they themselves are Concealed ... In short, one would think there was no Sin, according to them, but Fornication; or other Virtue besides keeping the Sabbath.[41]

In about 1710, public dancing was introduced into Edinburgh when a private association called 'the Assembly' was formed in imitation of the institution at Bath (which Queen Anne visited in 1703). It took over an old house on the lowest bend of the West Bow that had belonged to the Somerville family. This was a devout and Whiggish quarter of town, and according to one source the building was attacked by fanatics and its door pierced with red-hot spits.[42] Whatever actually occurred, the experiment was given up.

The Assembly reopened in 1723 in Patrick Stiell's Close, later Old Assembly Close, on the south side of the High Street (which burned in the fire of 1824). Margaret, Countess of Panmure wrote to her exiled Jacobite husband on 24 January 1723:

> There are not many Comppany here this Winter, but we have gott a new diversion which is an Assemblie, which I believe will take very well in spight of the Presbyterian ministers' railing att it I have been att one the young folks dances and the elder ones lookes one. They are to play att little games, I mean for little money but I am to be no

gamester and I believe I shall go but seldom. The president of the Session I am told was there this night to be an incouradgement to it.[43]

Anne Stuart, niece of the Earl of Moray, wrote to a friend in the country on 28 January: 'They have got an assembley at Edinburgh, where every Thursday they meet and dance from four o'clock till eleven at night. It is half a crown and whatever tea, coffee, chocalate, biscuit, etc, they call for, they must pay as the managers direct; and they are the Countess of Panmure, Lady Newhall, the President's lady, and the Lady Drummellier. The ministers are preaching against it, and say it will be another horn order.'[44] (The Order of the Horn, an aristocratic circle of acquaintances at the time of the Union, took its name on a whim from a horn spoon, but was thought by the Edinburgh citizens to have an immoral, even Satanic, association.[45])

Patrick Walker, in his hagiography *The Life and Death of Mr Alexander Peden*, printed at Glasgow in 1725, gave a half-deranged account of how such novelties as the theatre and Assembly struck old-fashioned Whigs: 'and now we have got a new Assembly and publick Meeting, called *Love for Love*, but more truly, *Lust for Lust*; all Nurseries of Profanity and Vanity, and Excitements to base Lusts; so that it is a Shame to speak of these Things that are said and done amongst them ... Now they [women] deform their Bodies with Hoops or Fardingales, Nine Yards about; some of them in Three Stories, very unbecoming Women professing Godliness, more fit for Harlots.'[46] What saved the Assembly from the zealots was a certain Edinburgh deference to social rank. Allan Ramsay senior, in his 'The Fair Assembly: A Poem' of 1723, reassured the Puritans in some truly awful verses:

> Sic as against the *Assembly* speak
> The rudest sauls betray,
> When *matrons* noble, wise, and meek
> Conduct the healthfu' play;
> When they appear nae vice daur keek...

Edward Burt thought the Kirk left the Assembly alone because it was for 'People of Distinction'.[47]

> I say it is not in my Memory there was any Riot at the first of these Meetings, but some of the Ministers published their Warnings and Admonitions against *promiscuous dancing*. And in one of their printed Papers, which was cried about the Streets, it was said, that the Devils are particularly busy upon such Occasions.[48]

But in the case of both the playhouse and the Assembly,

> The Ministers lost ground, for the most part of the Ladies turned Rebels to their Remonstrances, notwithstanding the frightful Danger.[49]

There were rumours that even students of divinity at the College were attending 'the dancing school'. *The Lord appear, and help, and pity his poor Church in time to Come!* Robert Wodrow prayed.[50]

The Assembly permitted women of the 'improving' class to take part in the new civic projects. From the sale of tickets the directresses made regular and substantial donations to the Royal Infirmary, the Charity Workhouse and their own charitable schemes.[51] In the National Archives of Scotland is an advertisement of 1728 in which the four directresses ask that only British lace and linen be worn to the dance and, twice a year, exclusively Scots garments, 'for the encouragement of the Manufactories of this country'.[52]

The formality and propriety were extreme, and Lady Panmure once sent her own nephew home for appearing from dinner a little 'flustered'. Goldsmith, who spent eighteen lonely months as a medical student in Edinburgh, left a description of the regime at Old Assembly Close in the 1750s. Writing to a friend from college in Dublin, he said:

> Let me say something of their balls which are very frequent here. When a stranger enters the dancing-room, he sees one end of the room taken up with the ladies, who sit dismally in a groupe by themselves. On the

other end stand their pensive partners, that are to be: but no more intercourse between the sexes than there is between two countries at war –
the ladies, indeed, may ogle and the gentlemen sigh, but an embargo is
laid on any closer commerce. At length, to interrupt hostilities, the lady
directress or intendant or what you will pitches on a gentleman and
lady to walk a minuet; which they perform with a formality that
approaches to despondence. After five or six couple have thus walked
the gauntlet, all stand up to country dances; each gentleman furnished
with a partner from the aforesaid lady directress; so they dance much
and say nothing, and thus concludes our assembly.[53]

Goldsmith was too honest to maintain this man-of-the-world
tone. In reality, he had almost no contact with women. 'An ugly
and a poor man,' he added, 'is society for himself; and such society the world lets me enjoy in great abundance.'[54] For all its stiffness, Burt noted that 'I think I never saw so many pretty women
of Distinction together as at that Assembly.'[55]

In about 1756 the Assembly moved to a new address in Bell's
Wynd to the east, sometimes called New Assembly Close. The
Jacobite laird in *Redgauntlet* claimed to have quartered three hundred men at the Bell's Wynd rooms during the Forty-five,[56] and
they later housed the Town Guard, but Arnot described them as
extremely cramped.[57] They were ruled by the sternest of the ballroom martinets, Lord Chief-Justice Mansfield's sister, The Hon.
Miss Nicholas Murray. 'Miss Nicky's' regime was, in Robert
Chambers's careful phrase, 'more marked by good manners than
good nature'.[58] 'Ye're a great idiot,'[59] she once said to Boswell, getting him in one. But with the building of the new districts, the
Bell's Wynd Assembly had competition from New Town hotel
keepers and the opening of the Buccleuch Place rooms in 1783–4.
In 1787 the Assembly moved to new premises in George Street in
the New Town, and dancing now had a temple as grand and
extravagant as any church in Edinburgh.[60]

Into this world of gaiety and improvement the Forty-five made
an eerie eruption. For Burns, the quintessential libertine, the
Rebellion was of course more about sex than divine right:

> He set his Jenny on his knee,
> All in his Highland dress;
> For brawly weel he ken'd the way
> To please a bonie lass.
> An' Charlie, &c.

Even so hard and dour a Whig as the author of the Woodhouse-lees MS somehow could not resist gazing after the Jacobite women as they rode off to England with the army on 1 November in 'huzare dress with furred caps, long swords or shabbers and limber boots', and above all 'the Secretary Murry's lady eqwiped in this dress with pistols at her syde sadle and her cape on distinguished with a white plumoshe fether'.[61] Lord President Forbes, gritting his teeth, wrote: 'All the fine ladies, if you will except one or two, became passionately fond of the young adventurer, and used all their arts and industry for him, in the most intemperate manner.'[62] Jacobites, it appears, like blondes, have more fun.

Likewise the French envoy, Boyer d'Éguilles, stranded in the snow in Inverness, with nothing to eat but mutton and oatmeal and for company only Irish officers in the French service whom he could not understand and did not like, was haunted by the image of Anne Mackintosh, '*l'intrépide ladi*' who brought six hundred men to the Prince against her husband's wishes, 'a pistol in her hand and money in the other'.

> She presented herself to [the prince] with the grace and nobility of a goddess, for nothing in the world can match this woman for beauty. She paraded for him the whole of the little army she had assembled, and having spoken to the soldiers of their duty in the crisis, and of the rights and virtues of their Prince, she swore in very categorical terms to break the head of the first man who turned his back, after having burned his house before his eyes and driven off his family.[63]

Dougal Graham, who served in the ranks of the Prince's army, reported that four women were taken at Culloden, one dressed as a captain of men.[64]

In later years, romantic Jacobites built a fantasy of chivalry

around the Prince's fifty days at Holyrood, a sort of imaginary repetition of the first Stuart restoration after the Cromwellian inter-
regnum when, as Defoe put it, 'the gay humour came on'.[65] The
scenes of gaiety at Holyrood in Scott's *Waverley* may be exagger-
ated. Lord Elcho, who admittedly did not much like the Prince in
those days, and after Culloden loathed him, wrote of Charles
Edward's first evening at the palace: 'At night their came a Great
many Ladies of Fashion to Kiss his hand, but his behaviour to them
was very Cool: he had not been much used to Women's Company,
and was always embarrassed while he was with them.'[66] Likewise,
the Jacobites Carlyle lived with in Edinburgh after Preston told
him the court was 'dull and sombre.'[67] Boyer d'Éguilles, who
thought such matters highly material to the interests of the French
court, wrote: 'It is not that he is a flirt, or a ladies' man – quite the
contrary: it is because he is not, that the Scotswomen, who are naturally serious and passionate, conclude of him that he is truly
affectionate, and will remain constant.'[68] Magdalen Pringle was
torn between her enthusiasm and her Whiggishness:

> O lass such a fine show as I saw on Wednesday last. I went to the
> Camp at Duddingston and saw ye Prince review his men. He was sit-
> ting in his Tent when I came first to ye field. The Ladies made a circle
> round ye Tent and after we had Gaz'd our fill at him he came out of
> the Tent with a grace and Majesty that is unexpressible. He saluted
> all ye Circle with an air of grandeur and affability capable of
> Charming ye most obstinate Whig and mounting his Horse which
> was in ye middle of ye circle he rode of to view ye men ... His horse
> was black and finely bred (it had been poor Gardners[69]) his Highness
> rides finely and indeed in all his appearance seems to be Cut out for
> enchanting his beholders and carrying People to consent to their own
> Slavery in spite of themselves. I don't believe Cesar was more engag-
> ingly form'd nor more dangerous to ye liberties of his country than
> this Chap may be if he sets about it ... Ye Principal Ladys were Lady
> Nithsdale Lady Ogilvie and all ye Traquair Ladies &c &c[70]

In fact, many Jacobite families had been attainted in the
Fifteen, and had learned to use the legal dependence of women to

protect their property: that is, vulgarly, to put it in the wife's name. One piece of evidence is a letter from Lord George Murray to his brother at Atholl, asking him to call in promises of money for the Rebellion: 'It is to their Ladys you will please to write, as they appear to do the thing, and not the Husbands.'[71] The Scottish women *appear* to be raising the tenantry for the Prince, just as in MacLaurin's diary they *appeared* to have bought all the bullets in Edinburgh. As Marx well understood about Scotland, the wildest romance often boiled down to real property.

Even so, the Forty-five has the character of a female Saturnalia. In conditions of emergency, traditional patterns of authority are demolished or suspended, and if the Prince could not find friends among the men, well, there were always the women. '*Après avoir parlé des hommes nos ennemis*,' wrote Boyer d'Éguilles, '*il faut parler de leurs femmes nos amies.*' It was just that the Prince himself should in the end have had to flee for his life disguised as a woman, done up, as Dougal Graham put it, 'like a Dutch frow'.[72] Of the legendary figures of the Rebellion, only Flora MacDonald is without any taint of weakness or opportunism, the only perfect Tory in the history of Scotland.

The Duke of Cumberland had no time for all that, and arrested the Duchess of Perth, Lady Ogilvy, Lady Strathallan and Lady Gordon and sent them to Edinburgh Castle. Old Sir John Clerk was embarrassed: 'This action was esteemed a little uncourtly for a young man like the Duke of Cumberland, but there was a necessity for this piece of severity, that women might understand that they might be punished for Treasone as well as others.'[73] Hume, who was at the time on the staff of General James St Clair and about to embark for France, reassured Margaret Ogilvy's father, Sir James Johnstone, that the general had heard from ministers that menaces and threatened prosecutions of women were 'not to proceed to execution; but only to teach our country-women (many of whom had gone beyond all bounds) that their sex was no absolute protection to them, and that they were equally expos'd to the law with the other sex'.[74]

Both Clerk and Hume sound uncomfortable. In that respect,

the Jacobite nostalgia of Scots women is a celebration not just of a moment of fantastic liberty but also of moral superiority. As for the political proprieties, well, that was men's business and they could do what they liked with them, or nothing. In a short story, 'My Aunt Margaret's Mirror', Scott sentimentalised that attitude in the person of Margaret Bothwell: 'My heart is full of plaids, pibrochs, and claymores ... The public advantage peremptorily demanded that these things should cease to exist. I cannot, indeed, refuse to allow the justice of your reasoning; but yet, being convinced against my will, you will gain little by your motion.'[75]

With the defeat of the Forty-five rebellion, Edinburgh women began to alter in habit and appearance. The veil or plaid which almost all women wore to cover their heads outside the house began to give way before the hats and head-dresses of London. The plaid, tartan or variegated, gained a short reprieve from an identification with the Highland dead at Culloden. Ramsay of Ochtertyre, who came up to the College in 1747, said that nine-tenths of Edinburgh women then wore plaids at least at church, but silk and velvet cloaks were coming into fashion. Returning in 1752, he found scarcely a woman in the national plaid, and 'in the course of seven or eight years the very servant-maids were ashamed of being seen in that ugly antiquated garb'.[76]

The advance of polite society created new occupations for women, both as consumers and as labourers. According to Ramsay, who displays a manly interest in the costumes of gentlewomen, there was just one milliner in town in the 1720s and no more than six in 1753.[77] Haberdashers, who sold a variety of wares including cloth, hats, gloves, linens and threads, were unknown in 1763, but twenty years later the trade 'was nearly the most common in town'.[78] Perfumers, unknown in the year of the North Bridge, in 1783 'had splendid shops in every principal street'.[79] These businesses, which offered semi-independent positions for women such as Effie Deans in *Heart of Midlothian*, coincided with the crumbling of the old masculine, corporate restrictions. Merchants complained to the Town Council about the number of 'young women' now keeping shops.[80] Anne Forbes, sent to Rome with her mother in 1767 to study painting under Gavin Hamilton, set up in George's Square as drawing teacher and portraitist in pastels and oils.

William Creech, the bookseller, reported that in 1763 there were only five or six brothels, and a 'person might have gone from the Castle to Holyrood-house ... at any hour in the night, without being accosted by a single *street-walker*'. By 1783, he said, there were a hundred times as many prostitutes in Edinburgh.[81] Boswell's diaries and letters reveal an astonishing variety of sexual experience open to Edinburgh men. One night in June 1767 he drank himself into a stupor 'saving' Miss Blair, passed the night in a bawdy-house, caught a venereal disease and was terrified lest he might have infected his pregnant mistress, Mrs Dodds.[82] Here, in a sort of parade, were the categories of embraceable women in Edinburgh: marriageable heiress, kept mistress, street-walker. (Boswell took his whores from Portsburgh and The Pleasance to the New Town, because the houses were further apart and there were masons' 'shades' or sheds for a little privacy.) Burns in Edinburgh migrated between his servant-girl 'pieces'[83] May Cameron and Jenny Clow, and the 'Clarinda' of his precious Arcadian correspondence, Mrs Agnes McLehose. When Kames's only daughter Jean Home was unfaithful to her husband, she was ostracised and sent to France: she merely said she despised her husband and could do without a maid.[84]

A boarding-school bill for a Forfarshire girl from 14 February to 11 August 1788 reveals the transformation of social life. She attends the theatre ten times, has dancing and riding lessons, hires coaches and sedan chairs, attends Leith races. Her bill, what with gowns, hair, ribbons and frills, amounts to no less than £47 3s. 9d. sterling; that, when a labourer's wage in Edinburgh was just a shilling a day.[85] Did Adam Smith have such an establishment in mind when he wrote in *The Wealth of Nations* that, because there were no public schools for girls, there was nothing 'useless, absurd, or fantastical in the common course' of women's education?[86]

By now men such as Dr Fordyce were promoting women's education, even in the 'severer studies', but there remained a strong prejudice in polite society, maintained by both old-fashioned men and uneducated women, against a 'learned lady', or what the English called a 'blue-stocking': a character that Lord Kames dismissed as 'ridiculous, and deserves to be so'.[87] John Gregory, Professor of the Practice of Medicine at the College, told his daughters: 'Be even cautious in displaying your good sense. It will be thought you assume a superiority over the rest of the company. But if you happen to have any learning, keep it a profound secret, especially from the men, who generally look with a jealous and malignant eye on a woman of great parts, and a cultivated understanding.'[88] Creech in his *Edinburgh: Fugitive Pieces*, the published collection of his Addisonian letters to the *Courant*, *Mercury* and *Gazette*, made fun of the poor Edinburgh merchant with his learned wife: 'If I set her down to mend my stockings, she is reading Locke upon the *Human Understanding* ... Tommy's breeches have hung about his heels all week, owing to the *Revolution in the Low Countries*; and Johnson's *Lives* have nearly starved my youngest daughter at breast.'[89]

Maria Riddell, the West Indian who wrote a fine memoir of her turbulent friendship with Burns, impressed William Smellie not with her verses but with her science. Aged eighteen, under cover of the usual pimping letter from Burns, she sent Smellie her account of a journey to the Caribbean, to which he replied on 27 March 1792, his prejudices rattling like ships' china in a high sea:

> I received your MS. of Travels and Natural History. When I considered your youth, and still more, your sex, the perusal of your ingenious and judicious work, if I had not previously had the pleasure of your conversation, the devil himself could not have frightened me into the belief that a female human creature could, in the bloom of youth, beauty, and consequently of giddiness, have produced a performance so much out of the line of your ladies works. Smart little poems, flippant romances, are not uncommon. But science, minute observation, accurate description, and excellent composition, are qualities seldom to be met with in the female world.[90]

As is well known, commercial markets tend over time to abolish or dissolve distinctions superfluous to their own functioning.

Walter Scott recognised that he must compete with Jane Austen at level weights – or rather, at a small handicap. Behind a veil of anonymity that was easily pierced, Joanna Baillie (born in 1762) became London's most successful dramatist of the late Regency, and Susan Ferrier (born in 1782) Edinburgh's own novelist of manners. It should be no surprise that the Edinburgh women, fifty years on, choose to enter history through the same door as their men: that of literature.

❧

Under the influence first of Montesquieu and then of Jean-Jacques Rousseau, Edinburgh spilled a certain amount of philosophy on women: God or nature's purposes for females, the proper education of young girls, the mental or sentimental differences between the sexes, the institution of marriage, and so on. If there was a development in the town's philosophical approach, it was this: women started as items of property and ended as items of emotion.

In his *Treatise of Human Nature*, Hume described sexual love as the origin not merely of the family but of society itself. A compound passion, it combined sensations of beauty and benevolence with the appetite of generation. But having begun so well, Hume soon lost interest in the whole topic. In Book III, 'Of Morals', he discussed female chastity, but did not proceed beyond a speculative anthropology. Female chastity was valued because men would not support their children unless they could be absolutely sure of their paternity. (There is in fact no particular reason to believe this to be the origin of the premium on female chastity.) Here, it appeared, Hume was in agreement with his enemy, Dr Johnson, who liked to argue that all property depended on women learning a simple lesson: 'the great principle which every woman is taught is to keep her legs together'.[91]

Amid such conventional thinking, Robert Wallace's unpublished essay 'Of Venery or of the Commerce of the 2 Sexes' is refreshing. In this manuscript, which is in the Edinburgh University Library, Wallace argues that sexual desire is as natural to women as to men: 'Every woman during certain seasons and a certain period of life is incited to lust.'[92] Women should be permitted to make proposals of marriage, and divorce should be simplified. 'Fornication should be Disscouraged ... [but] ought not to be accounted a very great blot even upon a woman's character.'[93] The essay includes a paean to masturbation, a practice – *autres temps, autres moeurs* – old Fletcher of Saltoun would have made a capital offence.

As a piece of salacious philosophising, Wallace's essay reads more French than Scots. It is hard to know to what extent his sexual free-thinking was shared by the Edinburgh *literati*. On the cover of the only copy in George's Square he has written: 'needless & by no means proper to publish att least att present ... [since its contents are] so contrary to our present notions & manners.'[94] In Hume's essays of the mid-century, such as 'Of Polygamy and Divorces', he effected to be considering all possible relations between the sexes but ended, as ever, at the *status quo*. He rejected polygamy as destroying 'that Nearness, not to say Equality of Rank, which Nature has established betwixt the sexes', and divorce as undermining a union that should be complete.[95]

Adam Smith, after leaving Oxford, lived his life with his mother or his maiden cousin, Janet Douglas. His experience of women was restricted, even monkish. His moral philosophy is nowhere more ingenious than where it manages to exclude sexual passion from the world. Women were capable of humanity, but only of that 'exquisite fellow-feeling' that requires no self-denial or self-control or exertion.[96] For true generosity, men were needed.

As for the feelings between the sexes, love was natural, and appropriate to a certain age of life, but also 'a weakness', which the Spartans and American Indians regarded as an 'unpardonable effeminacy' or 'sordid necessity'.[97] There was a 'grossness ... which mixes with, and is, perhaps, the foundation of love' which was offensive at close quarters.[98] Sexual love always appeared to outsiders 'entirely disproportioned to the value of its object', and all expressions of love to a third person, unless self-deprecating, were ridiculous.[99] If love between men and women possessed any grace or charm or propriety it was in the secondary emotions it gave rise to, such as humanity, generosity, courage and selflessness. In truth, Smith couldn't be doing with sex because it was sand in his great sympathetic machine: the last thing lovers need is an Impartial Spectator. (Henry Mackenzie reported that Smith was once 'seriously in love with Miss Campbell … a women of as different dispositions and habits from his as possible'.[100] However, Mackenzie also sentimentalised Hume, and is not reliable.)

Smith's ungallantry was in sharp contrast with the enthusiasms of his master at Glasgow, Francis Hutcheson, and his pupil, John Millar. Hutcheson saw sex as part of economy, in the old sense of domestic utility, yet in a passage of typically high-minded rapture he discovered in sexual passion a thousand opportunities for virtue and happiness:

> Beauty gives a favourable presumption of good moral dispositions, and acquaintance confirms this into a real love of esteem, or begets it, where there is little beauty. This raises an expectation of the greatest moral pleasures along with the sensible, and a thousand tender sentiments of humanity and generosity; and makes us impatient for a society which we imagine big with unspeakable moral pleasures: where nothing is indifferent, and every trifling service, being an evidence of this strong love and esteem, is mutually received with the rapture and gratitude of the greatest benefit, and of the most substantial obligation.[101]

Hutcheson went on to claim that even the most debauched men sought moral qualities in their mistresses, such as good nature and fidelity: even 'chastity itself has a powerful charm in the eyes of the dissolute'.[102] That might be straight from the pages of Samuel Richardson's *Clarissa Harlowe*.

But it was Millar, Adam Smith's favourite pupil at Glasgow, who elevated sexual attraction to the position it holds today as the very 'foundation of political society': the origin of the family, stimulus to taste and beauty, nurse of sentiment and gallantry, merciless destroyer of social form. Having civilised barbarian societies, sexual love had created a culture of politeness, but now threatened to demolish it. Millar's *Origin of the Distinction of Ranks*, first published in 1771, seems at first just another conjectural anthropology or, rather, what he called a 'natural history of mankind': applied this time to subordination and authority in society.[103] In reality, Millar reflected the growing feminism of intellectual Scotland by dedicating his first and most interesting chapter to women: 'Of the Rank and condition of Women in different ages'. What Montesquieu, Steuart, Ferguson and Adam Smith had attempted in jurisprudence, showing that differences in laws and customs and institutions were not isolated or adventitious facts but had deep-laid causes and effects, Millar tried to do in the realm of sex. 'Of all our passions,' the book begins, 'it should seem that those which unite the sexes are the most easily affected by the peculiar circumstances in which we are placed, and most liable to be influenced by the power of habit and education.'[104] In Millar, and not a moment too soon, women entered the realm of Scots philosophy.

Like Sir James Steuart, Millar saw sexual love as the dominant impulse in history, but preferred to examine its consequences not in population and prosperity but in manners and institutions. While sexual need is permanent, it is wonderfully subject to refinement and manipulation; and so, by extension, are the lives and manners of women in history. In a new mixture of sentiment and conjecture, in this and the *Historical View of English Government* of 1787, Millar attempted a political history of sex.

In the 'savage' phase of human existence, the search for food left little room for any refinements of sex. Only when society achieved the moderate stability of the 'pastoral' and 'agricultural' ages, and accumulated some property, could men and women indulge in 'indolent gratifications'. The spread of property and its unequal allocation gave women a new value. Marriages became alliances, attached with dowries and entangled with customs and regulations. Female chastity both frustrated and stimulated the masculine imagination.

Like Hugh Blair, Millar attributed the poems of Ossian to the

pastoral era. He sensed anachronism in MacPherson's 'translations', but drew precisely the wrong conclusion. In the poems, 'the
degree of tenderness and delicacy of sentiment ... can hardly be
equalled in the most refined productions of a civilised age.'[105]
Indeed. Millar viewed the Ossianic pastoral as a golden age of
sexual friendship, between the brute appetites of the savages and
the degeneracy of 'polished nations, who, being constantly
engaged in the pursuit of gain, and immersed in the cares of business have contracted habits of industry, avarice and selfishness.'[106]

Millar gave a sympathetic account of the feudal or gothic age,
in which a society based on alliance, trust and reciprocal service
set the chaste woman on an imaginary pinnacle. Ferguson, in Part
Four of his *Civil Society*, had derided the excesses of romance but
recognised the modern debt to chivalry.[107] Millar argued that
romantic love had created a respect for women that would have
been quite baffling to the Greeks and Romans, and had delivered
chivalrous sentiments even into the commercial age. Such mental
attitudes passed, by way of Henry Mackenzie and Walter Scott,
into the nineteenth century and beyond.

In the fourth or 'commercial' phase of social history, as trade
creates opulence unimaginable in previous ages, courtship becomes a pastime and women are brought out from seclusion and
'lay aside the spindle and the distaff'.[108] They are admired for their
'useful or agreeable talents'.[109] In short, 'women become neither
the slaves, nor the idols of the other sex, but the friends and companions.'[110] In the finest of Raeburn's double portraits, of Sir John
Clerk of Penicuik, 5th Bt, and his wife, that companionship can
be seen raised to sublimity. Sexual passion is mellowed and refined
in the brilliant Scottish light to an unwavering love to the grave and
beyond. (Raeburn, who was born in Stockbridge and later laid out
the beautiful set of streets around Ann Street, returned from Rome
in 1784 and set up as a portrait painter on the south side of
George Street.)

Millar was not Adam Smith's pupil for nothing. He saw the
modern woman as a participant in a sexual division of labour. Her
skill and dexterity, and her ability to rear children, fitted her for

the 'interiour management of the family',[111] while her delicacy and, yes, sensibility, 'whether derived from original constitution, or from her way of life', enabled her to please and solace her husband. Yet in Millar, and John Gregory, that same note of commercial pessimism that Smith and Ferguson sounded for men is heard for women. In relations between the sexes, as elsewhere, 'it would seem ... that there are certain limits beyond which it is impossible to push the real improvements arising from wealth and opulence.'[112] Already where it is most advanced, in France and Italy, the free intercourse of the sexes 'gives rise to licentious and dissolute manners' which are bad for public order, the welfare of society, and women themselves. 'The natural tendency, therefore,' Millar wrote sadly, 'of great luxury and dissipation is to diminish the rank and dignity of the women.'[113] The geologist James Hutton, writing at the end of his life, saw the tide of opulence sweeping away virtue and happiness, but women had the power to save society. 'Nothing will contribute more to the perfection of a state,' he wrote in his *Investigation of the Principles of Knowledge* in 1794, 'than to have the women employed as instruments in promoting human virtue; and to see them valued for those accomplishments which best can make men happy; instead of being considered as only fitted for domestic service, and for the idle entertainment of the little tyrant, in the thoughtless moments of his life.'[114]

Edinburgh's interest in female education was largely due to the prestige of Rousseau. Having quarrelled quite pointlessly with Hume in 1766, the French philosopher came no nearer to Edinburgh than Staffordshire in England. Boswell, however, spent a night with his mistress, and was terrified and bored in equal measures.[115]

In his didactic novels, *Julie, ou la Nouvelle Héloïse* of 1761, and *Émile* of 1762, Rousseau picked up the theme of women's education that had been more or less neglected since Fénelon in the seventeenth century. At the close of the later novel Rousseau created Sophie, a perfect mate for his perfect youth Émile. Whereas the boy's nature is to be given free rein by his tutors, the girl's – with its sensitivity and powerful sexual appetite – must be constrained to serve him.

Among the first Edinburgh men to attempt the creation of a Scottish Sophie was, as might have been expected, Lord Kames. In May 1764, while on the Southern Circuit, he met Katharine Gordon, the eighteen-year-old daughter of Thomas Gordon of Blair House, near Ayr. Rising seventy years of age, Kames fired off letters to her in the character of Mentor, the friend of Odysseus and companion of his son Telemachus, as used by Fénelon in *Télémaque*. Miss Gordon seemed reluctant to enter into this correspondence and had to be prompted by her mother. Kames was forever urging Miss Gordon to apply herself to French, to keep a common-place book, not to talk scandal and, in general, to correct her usual woman's faults. Hume's comment (as reported by Adam Smith to Boswell) comes to mind: 'When one says of another man he is the most arrogant Man in the world, it is meant only to say that he is very arrogant. But when one says it of Lord Kames, it is an absolute truth.'

Although Kames talked of making Miss Gordon a 'complete Creature', that creature had her limits: she was to 'be a companion to a man of merit; and next to educate your offspring so as to be a blessing to their Parents and an honour to the country they live in'.[116] If his virtue, when applied to women, looks suspiciously like a tool of control, it was not without some wider social purpose: women's superior 'sense and feeling improved by regular discipline' permitted 'an intimate correspondence of the two Sexes [that] will prove the chief support of virtue, instead of Being as at present a fruitful source of vanity and folly'.[117]

While maintaining the character of Mentor, Kames was quite clearly mad about Miss Gordon, teased her with invented rivals, and showed none of the self-control that he was urging on his female Telemachus. Edinburgh felt there was something not quite proper about an elderly Senator of the Court of Justice and his young 'favourites', and he was once publicly scolded for letting his thoughts stray from his pupils' heads and hearts to their legs.[118] As Robert Wallace remarked in his secret pamphlet 'On Venery',

'Seldom I believe can a man admire the good qualities of a fine woman's mind and conduct without a secret wish to be familiar with her person.'[119]

On 29 April 1765 Kames complained that Miss Gordon had said nothing about applying herself to French or keeping up her common-place book. Unfortunately, Kattie Gordon was 'desperately afraid of the character of a learned Lady',[120] a trip to Blair House during the Southern Circuit in September was not a complete success, and by the New Year the correspondence had all but dried up. On 28 July 1766 Kames took his leave with the argument that Miss Gordon was now 'so much improved'[121] as to be well able to act her part in life without further instruction. Miss Gordon was, after all, no different from a field or wood on his estates at Kames and Blair Drummond.

The Reverend Dr James Fordyce, a fashionable preacher who had moved to London to take over the Presbyterian church in Monwell Street and became a friend of Dr Johnson's, was more ambitious for women's education. In his popular *Sermons to Young Women* of 1765, he recommended that his listeners should not merely dance but apply themselves to history, biography, geography, astronomy, sentimental novels, *The Spectator*, moral and natural philosophy and even, for the most gifted, 'prosecuting severer studies to every prudent length' – that is, so long as study did not interfere with family duties or create a female pedant.[122] James Hutton also wanted women 'informed'. 'It is in the wisdom of men,' he wrote, 'to form women virtuous and intelligent; it is in the wisdom of women, to form men honourable and temperate; and, it is in the wisdom of philosophy, to consider both female virtue, and manly honour, as bulwarks of the state which they adorn.'[123]

Sex will derail philosophy, as every other human endeavour, and these Edinburgh lucubrations were nowhere profound, and often mere flirtations. Fordyce quoted the *Phaedrus* for the power of virtue in human form, but his virtue had a markedly more erotic character than Plato had in mind: 'Virtue exhibited without affectation by a lovely young person, of improved understanding and gentle manners, may be said to appear with the most alluring aspect, surrounded by the Graces, and that breast must be cold indeed which does not take fire at the sight.'[124] In Julia de Roubigné's presence, Henry Mackenzie's heroes felt the 'throb of virtue', which doesn't sound all that virtuous. In reality, sex entered polite society in the garb of virtue only to throw it off.

In later life, and particularly after his marriage, Mackenzie came to regret the cult of female sensibility he had done so much to promote. In a letter to his cousin Elizabeth 'Betty' Rose in January 1777 he wrote: 'I am every day more and more stumbled about the proper education of your sex. There is a bewitching sensibility we are apt to encourage in them which I begin to fear is often a very unsafe guide to them thro' life; and I sometimes am repentant myself for having done even the little that was in my power towards its encouragement.'[125] But it was the Englishwoman Mary Wollstonecraft, in her withering attacks on Fordyce and Gregory in A Vindication of the Rights of Women of 1792, who revealed the weakness of this strand of Scottish didactic thought. A sterner moralist than Fordyce or the medical doctors, she recognised that eighteenth-century modesty, chastity, virtue and education were not moral but erotic values. Their purpose was to make women attractive not to God (or morality) but to men: 'insignificant objects of desire – mere propagators of fools!' She could not tolerate women losing their self-control, let alone revelling in it. She despised the cult of sensibility.

Meanwhile, neither education nor sentiment addressed the problem of property. William Alexander, in the last chapter of his History of Women, catalogued the 'Rights, Privileges and Immunities of the Women of Great Britain', but began with their handicaps. In theory, a British wife of the eighteenth century could gain property through neither labour nor inheritance. In the inheritance of landed estates she came after all males, while in the absence of a specific marriage settlement her property passed to her husband for administration. (In marriage contracts, the groom's family sometimes paid or committed themselves to pay a jointure on which the bride, once widowed, would live.) By law, the wife's 'natural and proper province' was the domestic sphere.

Men grumbled that the eighteenth-century woman was giddy, thoughtless, idle and extravagant. Alexander saw those as constructive faults: 'for should she even toil with the utmost assiduity, she cannot appropriate to herself what she acquires; nor lay out any part of it without leave of her husband.'[126] She had been 'but improperly or slightly educated; and at all times kept in a state of dependence'. It was absurd that a British woman might govern the country, but nothing else, 'as if there were not a public employment between that of superintending the kingdom, and the affairs of her own kitchen, which could be managed by the genius and capacity of women'.[127] She became in effect a minor. At marriage, her husband controlled her body, residence, movable property, and debts.

Despite that, Alexander approved the *status quo*. 'Her political existence is annihilated, or incorporated into that of her husband,' he wrote, 'but by this little mortification she is no loser, and the apparent loss of consequence is abundantly compensated by a long list of extensive privileges and immunities.'[128] This list turns out to comprise a husband's duty, while he lives with his wife, to keep her fed and secure. She cannot be imprisoned for debt, and can demand security against being beaten. More radical was Hutcheson, who thought the 'powers vested in husbands by the civil laws of many nations are monstrous' and saw the tyranny of some women over their husbands as half-justified 'by the unequal condition in which the laws have placed them'.[129]

Women could not inherit an entailed estate, and that was the origin of the greatest civil trial at Edinburgh in this period, the Douglas Cause. Archibald, first and only Duke of Douglas, died in 1761 without direct heirs. Almost all his estate was claimed on behalf of the Duke of Hamilton, a minor, whose guardians included two close friends of David Hume, William Mure of Caldwell and Andrew Stuart of Torrance in Lanarkshire, Writer to the Signet. The claim was opposed on behalf of the dead man's sister, Lady Jane Douglas, who had married, at the mature age of forty-eight, Colonel John Steuart and, two years later in 1748, announced the birth of twins in Paris. One of the children died young, but the second, Archibald, survived and was duly served heir. The Hamilton trustees brought suit in the Court of Session to argue that Archibald and his brother were foundling French children. In 1763 the Court decided to hear further evidence, and did not convene to try the case until 7 July 1767.

On 14 July the justices divided equally, seven to seven, but Lord President Robert Dundas cast in favour of the Hamiltons. The Douglas camp appealed to the House of Lords – in itself a controversial action, since in the Act of Union no role had been envisaged for that court in Scottish law – where in February 1769 Lord Mansfield and the two other justices reversed the judgment. The case excited immense popular interest: not merely because it revived in civil guise the ancient baronial animosities between great houses that had once been the stuff of Scots politics, but for its peep into the secrets of aristocratic families and its intimate, even obstetrical character. There was a remarkable volume of pleading in connection with the size of Lady Jane's breasts.[130]

Most of the *literati* adhered to the Hamiltons. Boswell wrote pamphlets and songs in favour of the Douglasites, for he felt that to oppose them struck at the very root of aristocratical paternity (and he was never certain of his own). At the news from London, he led a mob from the Cross to break the opposing Senators' windows, not excepting those of the Lord President, and of his mentor Lord Hailes and his father Lord Auchinleck.[131] While at times Boswell showed a great understanding of the human mind, he could rarely fathom his own.

The French Revolution altered the public character of women, as of so much else, by first elevating and then discrediting the fashion for wild nature. The different women's personalities of Scotland of this period – romantic, spirited, sentimental, economical – survived as vestiges in the ministers' wives of the Kailyard School of Scottish domestic fiction and the pioneering Scotswomen of the British Empire. Jeanie Deans, the heroine of *Heart of Midlothian*, is no less a feminine ideal than the divinities of English fiction but is, in the words of the Englishwoman Lady Louisa Stuart, immeasurably more persuasive.[132] Jeanie will never tell a lie, is hard-working, spirited and thrifty: in truth, an authentic Scottish heroine, and all on a diet of vegetables and pure water.[133]

10

Earth to Earth

In his youth, David Hume had tried to make a revolution in thought by applying experiment to the processes of mind. In the most disreputable passage in his writings, he concluded his *Enquiry Concerning Human Understanding* of 1748 by consigning all but mathematics and empirical science to the fire: 'If we take in our hand any volume; of divinity or school metaphysics, for instance; let us ask, *Does it contain any abstract reasoning concerning quantity or number?* No. *Does it contain any experimental reasoning concerning matter of fact and existence?* No. Commit it then to the flames: for it can contain nothing but sophistry and illusion.'[1]

Yet if that was a call to experimental science in Edinburgh, Hume then revoked it by arguing in both the *Treatise* and the first *Enquiry* that no universal proposition can be derived from instances. Adam Smith, at least in his 'Astronomy', treated scientific theories not as realistic accounts of an objective universe but as mere pastimes or consolations, what the Englishman Addison called 'pleasures of the imagination'. He greatly distrusted medical doctors. Adam Ferguson fretted that the new scientific specialisms were destroying the unity and integrity of knowledge.

   Yet all the while, Edinburgh was making an outstanding contribution to medicine and to both natural and applied science. It was
not that the Edinburgh school retreated from Hume's 'capital or
centre' of the sciences, human nature. Scottish thinkers dominated English-language philosophy right up until the 1850s. But
from the 1760s onwards, the Edinburgh school indulged an
increasingly practical bent. Its passion for classification and
system, its ambition for improvement and ethical earnestness were
combined with what Johnson, talking about Dr Gillespie's treatment of his illness, called 'all solid practical experimental knowledge'.[2] Practical advances in both medicine and agriculture then
stimulated theoretical science.

   From being a mere dependency of the medical schools on the
Continent, Edinburgh became the most celebrated centre for
medical education in the world. Students flocked in from
Ireland, England, the West Indies and the American States to
hear a succession of fine teachers, including physicians such as
John Rutherford, John Gregory and William Cullen, the anato-
mists Alexander Monro *primus*, *secundus* and *tertius*, and
chemists such as Joseph Black. Monro *primus*, who began by
demonstrating anatomy to a handful of young men, had by the
1730s more than a hundred students, by the 1750s about two
hundred,[3] and nearly three hundred at his death in 1767.[4] By
1789, when the foundation stone was laid for the new College
buildings, forty per cent of the thousand-odd students were
medics.

   By the end of the eighteenth century Edinburgh had five professorships of medicine, as well as chairs of botany and chemistry.
There was a competitive extra-mural school clustered about the
Royal College of Surgeons in High School Yards. At the new Royal
Infirmary, the first institution of its kind outside London, young
men could witness surgical operations, attend the doctors at
patients' bedsides or assist as dressers. From 1731 there was a
medical society at the College (which became the Philosophical
Society), while botanic gardens off Leith Walk supplied medicinal
plants for studies in *Materia Medica* (Pharmacology) and

Chemistry. The demand for fresh cadavers was, as will become clear, insatiable.

Edinburgh graduates in both medicine and surgery came to predominate in the army and navy and the government of India, and helped found great medical schools in New York, Philadelphia, Montreal and the Antipodes. Enmired in rural Yorkshire, in 1761 the popular medical writer William Buchan badgered William Smellie for medical intelligence from 'Edinburgh ... the seat of the medical muses'.[5] Great Edinburgh physicians such as Andrew Duncan and Alexander 'Lang Sandie' Wood were well-known town characters, and even Dr Johnson, in his last illness in 1784, was not too proud to receive medical advice solicited by Boswell from the likes of Drs Gillespie, Hope, Cullen and Monro *secundus*.

If that all sounds reassuringly like a modern medical establishment, with its clear professional frontiers and its clinical base in a teaching hospital, it was also part of the familiar old Edinburgh, buried to its hips in nepotism, patronage, jobbery and the corporation spirit. Between 1720 and 1846 the chair of Anatomy was occupied by three Alexander Monros in direct succession, and their jealous control as of right was only breached with the establishment of new medical chairs at the College in the first years of the nineteenth century. The Gregory family supplied seven College professors. University, physicians and surgeons were often at one another's throats. William Cullen wanted universities to have a monopoly of medical education and provoked a powerful rebuke from Adam Smith. James Gregory, John Gregory's son, who succeeded Cullen as Professor of the Practice of Medicine in 1790, abused the surgeons working at the Infirmary in language scurrilous even by the depressed standards of medical dispute.[6]

Yet as part of the general improving enterprise, the Edinburgh (and Glasgow) medical schools created an atmosphere favourable to practical advances. John Pringle, MD at Leyden in 1730 and then absentee Professor of Moral Philosophy (to Hume's frustration) before the Forty-five, published in 1752 an outstanding work on military medicine, *Observations on the Diseases of the Army*.

He went on to dominate scientific research in London, becoming President of the Royal Society in 1772. James Lind, who became MD at Edinburgh at 1748, published the *Essay on Diseases incidental to Europeans in hot climates. With the method of preventing their fatal consequences* in 1768 and promoted lemon juice as a specific against scurvy.

By way of chemistry and botany, medicine stimulated industries quite remote from the sick trade. William Cullen, when he started as Professor of Medicine at Glasgow in 1751, saw (in John Robison's words) 'the great unoccupied field of philosophical chemistry open before him'.[7] In 1755 he transferred to Edinburgh, and eleven years later brought to the College his most brilliant pupil, Joseph Black. Searching at first for a treatment for the kidney stone, Black laid the theoretical basis for the use of alkalis in the glass, soap-boiling and bleaching trades. He applied vitriol to chalk, limestone and magnesia and called the gas that was released 'fixed air'. Carbon dioxide (as it is now called) transformed the understanding of respiration, mine safety and soil science, let alone soda water.[8] It also opened the way for other gases to be distinguished and described: Henry Cavendish's hydrogen, the oxygen of Priestley and Lavoisier, Daniel Rutherford's nitrogen.

Black's correspondence from the 1760s shows him answering queries from businessmen on the use of lime in bleaching, the manufacture of alkalis from kelp, all manner of water analysis, sugar-boiling, glazes for pottery, gilding, indigo dyes, smiths' forges, sheep dips and gold-panning.[9] Above all, his experiments in evaporation and latent heat were indispensable to James Watt's work on the steam engine at Glasgow University. As Robison commented, Watt's improved engine 'attracted the attention of all those engaged in the great business of making money. It was this more than all the love of knowledge, so boldly claimed by the eighteenth century, that spread the knowledge of the doctrine of latent heat, and the name of Dr Black.'[10]

James Hutton loathed medicine but used to say he chose it as a way to study chemistry. He farmed in Norfolk and Berwickshire for fifteen years, and became rich – not that he cared – from a sal ammoniac works established in the early 1750s, before devoting himself to geology and metaphysics. John Roebuck, who also studied medicine at Edinburgh, set up a sulphuric acid works at Birmingham, and another behind high and well-patrolled walls at Prestonpans. In 1759, with English capital, he founded the Carron Iron Works, where by the 1770s four immense furnaces belched out iron for cannon, sugar-boilers for the West Indies, stoves, grates, kitchen utensils, spades, hoes, firebacks, hinges and bolts. Both Erasmus Darwin and his son Charles studied medicine at Edinburgh.

Adam Smith had a glimpse of the scientific Promised Land he could not enter. In Glasgow he had met the instrument-maker James Watt, who in 1764 was working to repair and modify the Glasgow college's model of Newcomen's steam engine. While Smith could not have known that Watt's improved steam engine, once combined with the new spinning machinery being developed in Derbyshire and Lancashire, would transform Scotland and the world, he sensed a shift in the character of knowledge. 'It was a real philosopher only,' he wrote in the 'Early Draft' of *The Wealth of Nations*, 'who could invent the fire engine [steam engine], and first form the idea of producing so great an effect by a power in nature which had never before been thought of.'[11] To Scotland's loss, Roebuck was bankrupted and Watt moved to Birmingham in 1774.

One evening in the 1770s, Black demonstrated to a party of friends including James Hutton that he could raise a calf's allantois or foetal membrane to the ceiling. They searched in vain for pulleys and threads, until he explained that the balloon was filled with hydrogen, which is lighter than air. On 5 October 1785, watched by some 80,000 spectators on Heriot's Hospital Green, a young Italian named Vincenzo Lunardi ascended three miles over Edinburgh in a hydrogen balloon. He descended safely not far from his point of take-off. This inaugural moment for natural science as spectacle in Scotland was preserved for posterity by two great artists, John Kay, who was there to draw the scene, and Robert Burns. The brave Italian and his balloon were such sensations that a lacy lady's bonnet was christened a 'Lunardi'. Within months, a young girl went to Mauchline church wearing such a hat and the most famous head-louse in literature:

> But Miss' fine Lunardi! fye!
> How daur ye do't?[12]

The records of the Edinburgh Royal Infirmary (analysed with infinite pains by the modern medical scholar Günter Risse) reveal the illnesses and accidents that beset the Edinburgh poor in the second half of the eighteenth century. There are fevers arising from the stale or putrid 'exhalations' in tiny, windowless rooms; the effects of frosty air, cold drinks after dancing or 'disagreeable smells'; breathing difficulties exacerbated by climbing stairs; rheumatic complaints arising from anger or grief; excessive walking; contagious fevers; falling out of windows when drunk; being struck by falling stones or bricks, or coach- or carriage-wheels, or casks rolling down the steep hills; fights; brawls; the lethal fumes from lead, mercury and silver workings; chips in a stonemason's eye.

The bills of mortality published annually by *The Scots Magazine* add detail to the picture. Of the 1,463 persons buried in the city kirkyards and the West Kirk in 1745, no fewer than 290 died of consumption, 270 from fever, 141 from smallpox, 108 from 'chin-cough' (whooping-cough or diphtheria), 104 from being 'aged', 81 from 'teething', 98 from measles, and 56 from still-birth.[13] As the popular medical writer William Buchan noted, 'The whole atmosphere of a large town is one contaminated mass, abounding with every kind of infection.'[14]

All that was little changed from the Middle Ages. What distinguished the second half the eighteenth century was the gradual application to traditional medicine of piecemeal discoveries in anatomy, botany and chemistry. At the same time, the professions of physician and surgeon were fenced and defined, and men and women on the outside were denounced as quacks and mountebanks. Permanent institutions, beginning with the Royal

Infirmary and later the Edinburgh Asylum, took the sick out of the sole care of their families and into a relatively clean, ventilated and orderly house.

In 1505 the barbers and surgeons of Edinburgh had combined in a new incorporation which in 1657 also admitted apothecaries. What emerged was a sort of rough-and-ready general practitioner, capable of setting joints, drawing blood or applying leeches, and prescribing pills, purges and emetics.[15] With the growing prestige of anatomy, James Borthwick became a regular teacher of dissection: his tomb in Greyfriars Kirkyard is adorned with sculpted surgical instruments. In 1705 Robert Eliot was appointed by the Incorporation of Chirurgeons as teacher of anatomy, and then confirmed by the Town Council as Professor at the College at £15 sterling a year. He was expected to perform a public dissection at Surgeons' Hall.

Then, as now, physic was a separate profession. In 1681 the patronage of James, Duke of York overcame the opposition of the Surgeons-Apothecaries, the College and the bishops, and the Royal College of Physicians was established in Fountains Close. Already Dr Andrew Balfour and Dr Robert Sibbald had laid out a garden for medicinal plants, first at Holyrood in about 1670 and then, from about 1676, in the yard adjoining the church and hospital at Trinity Hospital. James Sutherland became first Professor of Botany at £20 sterling per annum. (The site was swampy, and in 1763 John Hope had the garden removed to the neighbourhood of Leith Walk. The present one in Inverleith Row was first planted in the 1820s.) In 1699 the Royal College of Physicians published the first *Pharmacopoeia Edinburgensis*, known in English as *The Edinburgh Dispensatory*.

Edinburgh was still a dependency of Leyden, but Archibald Pitcairne's appointment there in 1692 as Professor of Medicine confirmed the prestige of the Scots alumni. The greatest medical teacher of the new century, Hermann Boerhaave, who gave bedside lectures in the Caecilia Hospital in Leyden, had learned enough from his Scots pupils to describe oatmeal porage as '*nutrimentum divinum* [food fit for the Gods]'.[16] One of those pupils was a military surgeon named John Monro, who was to become deacon of the Surgeons-Apothecaries and marry a niece of Forbes of Culloden. He was ambitious to establish a Leyden in Edinburgh, where a chair of Physic and Chemistry had been created in 1713.

His son, Alexander Monro *primus*, was born in London on 8 September 1697. He studied at London, Paris and Leyden, and went to Edinburgh in 1719 to take over from Robert Eliot at the Surgeons' Hall. The following year he was appointed Professor of Anatomy, and the chair was made lifelong in 1722. The growing interest among students and apprentices in anatomical dissection could not be matched by a supply of fresh cadavers from the gallows; dreadful tales of body-snatching began to circulate. In the lowering storm, Monro retreated from the Surgeons' Hall to the sanctuary of the College yards. The result was that anatomy passed out of the control of the surgeons' Incorporation. (In a sign of the growth of specialisation, the barbers left the Incorporation in 1722.) In October 1726 the Town Council recognised a Faculty of Medicine at the College consisting of Monro, as Professor of Anatomy; Sir Walter Scott's maternal grandfather John Rutherford, Andrew St Clair, John Innes and Andrew Plummer as Professors of Medicine and Chemistry; Charles Alston, *Materia Medica*; and Joseph Gibson, who took over the old female province of midwifery.

But Monro was the chief attraction. As a teacher, Oliver Goldsmith wrote in 1753, the Professor of Anatomy 'launches out into all the branches of physic, when all his remarks are new and useful. 'Tis he, I may venture to say, that draws hither such a number of students from most parts of the world. He is not only a skilful physician, but an able orator, and delivers things in their nature so obscure in so easy a manner, that most unlearned may understand him.'[17] In comparison, Goldsmith found Leyden a disappointment. 'Physic by no means taught here so well as in Edinburgh,' he noted in 1754, and the cost of living was much higher.[18]

In 1764 the College built Monro a new anatomy theatre in the its garden. By then, Alexander *secundus* (born in 1733) was delivering all but the opening lecture of the course. In his fifty-year reign Alexander *secundus* taught no fewer than 13,000 students, over a third of them from outside Scotland. In 1808 he retired to his house in St Andrew's Square and his lectures were taken over by his son, Alexander *tertius* (born 1773). Alas, the Monro blood had thinned. Alexander *tertius* published nothing of importance and, according to a scurrilous portrait in *The Lancet*, was indolent, unpunctual, shambling, too stingy to use a cab from the High Street even in pouring rain, and so disorderly in his dress as to lecture with his trousers undone.[19] He was overshadowed by Robert Knox at the College of Surgeons.

In physic, William Cullen had been persuaded from Glasgow by Lord Kames and in November 1755 was appointed joint Professor of Chemistry. Two years later he began to give clinical lectures at the Infirmary, and by the early 1760s was also lecturing on *Materia Medica*. Passed over in favour of John Gregory to replace John Rutherford as Professor of the Practice of Physic in 1766, he succeeded Robert Whytt as Professor of the Theory of Physic (that is, physiology). From 1768 the two men taught each class in alternate years. This rearrangement of the academic furniture left the chair of Chemistry free for Joseph Black. The chair of Botany and *Materia Medica* was held from 1761 by John Hope, beautifully portrayed by John Kay in conversation with his head gardener. His successor from 1786, Rutherford's son Daniel, had already made his name by distinguishing Black's carbon dioxide from nitrogen.

At the beginning of the eighteenth century there was no hospital exclusively devoted to the sick, though the Royal College of Physicians treated the poor *gratis*. The problem, in the words of a pamphlet of 1729, was that of control of patients and 'want of proper diet and lodging'.[20] In 1721 'some gentlemen', led by Monro, issued a pamphlet for a voluntary hospital on the pattern of the new charity hospitals being established in London, the Westminster and Guy's. The arguments are skilfully marshalled, and express in correct order the charitable and economical impulses of the eighteenth century in Edinburgh.

First, 'as Men and Christians we have the strongest Inducements, and even Obligations to this sort of Charity, as it is warmly recommended and injoyned in the Gospel as one of the greatest Christian duties.' There follows a balancing argument from Nature: 'Humanity and Compassion naturally prompt us to relieve our Fellow-Creatures when in such deplorable Circumstances as many are reduced to, Naked, Starving, in the outmost Distress from Pain and Trouble of Body and Anguish of Soul.' Then the economical: 'That the Relief of these miserable objects is a Duty, so it is no less Advantage to the Nation, for as many as are recovered in an *Infirmary* are so many working Hands gained to the Country.' Finally, both students and their money will remain in Scotland: 'That Students in Physick and Surgery might hereby have rather a better and easier Opportunity of Experience, than they have hitherto had by studying abroad, where such Hospitals are, at a great charge to themselves, and a yearly Loss to the Nation.'[21]

Nothing came of the pamphlet until George Drummond, in his first period as Lord Provost, proposed that a fishery company then in the process of liquidation should transfer its stock to a voluntary hospital. In the course of 1728 subscriptions were opened at the Bank of Scotland for additional funds, and with contributions from the Surgeons, the Physicians and public-spirited individuals, £2,000 was raised. That was thought inadequate, and in May the General Assembly of the Kirk resolved to seek money from parishes outside Edinburgh, as did the Episcopalian clergy, while 'the Honourable the Ladies of the Assembly of Edinburgh' devoted the proceeds from two dances. In this great civic exercise there were more than 350 subscribers, including the entire faculty at the College, Drummond, the secretary to the Duke of Argyll, the Duke of Hamilton, the Duchess of Gordon, the Marquess of Lothian, the Earls of Buchan and Hopetoun, the President of the Session, ministers, merchants, advocates and Writers to the Signet, and even some physicians and merchants from London. In the list of by now familiar names – Duncan Forbes, Robert Dundas, Dr John Clerk, Archibald Stewart – is that of the unfortunate Captain Porteous of the City Guard.

In December a meeting was held to establish the rules of management of the hospital. There were to be twenty extraordinary managers, including Drummond, nine physicians, Monro and two other surgeons, and David Spence, secretary of the Bank of Scotland, as Treasurer. Of these, twelve were appointed ordinary managers to oversee the day-to-day running of the institution, including two 'visiters' who were to 'visit the Hospital once a Month at least, and inquire into the Management thereof, the Treatment and Dyet of the Patients, the Conduct of the Housekeeper or Mistress, the Behaviour of Servants and Patients, and whole Oeconomy of the House'.[22] (If that sounds unwieldy, in 1740 one hundred managers were the rule for the new Charity Workhouse.) From the College the managers took a lease of nineteen years on a house in Roberston's Close, 'made more agreeable and convenient, by the Professors of Medicine granting Liberty to the Patients to walk in a Garden immediately adjacent'.[23] The Physicians promised that at least one of their number would attend *gratis*, while the Surgeons-Apothecaries authorised six members to attend, and to dispense medicines. To keep contributions flowing, 'a Box with two Locks, having a Hole in the Top, is set up in the Passage to the Patients' Chambers'.[24]

The Infirmary received its first patients on 6 August 1729. In the year that followed it treated thirty-five men and women for chlorosis or green sickness, pain in the thigh and looseness, cancers, inflammation of the eyes, liver pains, scorbutic tumors, 'hysterick' disorders, blood flux, agues, melancholy, consumption, palsies, phthisis, vertigo, ulcers and tympany. According to the end-of-year assessment, nineteen were cured, five were recovering, five had been dismissed as either incurable or incorrigible, five were still being treated, and one had died.

The new building to accommodate two hundred patients, to which Drummond applied such energy, was started in August 1738 on a site of two acres between the College and the Surgeons' Hall, looking north over the Cowgate valley. William Adam's design for a rectangular, four-storey building with two wings was based on the standard Ordnance Board pattern for barracks, but enlivened by an amalgam of columns, pilasters and pediments, a niche for statuary and a dome to decorate the central façade. In 1759 a statue of George, Prince of Wales in Roman costume was placed in the niche, and that alone has survived to adorn the present building (erected in 1879).

According to surviving plans, the ground floor was given over to a large reception hall, kitchens, a laundry, an apothecary's shop, and living quarters for the matron, nurses and keepers. The wings accommodated store-rooms, a 'dead-room' or mortuary, a physician's consulting room and the so-called 'Governor's room' for the management of the house. The upper storeys were divided each into four wards of twenty-four beds. Surgical cases were housed in the attic floor on each side of the amphitheatre, which was lit from the dome. Seats were arranged in banks for spectators. In the wings of the attic were venereal wards for women. The floors were paved throughout with Dutch unglazed tiles, and the beds were spaced six feet apart. Tobacco-smoking was not permitted.

The first patients were transferred to the new building in 1741. During the Rebellion a ward of twenty-four beds was established for sick and wounded soldiers. The first students' tickets had been issued as early as 1730, but it was not till 1748 that John Rutherford began to give clinical lectures in the wards of the Infirmary. He was joined in 1756 by Monro *secundus* and Whytt, and the following year by Cullen. In 1751 the Infirmary hired two full-time physicians, at salaries of no less than £90 and £50 per annum.

In the poem 'Death and Dr Hornbook', Death complained to Robert Burns that Dr Hornbook 'wi' his skill and art' made all his reaping and scything 'not worth a *f—t*'. In reality, the medical revolution in Scotland seems to have brought only moderate improvements in public health. In 1745 there was a mortality of 1,463; in 1789, according to the bills published in *The Scots Magazine*, there were 2,075 burials in the city churchyards, the West Kirk, Canongate and the new Calton burying ground. Of those buried, 518 had died from consumption, 285 from fever, 328 from smallpox, 246 from 'age', 159 from teething, and 77 from still-birth.

The victims of chin-cough (35) and measles (16) were much fewer, but 'bowelhive' (general intestinal disorder) killed no fewer than 205 children.[25] As the population of Edinburgh had more than doubled since the Rebellion, there had apparently been a reduction in the death rate. At the very least, there were no battle casualties.

Yet the mortality rate at the Infirmary, while comparatively low for the eighteenth century, actually increased over the period, from one in thirty-five admissions in 1730 to 3.1 per cent of 840 admissions in 1760, and 4 per cent of 1,977 admissions in 1789.[26] For all its charitable intentions, and the civic energy of its founders, the Infirmary served only a fraction of the city. Yearly admissions to the Infirmary in the 1780s were just a couple of per cent of the population of Edinburgh and Leith. Income – from students' tickets, legacies including a property in Jamaica, government contracts and the dancing Assembly – was much the same in 1790 as it had been in 1770.[27] Much of the expense of the house was paid by the government in the form *per diems* for sick servicemen from the army, and later the Royal Navy. Unsurprisingly, venereal disease (15 per cent) was the most common cause of admission noted in the Infirmary General Register between 1770 and 1800.

Nurses were ill-paid and sometimes harsh in their treatment of patients, and there was little physical examination beyond taking of pulse and temperature (with inaccurate thermometers). Very little at all was known about any disease. In his best-selling *First Lines of the Practice of Physic* of 1778–84, the great production of the Edinburgh medical school, Cullen looked carefully into the remote causes of fever (cold, et cetera), but there is no evidence of any idea that symptoms were a bodily response to invasion by bacteria. Reproductive biology, and indeed almost anything to do with the inner operations of women, was an utter mystery.

The *Edinburgh Pharmacopoeia* became more, not less, disgusting over time. While the first edition of 1699 standardised medieval herbal remedies, the next three (1722, 1735, and 1744) added such animal substances as *sanguis hyrci* (dried goat's blood), *album graecum* (dog's excrement), *apes praeparatae* (dried bees), *vinum millepedatum* (bruised millipedes in Hock), *buso* (live toads killed and dried in an oven), *trochisci viperini* (viper's flesh with biscuit-crumbs), *bezoar* (a deposit found in the stomach of a species of eastern goat), as well as human and cattle urine, to the *materia medica*.[28] Presumably the School of Medicine and the College of Physicians felt a need to compete with the mountebanks, and with such popular manuals as John Moncrief of Tippermalloch's *The Poor Man's Physician*, printed in Edinburgh in 1731. Burns's Dr Hornbook had an impressive armoury:

> Forbye some new, uncommon weapons,
> Urinus spiritus of capons;
> Or mite-horn shavings, filings, scrapings,
> Distill'd per se;
> Sal-alkali o' midge-tail clippings,
> And mony mae.[29]

William Buchan's *Domestic Medicine*, beautifully edited and printed by William Smellie in 1769, recommended a household cupboard of only the simplest of drugs. With their powerful emphasis on children's health, diet, exercise, good air, cleanliness and engagement on the part of parents, Buchan and Smellie probably did as much for the health of Edinburgh and Scotland as all the professors put together. At the Infirmary, Cullen had simplified prescription. 'I have a rule', he told students in 1773, 'of sticking to a single remedy[,] for otherwise in the case of a single change we do not know to what it is to be imputed.'[30] But as late as 1795 Dr Andrew Duncan was prescribing opium in various dosages to four out of five of his patients, and cantharides – a preparation of powdered flies, melted suet, wax and resin as a blister or poultice – to every other one. The non-drug treatment most practised was, of course, bleeding.[31] Most treatments still harked back to the old Galenic pathology based on the Doctrine of Humours and, by depleting the patient's body of physiological components like blood or dosing it with toxic heavy metals, risked doing as much harm as good.

For all their dissections and observations, the Edinburgh doctors had no idea whether blood-letting, cold effusions, purging, blisters and tartar emetic did or did not contribute to recovery. These were treatments associated with recovery that arose from thousands of years of observation of the symptoms of illness and of reproducing nature's remedies of nausea, vomiting and diarrhoea. Hume, who had argued precisely that custom was the basis of all common-sense notions of causation, was no doubt smiling from beyond the grave.

In the search for an over-arching theory, the old pathology in which the body was ruled by four humours (blood, black and yellow bile, gall) had given way to mechanical explanations of disease. Cullen saw diseases as possessed of a sort of life of their own, and amenable to being laid out in their genera and species as Linnaeus did plants.[32] Life was a function of nervous power, and he invented the term 'neurosis' to describe disruptions in the nerves: 'In a certain view, almost the whole of the diseases of the human body might be called nervous.'[33] Here was a counterpart, in the hospital, of the cult of drawing-room sentiment that his pupil John Brown took to romantic lengths, but it was all mere speculation. All the theories were made obsolete by changes in conceptions of biology and chemistry, and as comprehensively as the contemporary theories of heat. As James Gregory commented to the managers of the Infirmary in 1800: 'Much more than ninety-nine parts in the hundred of that which has been written in the theory and practice of physic for more than 2000 years is absolutely useless.'[34]

Cullen had the confidence and authority to accept the role of trial and error. Walking the wards of the Infirmary, he told his students:

> In these lectures ... I hazard my credit for your instruction, my first views – my conjectures – my projects – my trials – in short, my thoughts – which I may correct, and if necessary, change; and whenever you yourselves shall be above mistakes, or can find any body else who is, I shall allow you to rate me as a very inferior person.[35]

'A dogmatic physician', he said, 'is one of the most absurd animals that lives.'[36] The students must depend on 'EXPERIENCE alone; always, however, aware of the hitherto incompleat and fallacious state of Empiricism'.[37] If there was an advance in medical practice, it was often quite adventitious, as in the treatment of smallpox, for which inoculation had arrived by way of London and Constantinople by the mid 1720s, vaccination after Edward Jenner's experiment of 1796. As Gregory told the managers in 1803, 'If all the Physicians in Europe were put to the rack, they could not assign even the shadow of a reason why the cow-pox ... should secure a person for ever afterwards from the small-pox.'[38]

The virtue of the medical school was that it attracted money, students and royal patronage to Edinburgh, and set in train speculations that were to play a role in the establishment not only of modern medicine but of many industrial processes. The Infirmary showed that Edinburgh could maintain, through charity, an institution that could match anything in the English capital. Together, the school and the Infirmary brought to Edinburgh the most important attribute for a city – fame.

They also brought infamy. The primacy of anatomy, the superstition of the town and the macabre temper of medical students gave Edinburgh a sinister reputation which it has never quite shaken off.

The original Seal of Cause, the petition for the incorporation of the barbers and surgeons in 1505, requested, besides a craft monopoly, 'that we may have anis in the yeir ane condampnit man efter he be deid to make anatomell, quhairthrow we may haif experience, ilk ane to instrict utheris, and we sall do suffrage for the soule.' It was granted by the town on 1 July 1505.[39] In the next century, the College was given rights to the corpses of the workhouse poor, foundlings, stillborn children and suicides, but even those were inadequate, and by 1711 the College of Surgeons was protesting at violations of Greyfriars Kirkyard.

The flood of students and apprentices to the demonstrations of Monro *primus* led to the first crisis. Some time in the summer of 1724 a woman from Inveresk was executed in the Grassmarket for concealing a pregnancy. As her relations were carrying the body out through the Society Port, they were ambushed by surgeons' apprentices. In the fracas the corpse revived, and lived long and respectably as a salt-crier known as 'Half-hangit Maggie Dickson'.[40] On 17 April 1725 the Incorporation of Chirurgeons circulated what must be the most bizarre advertisement in the town's history:

> The Incorporation understanding that country people and servants in town are frightened by a villainous report that they are in danger of being attacked and seized by Chirurgeons' apprentices in order to be dissected; and although this report will appear ridiculous and incredible to any thinking person, yet the Incorporation, for finding out the foundation and rise thereof, do promise a reward of five pounds stg. for discovering such as have given just ground for this report, whether they be Chirurgeons' apprentices or others impersonating them in their rambles or using this cover for executing their other villainous designs.[41]

It was the unrest arising from these rumours that caused Monro to seek sanctuary within the precincts of the College, where he was allowed to occupy the lecture theatre.

As the number of students increased, so did the Resurrection movement. Alexander Carlyle, who went up to the College in November 1735, roomed with two Irish medical students, Lesly or Leslie and Conway, who had been attracted by Monro's fame. 'They had almost induced me to be a doctor, had not the dissection of a child, which they bought of a poor tailor for 6s., disgusted me completely. The man had asked 6s. 6d., but they beat him down the 6d. by asserting that the bargain was to him worth more than 12s., as it saved him all the expense of burial. The hearing of this bargain, together with that of the dialogue in which they carried it on, were not less grating to my feelings than the dissection itself.'[42] In 1742 the town was convulsed by riots. On 9 March a body that had been exhumed from the West Kirkyard was found in a house by the shop of the surgeon Martin Eccles. A crowd gathered and demolished the shop early the next day. On the 16th there were reports that the beadle at the West Kirk,

George Haldane, had been paid to look the other way, and his fine house – 'Resurrection-hall', as it was called, as if built with the profits of exhumation – was burned down. On the 18th the same fate befell the house of Peter Richardson, a gardener at Inveresk accused of lifting bodies from that churchyard. On the 26th, a sedan-chair that had been stopped at the Netherbow Port and found to contain a corpse was burned at the Cross and two Highlanders, John Drummond (chair-master) and John Forsyth (chair-carrier) were banished the town. On 6 April a gardener named John Samuel was caught at the Potterrow Port with the body of a child buried the previous Thursday at Pentland. (On 28 July he was whipped through the streets and banished Scotland for seven years.)[43]

A doggerel pamphlet in the British Library, printed that dreadful spring and entitled 'Groans from the Grave: or a Melancholy Account of the New Resurrection, practised In and about Edinburgh', evokes the pain and grief of the relations:

> Incens'd with Rage, immediately they streach
> Ther Power, to get a Warrant and to search,
> And in some Vault, or in some private Place,
> They find the Body lying in Disgrace,
> Mangled and cut, and stinking like a Beast.
> Ill smell'd, curroupt, that anes was blooming fair.
> Enough to putrefy the common Air...[44]

A gravedigger at the West Kirk was later heard to say, 'there will be a hantle folk fund missin' there on the resurrection mornin'.'[45]

By the early nineteenth century, when there were some eight hundred medical students at the College alone and Dr Robert Knox (1791–1862) was performing dissections at the Surgeons' Hall, there was what the liberal lawyer Henry Cockburn termed 'a constant premium for murder'.[46] The stage was set for the so-called West Port murderers Thomas Burke and William Hare, two Irish navvies who in 1827–8 lured at least twelve women and three men to Burke's house in Tanner's Close, Wester Portsburgh and suffocated them, for between £7 and £10 a body. According to

Hare, who turned King's Evidence and thus saved his life, their client was Dr Knox:

> Burke an' Hare
> Fell doun the stair
> Wi' a body in a box,
> Gaun to Doctor Knox.

On the day of Burke's execution Walter Scott watched from a bookbinder's window in the Lawnmarket, and heard the crowd hooting for Dr Knox's neck and blood. A committee of inquiry that included Robison and Sir William Hamilton white-washed him.[47] Three years later, the Anatomy Act permitted the use of the unclaimed bodies of paupers for dissection. The iron hoops across the graves in Greyfriars Kirkyard – surely the most haunted place in all the British Isles – still attest to this grisly chapter in Edinburgh's history. As Buchan and Smellie began their *Domestic Medicine*, 'It is astonishing, after medicine has been so long cultivated as a liberal art, that philosophers and men of sense should still question whether it is more beneficial or hurtful to mankind.'[48]

The greatest advance in the Scotland of the eighteenth century was in agriculture. The old farms of starveling cattle, bad ploughing and inadequate manure were vanishing from the Lowlands. Everybody, but everybody, in this story was farming in the new style, pioneered in Norfolk, of rotating cereals and fodder crops. Lord Kames, when he began managing his wife's estate at Blair Drummond near Stirling in 1766, moved up from the Borders to find standards of husbandry that, as in old age he told the agricultural journalist Andrew Wight, who reported on Scottish husbandry between 1778 and 1784, were a 'miserable sight to an improver'.[49] The roads were so bad that the men used sledges to carry corn to the stack-yard. There were no sown grasses, no red clover for summer fodder nor turnips for winter, no summer-fallow, and both infield and outfield were overrun with weeds. It sounds a little like parts of Caithness, say, today.

Meanwhile, women spun by candlelight in the long winters' evenings, but the men did nothing at all. In the next years Kames bridged the Forth at the cost – almost half to himself – of £900, improved the roads, began draining the moss, and set up a flax-dressing operation to provide winter work. At first, in the style of the Select Society or the Committee for Scottish Manufactories, he offered premiums to his tenants to improve their lands, but when that did not work he raised his rents, which did. None the less, Ramsay of Ochtertyre was quite scornful of Kames's 'rural operations'.

Wight visited another old metaphysician, Sir James Steuart, at Coltness. The farm, of about three hundred acres, was too boggy for corn or turnips, but was successfully sown for pasture. Wight was greatly impressed with Steuart's 'genius and zeal to promote agriculture'.[50] In Kincardine, Lord Monboddo bred Lancaster cattle and Suffolk draught-horses and employed twenty-two families of labourers. Lord President Robert Dundas farmed at Arniston near Musselburgh, and his brother Henry, Lord Advocate, at Melville; Robert Macqueen, Lord Braxfield at Broughton in Peebleshire; Sir Gilbert Elliot at Minto in Roxburghshire; John Hume at Ninewells in Berwickshire, and John Home at Kilduff in East Lothian. Alexander Carlyle's work on the land he held on lease from the Duke of Buccleuch at Dalkeith was an 'example to the most skilful tenants in the neighbourhood',[51] while Dr Cullen's turnip and wheat made 'a figure'[52] at Ormiston Hill.

Among these experimental farmers was James Hutton, who turned his agricultural experience into the last great philosophical masterpiece to come out of Edinburgh, *The Theory of the Earth*. Born in 1726, Hutton had some difficulty choosing a career. His passion was chemistry, and to that end he studied medicine at the Edinburgh College, in Paris and at Leyden, where he received his MD in 1749 with a thesis on the circulation of the blood.[53] Yet he could not settle. Having inherited a small property at Duns in Berwickshire, he turned to agriculture, travelled to Norfolk and lodged with a farmer to study 'rural economy' in its most productive British territory. The plain Norfolk society was much to his taste. Tramping round England and the Low Countries to study improvements, he also was drawn to mineralogy and geology: he wrote from Yarmouth in 1753 that he had looked into every pit, ditch and river-bed, and sometimes suffered the fate of Thales, the pre-Socratic philosopher who, intent on his observations, fell down a well.[54]

Returning to Scotland in 1754, Hutton brought a Norfolk plough and ploughman with him, conducted experiments on the effect of heat on plant growth, and farmed successfully until 1768. His neighbours included Hume's elder brother John at Ninewells, and Lord Kames. In 1764 he made his first excursions – accompanied by George Clerk-Maxwell, son of old Sir John Clerk – to the north of Scotland, travelling to Caithness by way of Dalwhinnie and Inverness, and returned by Aberdeen. According to a long and important letter Dr Joseph Black wrote some twenty years later to Princess Catharina Romanovna Dashkova, President of the Imperial Academy of Arts and Sciences in St Petersburg, Hutton had begun even then to develop the rudiments of what was to be his Theory of the Earth.[55]

The little sal ammoniac works he had set up with John Davie in the 1750s was incorporated as a co-partnership in 1765, and Hutton had the leisure and fortune to set himself up in Edinburgh. In 1770 he built a house for himself and his three sisters on St John's Hill in The Pleasance. From here he could contemplate the gap-toothed rampart and striations of Salisbury Crags, where quarrymen dug out whinstone for street cobbles, and the top of Arthur's Seat. His book made this one of the most famous geological landscapes in the world, one which gives even the teeming modern Edinburgh an aeonian character: not so much Cockburn's *rus in urbe*[56] as a metaphysical *mons in urbe*.

The Salisbury Crags are now held to be a sill of igneous rock left as the debris of a volcanic eruption which took place some three to four hundred million years ago. Behind the Crags, and tilted to the east by unimaginable forces, are the ruins of the volcanic pipe through which the lava flowed, known whimsically as Arthur's, or Arthur, Seat. The volcano sits on the ruins of sedimentary rocks that were deposited on the bed of a shallow sea, and still expose the fossilised remains of fish and plants. In the King's Park, and especially in what is now known as 'Hutton's Section' at the base of Salisbury Crags, he had a ready-made laboratory in which to test a theory which was both Biblical and astonishingly modern: that time and the earth were the same.

From 1767 until 1774 Hutton served on the management committee of a scheme to link the Forth and Clyde rivers (and also the Carron works near Falkirk) by a canal (completed only at the end of the century, at a cost of £300,000). In 1774, together with James Watt, now removed to Birmingham, he made a tour of England and Wales on foot. In 1777, at the ripe age of fifty-one, Hutton first entered print. At that time, the stony Scottish coal known as culm was subject to the same excise as high-grade English coal, which was a source of grievance to Scots colliers, including his friends the Clerks of Penicuik. Hutton's *Considerations on the Nature, Quality, and Distinction of Coal and Culm*, in conjunction with evidence presented by Black to Adam Smith's Customs Board the following year, persuaded the ministry to lift the duty.

Like David Hume, James Hutton never occupied an academic post in Edinburgh. In a 'Biographical Sketch' of Hutton, John Playfair described his pleasant bachelor life in Edinburgh thus: he rose late, studied until dinner, which he generally took at home, ate sparingly, drank no wine. After dinner, he read or in fine weather walked – study of another kind – for an hour or two and spent the evening in the society of such friends as Black, Ferguson, the Professor of Natural Philosophy James Russel, the younger James Lind (friend of Shelley, and putative model for Dr Frankenstein) and Adam Smith.

A short paper Hutton read to the Philosophical Society in 1778, entitled 'Of Certain Natural Appearances of the Ground on the Hill of Arthur's Seat', captured the tone of his life and friendships. He related how Ferguson, walking on the summit of Arthur's Seat in the summer of 1776, discovered curious concentric tracks of withered grass, and 'carried' himself and Blair up to investigate. Hutton's interest was not so much in the phenomenon itself as in the scientific approach to unravelling it. One could, he wrote, hypothesise certain causes (lighning, insects), but those were not the only possible causes, nor did they explain all the phenomena, and they should not be admitted as proved.[57]

In this companionable obscurity Hutton might have lived out his life. When the French lithologist Barthélemy Faujas de Saint-Fond, who had just produced a sumptuous folio on the extinct volcanoes of the Vivarais, visited Scotland in 1784 he took little account of this *savant modeste* except to cavil at the arrangement of his collection of minerals.[58] (Faujas, poor man, was fleeced at the aptly-named Dunn's Hotel in the New Town, shocked to the core by the lavish punch bowl at Dr Cullen's, and subjected by Commissioner Smith to an excruciating morning at the annual competition of Highland pipers: Adam Smith is a witty writer, but this is his only known *practical* joke.)

By then the Philosophical Society had become the Royal Society of Edinburgh. Anxious to support the infant society and his friends, as Playfair put it, Hutton was persuaded to present his long-considered theory. On 7 March 1785, according to the first volume of *Transactions of the Royal Society of Edinburgh*, his alter-ego 'Dr Black, in the absence of the author, read the first part of Dr Hutton's *Theory of the Earth*'.[59] On 4 April Hutton himself read the second part.

At that time, what became the science of geology had still fully to break its Biblical limits. The Book of Genesis taught that the earth was more or less co-eval with mankind, or about six thousand years old, calculating by Biblical generations. Created for Man's use, the earth of Moses was a static entity shaken by miracles and catastrophes. In the course of the seventeenth century, mining, scientific tourism and the examination of strata and fossils suggested there might be gradual changes over time. While the scale and remote epoch of these operations were not suspected,

Colin MacLaurin among others was well aware even in the 1740s that 'great revolutions have happened in former times on the surface of the earth, particularly from the phaenomena of the *Strata*; which sometimes are found to lie in a very regular manner, and sometimes to be broken and separated from each other to very considerable distances, where they are found again in the same order; from the impressions of plants left upon the hardest bodies dug deep out of the earth, and in places where such plants are not now found to grow; and from bones of animals both of the land and sea, discovered some hundreds of yards beneath the present surface of the earth, and at very great distances from the sea.'[60] A single Biblical deluge could not explain this variety of phenonema.

Volcanoes and earthquakes were no longer viewed solely as catastrophes but also as positive forces in the shaping of nature: evident in the present, but also in the past. On his trip to Scotland Faujas correctly identified the sensational basaltic columns of 'Fingal's Cave' on Staffa and the Castle rock, Calton Hill, Arthur's Seat and the Salisbury Crags as being of volcanic origin.

In his paper read by Black on 7 March 1785 Hutton set himself to discover how long the earth had existed and how long it would survive, based on the evidence not of human history (as in the Bible and the annals of the ancient world) but of 'natural history'.[61] It was clear to him that the earth he and his contemporaries saw was not the earth as originally created, but a composition over time of different ancient bodies in which what was now land had evidently once been sea, and *vice versa*. The present rocks had clearly been consolidated from debris, and then raised by some force in masses from the depths of the sea – witness the fossilised remains of fish at high altitudes. That force, Hutton pronounced, was the subterranean heat of the earth's core under the pressure of a weight of ocean. The veins of minerals in porphyry, granite and whinstone, forced into fissures in existing sedimentary rocks, were further evidence of this process.

What was happening was that continents were continuously being wasted away by processes that were visible in the present and

had similarly operated in the geological past. Over time mountains would be worn away by the action of rivers and become the
basis of new strata on the sea-bed, strata which would in time
themselves be raised up as new continents. As for the time elapsed,
it was both indefinite and beside the point. Unlike man or his institutions, the earth did not show signs of infancy or senility. 'With
respect to human observation,' Hutton said, 'this world has neither a beginning nor an end.'[62]

In their abstract speculations, human beings employ the tech-
nologies to hand. If Sir James Steuart imagined the world as a fine
French watch, James Hutton saw it as an immense machine fired
by heat: as Smith had predicted, a real philosopher had indeed
invented a 'fire engine'. Yet Hutton's theory, though it rejected
scripture and miracles, also accepted natural religion more or less
in the state that Newton, Boerhaave and MacLaurin had left it.
For Hutton – as for Smellie in the realm of what is now called zo-
ology[63] – a belief in a wise and good Nature would either be confirmed by the elegance of the Theory or, 'in supposing the theory
to be just, an argument may be established for wisdom and benevolence to be perceived in nature'.[64]

In other words, the old catastrophic view of Nature, which had
plunged the superstitious Dr Johnson into gloom in the
Highlands, was replaced in Hutton by a Smithian Stoicism. *The
Theory of the Earth* was a geological *Wealth of Nations*. Terrible
events such as the earthquake that destroyed Lisbon in 1755 and
the eruption of Vesuvius in the 1750s and 1760s were as integral
to Nature's benign operations as the greed of merchants was to
those of society: 'as forming a natural ingredient in the constitution of the globe'.[65] The world sustains itself and repairs the injuries of its own making.

Hutton's friends were captivated by the Theory, for it was
grand, systematic, elegant in the extreme, non-atheistic and – to
use Hugh Blair's favourite word – *sublime*.[66] In his letter to
Princess Romanovna Dashkova in St Petersburg, Black made all
these points: 'The boundless pre-existence of time & the operations of Nature which he brings into our view; the depth and

extent to which his imagination has explored the action of fire in the internal parts of the Earth, strike us with astonishment.' Moreover, and this was the beauty of the Theory, 'it is founded on the efficacy of powers which we see actually existing, and daily producing similar effects.'[67] None the less, the Theory in its first version made little impression. Too brief, abrupt and abstruse, according to his biographer John Playfair, it was ignored. A second paper by Hutton published in the first volume of the Society's *Transactions*, 'The Theory of Rain', fared better. It set off a European controversy that was fought, in Playfair's phrase, 'with more sharpness, on both sides, than a theory in meteorology might have been expected to call forth'.[68]

Throughout the 1780s Hutton criss-crossed Scotland, searching for granite outcrops and such unconformities as might reinforce his picture of continuous decay and renewal. In 1785, surveying at Glen Tilt in Blair Atholl in the shooting season, he displayed such delight at finding red granite veins in the schist that the Duke of Atholl's gillies thought he had struck silver or gold. His friend John Clerk of Eldin, an accomplished draughtsman and etcher who went on to great fame as an armchair naval tactician, sketched him at work. In 1786 he was in Galloway, the following summer in Arran and Jedburgh, and in 1788 at Siccar Point in Berwickshire. Young John Playfair (who was teaching Adam Ferguson's mathematics class at the College) was present at Siccar, and in his 'Biographical Account' he captured the tremendous mental thrill of these excursions:

> On us who saw these phenomena for the first time, the impression will not easily be forgotten ... We often said to ourselves, What clearer evidence could we have had of the different formation of these rocks, and of the long interval which separated their formation, had we actually seen them emerging from the bottom of the deep? ... The mind seemed to grow giddy by looking so far into the abyss of time; and while we listened with earnestness and admiration to the philosopher who was now unfolding to us the order and series of these

wonderful events, we became sensible how much farther reason may sometimes go than imagination can venture to follow.[69]

As Playfair wrote, the implications of Hutton's geological discoveries 'were matter, not of transient delight but of solid and permanent happiness'.[70] Raeburn's beautiful portrait of Hutton of about 1790 shows a thin, plain, modest man in homespun clothes, his face lit up by an intense sensibility to intellectual pleasure. All his life, said Playfair, Hutton had an 'exquisite relish for whatever is beautiful and sublime in science'.

The death of Adam Smith, whose reluctant executor he was, and a severe illness of his own in 1793 may have turned Hutton to recording his general system of physics and metaphysics. Always curious, like Hume, as to how mankind forms his conceptions of the external world, he concluded, somewhat in the manner of Bishop Berkeley, that colour or shape or magnitude were mere conceptions of the mind – not faint approximations of real objects (as in the famous cave passage in Plato's *Republic*), but wholly dissimilar. In Playfair's summary, external nature bore the same relation to man's perceptions as 'opium to the delirium which it produces'.[71] In Hutton's picture, death was just a passage from one condition of mind to another, and thus the mind survived to enjoy quite conventional doctrines of psychology, morality and natural religion.

A Continental attack on his 'Theory' caused Hutton to issue a much expanded version in two volumes in 1795. (A third volume was finally printed in 1899.) The earth has become a machine whose engineer has departed:

> When we trace the parts of which this terrestrial system is composed, and when we view the connection of those several parts, the whole presents a machine of a peculiar construction by which it is adapted to a certain end. We perceive a fabric, erected in wisdom, to obtain a purpose worthy of the power that is apparent in the production of it.[72]

Or:

We are thus led to see a circulation in the matter of this globe, and a system of beautiful oeconomy in the works of nature. This earth, like the body of an animal, is wasted at the same time that it is repaired. It has a state of growth and augmentation; it has another state, which is that of diminution and decay. This world is thus destroyed in one part, but it is renewed in another; and the operations by which this world is thus constantly renewed, are as evident to the scientific eye, as are those in which it is necessarily destroyed.[73]

History was almost certainly infinite. The earth showed 'no vestige of a beginning – no prospect of an end'.[74] His last years Hutton devoted to a compendious 'Elements of Agriculture', which has still to be published. In the winter of 1796–7 there was a return of his old illness, and he died on 26 March 1797. In 1802 Playfair faithfully issued his *Illustrations of the Huttonian Theory*, which finally brought Hutton's work to the scientific world.

The Theory of the Earth was the swan-song of the Scottish intellect in its heroic phase. A mental enterprise that began with Hume's *Treatise*, passed through Steuart's and Smith's *Inquiries*, Ferguson's *Civil Society* and Millar's *Origin of Ranks* ended in a timeless chaos of shattered rock. The universe is now held to have had an origin, and to be doomed to be destroyed, but at epochs so remote as to make Hutton's contention all but indistinguishable from the truth. Time and the universe are co-terminous. Yet the conclusion now drawn has lost its blithe Huttonian optimism. In Hutton's *Theory of the Earth*, God has departed but has left His earth in perfect working order. It is that optimism that mankind today, as he surveys the ruins of his own making, cannot share.

11

The Man of Feeling

Henry Mackenzie, a lawyer in Edinburgh specialising in Exchequer practice, was a figure with a foot in two centuries. Born on the day of Charles Edward's landing in the West, he lived to see George IV hold court at Holyrood in billows of tartan and ceremony. Antediluvian, he knew both David Hume and Walter Scott, and unites them: he was, said Scott, 'the last link of the chain which connects the Scottish literature of the present age with the period when there were giants in the land'.[1] He had shot hare and wild-fowl in fields now covered with the houses, churches and assembly rooms of the New Town and 'witnessed moral revolutions as surprising'.[2]

Mackenzie sentimentalised not merely Edinburgh society, but society itself. With his His-and-Hers short novels, *The Man of Feeling* and *Julia de Roubigné*, he turned the social theories of Adam Smith and David Hume into moral fables for the New Town and George's Square to read. The classical drama of Hume's deathbed was turned in 'The Story of *La Roche*' into an improving tale of sex and salvation. In Mackenzie, the brawling philosophical Edinburgh disappeared behind the respectable 'feeling' Edinburgh.

Yet as Scott and Cockburn were at pains to record,[3] Mackenzie was no lachrymose exquisite such as John Kay might have etched in The Meadows walks or on the New Bridge, but a successful revenue lawyer, a ruthless Tory pamphleteer, a suave writer and critic who helped launch the Romantic movement in literature with his ground-breaking lecture on the German stage given to the Royal Society of Edinburgh in 1788, settled the business of the antiquity of MacPherson's Ossian, set Walter Scott on his literary career and as 'our Scottish Addison' received the dedication of *Waverley*, befriended the novelists John Galt and Susan Ferrier. Above all, his review of Burns's *Poems, Chiefly in the Scottish Dialect* in the winter of 1786 established the 'heaven-taught ploughman' for the Scots to love and the English to despise.

In his rhetoric lecture given at Glasgow on 12 January 1763 Adam Smith made a characteristically brilliant analysis of the fashion for sentiment. Though he was speaking of the Roman historian Tacitus, and the unexampled tranquillity and security of the age in which he lived under the Emperor Trajan, Smith must have had the modern European capital in mind.

> They who live thus in a great City where they have the free Liberty of disposing of their wealth in all the Luxuries and Refinements of Life; who are not called to any publick employment but what they inclined to and obtained from the favour and Indulgence of the prince; Such a people, I say, having nothing to engage them in the hurry of life would naturally turn their attention to the motions of the human mind, and those events that were accounted for by the different internall affections that influenced the persons concerned, would be what most suited their taste.[4]

In short: 'Sentiment must bee what will chiefly interest such a people.' That, indeed, is what happened in the literature and society of France, Germany, Italy, North America, England and Scotland in the second half of the eighteenth century, with a force that has not dissipated.

Sentiment, not reason, was the great social invention of the second half of the eighteenth century. Nowadays, the word has come to mean emotion enjoyed purely for the sake of its pleasurable sensation. In eighteenth-century Europe and America, and most notably in the Edinburgh of Henry Mackenzie, sentiment was nothing less than modernity. In the guise of a revolt against aristocratic cynicism and licentiousness, the corrupt political managements of Walpole and his successors, and the mistreatment of women and children, sentiment and sensibility were code for all that was modern, progressive and companionable in polite society; for what the novelist John Galt called 'that remarkable suavity'[5] that pervaded manners and social life until the political controversies of the 1790s.

In the end, like all philosophical fashions, sentiment became disgusting, though not before it had driven Henry Cockburn, in recalling the manners of the George's Square of his youth, into a passion:

> Worst of all, there were 'sentiments'. These were short epigrammatic sentences, expressive of moral feelings and virtues, and were thought refined and elegant productions. A faint conception of their nauseousness may be formed from the following examples, every one of which I have heard given a thousand times, and which indeed I only recollect from their being favourites. The glasses being filled, a person was asked for his, or her, sentiment, when this or something similar was committed – 'May the pleasures of the evening bear the reflections of the morning.' Or, 'May the friends of our youth be the companions of our old age.' Or, 'Delicate pleasures to susceptible minds.' 'May the honest heart never feel distress.' 'May the hand of charity wipe the tear from the eye of sorrow.' 'May never worse be among us.'[6]

The modern West is a creation of sentimentality. Many of the twenty-first century's most precious notions of propriety have their origin in this movement. In attributing the capacity for feeling to unenfranchised fractions of society – girls, married women, country people, African servants, children, domestic and wild animals – it created the urgent political programmes of the modern era. Henry Mackenzie was in the van. Savillon, whom the heroine loves in *Julia de Roubigné*, disappears to the West Indies like Sir Thomas Bertram in Jane Austen's *Mansfield Park*, but he goes there not simply to give the metropolitan characters freedom of action but to emancipate a plantation slave work-force.

Already in the first half of the eighteenth century there existed in England, France and Germany a new reading public, solemn, earnest, puritanical, non-aristocratic, practical, warm-hearted, class-conscious, feminist, optimistic. What these new readers wanted, they received – from *The Tatler* and *The Spectator*, from Samuel Richardson, Laurence Sterne and Oliver Goldsmith, Jean-Jacques Rousseau, Jean-François Marmontel and, in the end, Johann Wolfgang von Goethe – glorifications of virtue in a domestic and often commercial setting.

Edinburgh, as the burgess town *par excellence*, took to sentiment like a duck to water. Being Edinburgh, it did so philosophically. Frances Hutcheson tended to mingle the moral and aesthetic senses. For him, good and evil were distinguished by the moral sense, counterpart of the supposed aesthetic sense which distinguished between the beautiful and the ugly. In Hutcheson, as in the novels of Richardson, Virtue was a *sensation*. Mankind is as sensitive to virtue as to a pretty face, and it is a feature of the sentimental novel, as it developed, that the one became code for the other. Richardson's purpose with his beautiful Pamelas and Clarissas was edification. His successors had other interests. By 1783, the Edinburgh Pantheon Society in the town was asking its members to debate 'Does reading Novels tend more to promote or injure the Cause of Virtue?'

Hume's *Treatise of Human Nature*, which argued that belief itself was a *feeling* rather than a rational process,[7] and Smith's *Theory of Moral Sentiments*, with its sociable burrowing into other people's consciousness, were the philosophical phantoms behind these literary exercises. Indeed, as one English critic put it, '*The Man of Feeling* is *The Theory of Moral Sentiments* in action.'[8]

But sensibility could just as easily be anti-social. When Rousseau was expelled from France by the Parlement on the ground of his atheism in early 1766, David Hume found him sanctuary in Staffordshire. In the isolation of the Peak district, the exile fell prey to a brooding paranoia and came to believe that he was victim of an international conspiracy – manipulated by David Hume. Hume wrote in exasperation to Hugh Blair on 25 March: 'He has reflected, properly speaking, and study'd very little; and has not indeed much Knowledge; He has only felt, during the whole Course of his Life; and in this Respect, his Sensibility rises to a Pitch beyond what I have seen any Example of: But it still gives him a more acute feeling of Pain than of Pleasure. He is like a Man who were stript not only of his Cloaths but of his Skin.'[9] Sensibility became in the end a prison, from which the exits were paranoia (Rousseau), consumption (*The Man of Feeling*) or suicide (*Werther*). As William Smellie wrote in the fine passage that closes his *Philosophy of Natural History* of 1790, human beings have been given precisely as much sensibility as is good for them, and should be content.[10]

Henry Mackenzie was born on 26 July 1745, supposedly in Libberton's Wynd, now obliterated under the George IV Bridge. His father, Dr Joshua Mackenzie, had been an army surgeon in Ireland, and set up in practice in Edinburgh. His mother was Margaret Rose of Kilravock, of a well-known Nairn family famous for its hospitality – embodied in a sixteen-quart punchbowl – and for its political discretion. Margaret's father Hugh, 16th of Kilravock, managed to entertain both Charles Edward and the Duke of Cumberland within days of each other before Culloden. 'You have had my cousin here,' Cumberland is alleged to have said.

At some date the Mackenzie family moved to the Cowgate head, where it meets the Grassmarket. Henry went through the High School and by 1758 at the latest had entered the College.[11] Though he showed an interest in literature, and wrote verses from the time of his first class at the College, he left before the regime of Blair and Robertson was fully established and in November

1761 was apprenticed to George Inglis of Redhall in the business of the Exchequer, the Scots court that tried cases relating to customs, excise and other branches of the Crown revenue. It was a judicial arena less competitive than the Outer House. As a shy lawyer's clerk of eighteen Mackenzie wrote a romantic tragedy and contributed poems to *The Scots Magazine*, delivering them after nightfall to Sands's shop in Parliament Close where the magazine was sold.[12] Starting with English odes – 'Happiness' in November 1763, and 'Ode to Melancholy' – and a blank-verse account of the High Street, 'The Street', he passed on, amid the fashion for antiquities established by MacPherson's *Fingal*, to imitations of old Scots ballads, 'Duncan' and 'Kenneth'. 'Duncan', printed by Smellie in *The Scots Magazine* of April 1764, was labelled, in MacPherson style, 'From an old Scots manuscript'.[13] If here Mackenzie blazed a romantic trail for Walter Scott, it was one he soon abandoned.

The Exchequer court followed the forms of English law, and in 1765 Mackenzie went to London to study the workings of the court there. Freed from the restraint on his imagination of his family and his native town, Mackenzie conceived the plan for a novel which would sketch 'the life and sentiments of a man of more than usual sensibility'.[14] This man would be a martyr to shyness like himself, warm-hearted, revolted by much to do with the legal profession, and exposed to the same states of consciousness and the same sensations Mackenzie himself had experienced. 'The *Man of Feeling* a real picture of my London adventures – Palpitation of heart walking along the pavement of Grosvenor Square to a man of high rank to whom I had a letter of introduction ... My being urged to remain in London; probable success if I had; but shy ...', as Mackenzie noted in his characteristically fragmentary memoir *Anecdotes and Egotisms*. In a letter to his cousin Elizabeth Rose of Kilravock in 1769, Mackenzie said he had a notion of 'introducing a man of sensibility into different scenes where his feelings might be seen in their effects, and his sentiments occasionally delivered without the stiffness of regular deduction'.[16]

In that he succeeded. He returned to Edinburgh, probably in 1768, and the next year appeared as an attorney at the Court of Exchequer, at a salary of £50 a year. *The Man of Feeling* was published anonymously in April 1771, had sold out by June,[17] and was so well-known that four years later, in his play *The Rivals*, staged at Covent Garden, Richard Sheridan had Lydia Languish scoop the book out of her aunt's sight, leaving Fordyce's unctuous *Sermons to Young Women* open on the table.[18] An Irish clergyman living in Bath, John Eccles, claimed authorship of *The Man of Feeling*. He was drowned in 1777 in the river Avon trying to save a child, an act of selflessness, said Mackenzie in true sentimental style, which vindicated his title to *The Man of Feeling*.[19]

Like MacPherson's *Fragments*, the book is a set of glimpses. As Mackenzie told Miss Rose, he had 'fallen into some detached essays, from the notion of its interesting both the memory and the affections deeper than mere argument or moral reasoning'.[20] To this end, he introduced the device of a sporting curate who used the pages of the manuscript for wadding for his gun, so that connecting incidents are missing. *The Man of Feeling* is not a novel with 'real' characters – in the sense that Goldsmith's Wakefield Vicar is a half-way recognisable social type – but a series of short-winded episodes in which successive incidents in town and country act on the super-sensitive feelings of the hero Harley, rather as incidents act on the delusions of Don Quixote. In other words, it is a *picaresque* novel of the heart. In each of his confrontations with modernity – the London Bedlam, a prostitute, a press-gang, a sharp, a stage-coach, agricultural improvements, a ruined village – Harley responds with not rage or ridicule, but incessant floods of tears. In a short novel, there are forty-three separate incidents of weeping.[21] The benevolence propounded by Hutcheson as the natural and instinctive basis of morality was a dead-end. In modern commercial society, benevolence leads but to sorrow and the grave. As he lies dying – evidently, of sensitivity – Harley seems to be saying that virtue is to be found only in Heaven.

There are some feelings which perhaps are too tender to be suffered by the world. The world is in general selfish, interested, and unthinking, and throws the imputation of romance or melancholy on every temper more susceptible than its own. I cannot think but in those regions which I contemplate, if there is any thing of mortality left about us, that these feelings will subsist: they are called – perhaps they are – weaknesses here; but there may be some better modifications of them in heaven which may deserve the name of virtues.

Still, he does not, unlike Goethe's Werther, kill himself to get there.

Encouraged by his success, Mackenzie attempted a triple-decker, a longer novel to be bound in three volumes. *The Man of the World* (1773) extols the society of the Cherokees with the primitivist enthusiasm of Rousseau or Lord Monboddo: 'On this side [i.e., Europe] fraud, hypocrisy, and sordid baseness; while in that seemed to preside honesty, truth, and savage nobleness of soul.'[22] *Julia de Roubigné* (1777) was prompted by a suggestion of Lord Kames that a catastrophe might arise not from evil intentions but from an excess of feeling. A tale of baseless matrimonial jealousy, it is a sort of sentimental *Othello*. In the magazine he founded with other literary lawyers in the George's Square district, *The Mirror*, Mackenzie continued to investigate the pleasures and perils of sentimentality. In the 'Story of *Emilia*' in the paper of 25 April 1780, he wrote: 'I found that *sentiment*, like religion, had its superstition and its martyrdom.'[23] But his chief essay sentimentalised the town's most famous citizen.

When Hume lay a-dying without benefit of clergy in 1776, Boswell was overwhelmed with religious panic, scurrying for security to the half-remembered lectures of his mother and Dr Johnson.[24] Dr Black reported that when he examined Hume and told him that there was no hope, Hume shook his hand. 'I thank you', he said, 'for the news.'[25] Adam Smith converted his friend from an anti-Christian to an ante-Christian. His letter of 9 November 1776 from Kirkcaldy to the publisher Strahan imitated the Attic restraint of the obituary for Socrates with which Plato closed the *Phaedo*.[26] The identification of Hume with Socrates gave great offence to Dr Johnson and Dr James Beattie. The English Christians were livid.

In Mackenzie's best-known story, 'Importance of religion to minds of sensibility; story of *La Roche*', which appeared in three issues of *The Mirror* for June of 1779, he substituted for Hume's antique restraint a decorous Christianity in which the stock characters of sentimental fiction – the Noble Old Man, The Dutiful but Doomed Daughter, the Good Swiss – were enlisted to save *le bon David* from the fires of hell.

The story recounts how a philosopher, living in a small French town very much of the nature of Hume's La Flèche, meets and restores to health a pastor from Switzerland, land of innocence, Alps and, for some of the time, Jean-Jacques Rousseau. The philosopher accompanies Pastor La Roche and his beautiful daughter home to the Bernese, attends divine service with them where the young woman makes a chaste figure at the organ, breaking down in tears during the psalm. Three years later the philosopher returns to Switzerland and, hurrying to their village, learns that Mlle La Roche has died of a broken heart at the death of her lover in a duel. The sight of La Roche, firm in his faith as he speaks to his congregation, leaves a deep impression on the philosopher.

As Mackenzie himself recognised, this was not religion, merely a religious feeling, 'as appealing to the sentiments of the heart, not to the disquisitions of the head':[27] but that, according to Ramsay of Ochtertyre, was now the faith of polite Scotland, 'which was disposed to make the yoke of duty more light, by reducing religion to matter of *sentiment*'.[28] As for the French setting and the English philosopher, they were merely the key or clue to the Hume connection.

The story ends with a Sentiment:

Mr ——'s heart was smitten; and I have heard him, long after, confess that there were moments when the remembrance overcame him even to weakness; when, amidst all the pleasure of philosophical discovery, and the pride of literary fame, he recalled to his mind the venerable figure of the good *La Roche*, and wished that he had never doubted.[29]

That is a scurrilous rewriting of philosophical history. It was society, not piety, that saved the philosopher of the *Treatise* from monstrosity: 'I dine, I play a game of backgammon ...' Mackenzie went so far as to claim that Smith himself found that last Sentiment 'so natural that in his usual absence of mind he told me he wondered he had never heard the anecdote before'.[30] His manipulation of both Smith and Hume shows how ruthless Mackenzie could be in the cause of sentimentality.

In his essay on Mackenzie for a series of prefaces to Ballantyne's *Novelists' Library*, later collected as *Lives of the Novelists*, Scott identified his townsman's strength and his weakness as a writer. Mackenzie's career showed him to be a skilful satirist, capable of caricature of the milder sort, an enthusiast of nature, a fine critic, a witty talker:

> But all these powerful talents, any single one of which might have sufficed to bring men of more bounded powers into notice, have been by Mackenzie carefully subjected to the principal object which he proposed to himself – the delineation of the human heart ... In short, Mackenzie aimed at being the historian of feeling ...[31]

The sentimental pitch had to be lowered (to put it mildly) before a novel could be produced that would outlive the fad in sentimentality. It was not until 1811 that from the flotsam of Mackenziean sentimental titles washing around London and Edinburgh – *The Man of Sorrow*, *The Man of Nature*, *The Man of Sensibility*, etcetera – there bobbed up a book which, in introducing social menace to this stew of virtue and tears, tipped every other novel of the age (not excluding Scott's) towards oblivion: Jane Austen's *Sense and Sensibility: A Novel, By a Lady, in three volumes*.

Robert Burns read *The Man of Feeling* (along with Sterne's *Sentimental Journey* and *Tristram Shandy*) in the mud of Lochlea in about 1780, and was entranced. He called it a 'bosom favorite'[32] and 'a book I prize next the Bible'.[33] For a while, Burns *was* Harley.

Writing just after Christmas of 1781 from Irvine, where he had
tried and failed to set up a flax-dressing business, the young Burns
struck Harleian poses even for his father: 'I am not formed for the
bustle of the busy nor the flutter of the Gay.'[34] (One wonders what
the guidman cottar made of this.) Yet it was these very sentimental postures that kept Burns's personality intact amid the hard-
scrabble life at the farms of Mount Oliphant and Lochlea and the
frustrations and religious bigotry of village Ayrshire. ''Tis absurd
to suppose', he wrote in an attempt at the elevated style to his old
schoolteacher in January 1783, 'that the man whose mind glows
with sentiments lighted up at their [Mackenzie, Sterne, Ossian]
sacred flame – the man whose heart distends with benevolence to
all the human race – he "who can soar above this little scene of
things" – can he descend to mind the paultry concerns about
which the terrae-filial race fret, and fume and vex themselves?'[35]

In 1772, on 2 January, the saddest day of any year, there appeared
in *The Weekly Magazine or Edinburgh Amusement* a poem called
'The Daft Days'. The magazine, started in 1768 by Walter Ruddiman, nephew of the Jacobite antiquarian Thomas Ruddiman who
had preceded Hume as Keeper of the Advocates' Library, ran until
1784 and would be forgotten but for the poems of Robert
Fergusson. 'The Daft Days' are the holidays between Christmas
and the New Year, celebrated to this day in Edinburgh:

> Now mirk December's dowie face
> Glours our the rigs wi' sour grimace
> While, thro' his *minimum* of space,
>   The bleer-ey'd sun,
> Wi' blinkin light and stealing pace
>   his race doth run.[36]

Poets must select the language they write in, and it is never the
language of conversation. The language of 'The Daft Days', with
its intense Scots words – 'mirk', 'glours' – sprinkled with elegant
Latin, is so assured that it is as if a door had suddenly opened for

a tourist onto a prosperous and hospitable town society of whose existence, up to that moment, he had had no information. For Fergusson's twentieth-century editor Matthew McDiarmid, 'The Daft Days' comes 'with the startling effect of a stone breaking the windows of the hot-house'.[37]

Fergusson's stanza is an old Troubadour measure employed in Scotland since the Middle Ages, made popular in the seventeenth century by Sir Robert Sempill of Beltrees in his 'Epitaph of Habbie Simson, Piper of Kilbarchan', and printed in Watson's *Choice Collection* of 1706–11:

> Now who shall play *The Day it Dawis*
> Or *Hunts Up*, when the cock he craws?
> Or who can for our kirk-town cause
>   Stand us in stead?
> On bagpipes now nobody blaws
>   Sen Habbie's dead.[38]

Allan Ramsay christened this form of stave or stanza the 'Standart Habby'. Fergusson's poem draws on the same melancholy as the original, but is also authentically mid-century Edinburgh: subterranean, classical, townee, learned, alcoholic, parochial, intimate:

> And thou, great god of *Aqua Vitae*!
> Wha sways the empire of this city
> When fou we're sometimes capernoity,
>   Be thou prepar'd
> To hedge us frae that black banditti
>   The C[ity]-G[uar]d.[39]

For a moment we see John Kay's picture of that 'lumbersome biggin' the City Guard-house, with John Dho lowering ferociously over the transom.[40]

Robert Fergusson, who died at the age of twenty-four in the Edinburgh Bedlam, wrote substantially all his poems in the space of two years. As Stevenson observed, 'had Burns died at the same age as Fergusson, he would have left us literally nothing worth remark'.[41] Fergusson added a new tone to British poetry which

Burns later played out. Without Fergusson, it is open to question whether Burns would have had the confidence to write in Scots. In an important autobiographical letter Burns told the physician Dr John Moore that he had all but given up writing verse in the early 1780s, till 'meeting with Fergusson's Scotch Poems, I strung anew my wildly-sounding, rustic lyre with emulating vigour'.[42]

Fergusson was born on 5 September 1750 in an apartment in Cap-and-Feather Close, an alley on the north side of the High Street that vanished in the next decade in the building of the approaches to the New Bridge. His father, from the Jacobite country in Aberdeen, had a precarious living as a clerk, with an income – according to a household account printed by Fergusson's first biographer, the Reverend A.B. Grosart – of under £20 per year to support a family of five.[43] His mother, Elizabeth Forbes, came from a landed family. He was what used to be called a sickly child, and seems to have lost a year at the High School in 1760. With the help of his uncle John Forbes he received a bursary to the Grammar School in Dundee, a fateful move since he passed from there not to the College – where he would have met Hugh Blair – but to St Andrews, in 1764. As a seat of learning, St Andrews had fallen on hard times, and drew some caustic comment from Dr Johnson during his visit in 1773; but William Wilkie, author of *The Epigoniad*, was Professor of Natural Philosophy and took a friendly interest in Fergusson, interceding to restore him when he was rusticated for causing a riot at prayers.[44] Robert was already a fine singer, and was writing verse. The question, as ever in Scotland, was which language to use.

The famous *makaris* – the 'makers', or poets – of Chaucer's age had written in an international idiom made possible by the classical culture shared by all Europe, but by the time Allan Ramsay senior came to prepare the preface to his anthology of Scots poetry, *The Ever Green* (printed by Ruddiman senior in 1724), 'the good old bards of Scotland' were as remote as Sappho and Alcaeus. 'It was intended that an account of the Authors of the following Collection should be given,' Ramsay wrote, 'but not being furnished with such distinct Information as could be wished

for that End at present, the Design is delayed.'[45] Demotic Latin
verse had died with Archibald Pitcairne in 1713, while the tradition of Scots vernacular verse had been disrupted by the use of
southern English in the Kirk and colleges and at the bar. (In *Tam
o' Shanter*, Burns breaks into southern English when he wants to
moralise.) Ramsay had written low-life poems in Edinburgh
dialect and also, in *The Gentle Shepherd*, a pioneering Scots pastoral. But the best-known Scots poet before Burns, James
Thomson, had moved to London and written in a stately if somewhat languid southern English. Most of the poems printed in the
Edinburgh periodical press were provincial imitations of Gray or
Shenstone. Fergusson could write in southern English as well as
any of these, but he also wanted to write in Scots. His way in was
Habby Simson.

The 'Elegy on the Death of Mr David Gregory, late Professor of
Mathematics in the University of St Andrews' was printed in
1773, but probably written soon after Gregory's death in 1765:

> He cou'd, by *Euclid*, prove lang sine
> A ganging point compos'd a line;
> By numbers too he cou'd divine,
> > Whan he did read,
> That *three* times *three* just made up nine
> > But now he's dead.
>
> In *Algebra* weel skill'd he was
> An' kent fu' well *proportion's* laws
> He cou'd make clear baith B's and A's
> > Wi' his lang head;
> Rin owr surd roots, but cracks or flaws;
> > But now he's dead.

Habby has been washed and brushed, become learned of a
sudden, and polite. The vulgarity of Ramsay's low-life elegies is
gone. If there is anything sophomoric about Fergusson's poem, it
is only the use of 'but' in the same last couplet to mean first 'without' and then 'however'.

His father died in May 1767 and Robert Fergusson went down from St Andrews the next year to join his mother and family living in very tight circumstances in Jamieson's Land in Bell's Wynd. For a while he stayed in Aberdeenshire with his uncle John Forbes, but John objected to his nephew's shabby dress and Robert returned to Edinburgh and took a clerking job at the Commissary Office, copying documents relating to matrimonial cases and wills at a penny a page. He wrote verses and attended concerts and the theatre, where he contributed some songs. He made friends with the actor West Digges, and had the habit of occupying the central box at the Canongate theatre and applauding by smashing his right hand down on the front of the box. He joined the Pantheon Society, and on 10 October 1772 the Cape Club, a collection of able if not quite genteel men who included Digges's leading actor William Woods and, in time, Deacon William Brodie, Town Council member by day and burglar by night, whose double life inspired Stevenson's *Dr Jekyll and Mr Hyde*. (After a botched break-in at the Excise Office in Chessel's Court, just opposite where Adam Smith was living, Brodie was tried and executed.) Like many other Edinburgh clubs, the Cape boasted a facetious name and elaborate rituals of the kind that now survive only at old-fashioned British and American universities: each member was designated a knight and Fergusson, because of his fine voice, Sir Precentor.

From February 1771 Fergusson's verses began to appear in the Ruddimans' magazine – pastorals, elegies in the Shenstonian style, and mock heroics in standard English. In the new year came 'The Daft Days' and then a succession of Scots poems: 'Elegy on the Death of Scots Music' on 5 March, 'The King's Birth-Day in Edinburgh' on 4 June, 'Caller Oysters' on 27 August, 'Braid Claith' on 15 October, 'Geordie and Davie: An Eclogue to the Memory of Dr William Wilkie' on 29 October, 'Hallow Fair' on 12 November and 'To the Tron-Kirk Bell' on 26 November.

'Is Allan [Ramsay] risen frae the deid?' asked a verse correspondent, J.S. of Berwick, in the magazine of 3 September 1772. The

Ruddimans must have thought so, because in the new year they issued his first collection, *Poems by Robert Fergusson*. These included the eight Scots poems already noted, 'Sandy and Willie: An Eclogue' and twenty-seven poems in southern English. Alexander Grosart reported that 500 copies were sold, and that the poet earned £50, which is more than Burns received from the Kilmarnock edition of *Poems, Chiefly in the Scottish Dialect*. In the course of 1773 *The Weekly Magazine* printed a new poem almost every fortnight.[46] In addition, the Ruddimans issued the first canto of what was set to be a full-scale evocation of Edinburgh in octosyllabic couplets, *Auld Reikie*. The canto ends with a sensational *envoi*:

> Aft frae the *Fifan* coast I've seen
> Thee tow'ring on thy summit green;
> So glour the saints when first is given
> A fav'rite keek o'glore and heaven
> On earth nae mair they bend their ein
> But quick assume angelic mein;
> So I on *Fife* wad glour na more,
> But gallop'd to EDINA'S shore.

No second canto was printed. Grosart reported seeing a copy with a dedication to Sir William Forbes of Pitsligo, but the banker disdained to accept it, and Fergusson became discouraged.[47] (Certainly, there is no such dedication on the British Library copy or in any copy that has appeared in the modern book trade.)

Fergusson is known to have been planning to follow Gavin Douglas in translating the *Eclogues*, *Georgics* and *Aeneid* of Virgil into the Scots vernacular: in other words, to undertake for the Ruddimans the utterly doomed and hopeless task of preserving in post-Union Scotland the Scots humanist tradition of Buchanan and Douglas. What survives instead is a translation of Horace, *Odes* 1:11, and, more important still, 'Dumfries', which commemorates Fergusson's visit to the town in the summer of 1773. 'Dumfries' is not so much an imitation of Horace as Horace himself:

> O Jove, man, gie' some orrow pence,
> Mair sillar, an' a wie mair sense,
> I'd big to you a rural spence,
>   An' bide a' simmer,
> An' cald frae saul and body fence
>   With frequent brimmer.

'Dumfries' is one of the last authentic reverberations of antiquity in British literature: truly, the end of an old song.

Towards the end of 1773 Fergusson fell prey to severe depression. His last two poems to appear in the Ruddimans' magazine, the Villonades 'Rob. Fergusson's Last Will' on 25 November, and the 'Codicile to Rob. Fergusson's Last Will' on 23 December, told their own story. After 30 December, Fergusson does not seem to have attended the Commissary Office. Early biographers report that he had contracted syphilis, and that he had joined a party 'interested in an election business' in eastern Scotland. 'The riotous enjoyments incident to such occasions' produced a 'feverishness and decrepitude of mind'.[48] He became troubled by religious doubts. One account said that he was woken one night by the cries of a starling being tortured by a cat: the fate of the bird 'wrought his mind up to a pitch of remorse that almost bordered on frantic despair'.[49] The summer brought some relief and _The Caledonian Mercury_ of 7 July printed some verses by William Woods, 'To Mr R. Fergusson: On his Recovery'. But on the 28th, he suffered a bad fall downstairs. Dr Duncan decided that the poet was too much for his old mother to look after in the cramped, poverty-stricken and insanitary apartment in Bell's Wynd. 'After several fruitless attempts to have him placed in a more desirable situation'[50] Duncan had him removed to the Bedlam, a set of damp and unheated cells attached to the Charity Workhouse in a kink of the walls south of the College. There Duncan attended him, as did the popular surgeon Alexander Wood, who treated Burns when he hurt his foot in 1787. John Pinkerton, an antiquarian who printed a collection of late medieval Scots poetry in the year of Burns's visit to Edinburgh, recalled:

I was told by a person who was his most intimate friend, and who
went to see him lodged there, as otherwise force alone could not
have carried him, that it was about nine o'clock at night when they
went; and that the dismal habitation was quite silent: but upon
Fergusson's entering the door he set up a strange halloo, which, in
the instant, was repeated by the miserable inhabitants of all the cells
in the house. My informant says, the sound was so horrible that he
still hears it.[51]

Fergusson died on 17 October 1774 and was buried in the
Canongate Churchyard, on the west side, where he lay without a
monument – in all senses – till the arrival of Burns. Greatly distressed by the manner of his death, Dr Duncan went on to found
the Royal Public Dispensary in 1776, where the poor were treated
free of charge, and to lay the foundation for the Royal Edinburgh
Asylum, which was opened in 1813.

In 1779 the Ruddimans issued Fergusson's uncollected poems
from *The Weekly Magazine*, but the *literati* continued to hold
themselves aloof. Fergusson's fault was that he was not a sentimental author but a satirist of Latin character, a man so at ease
with the classical world that he could rhyme 'dinna fash us' with
'Parnassus'; not a gentleman nor yet a savage, but a precocious
bursary boy and tavern rat. Though he died young and obscure it
was in Bedlam, not in a cottage trained with honeysuckle like
Michael Bruce, the darling of the *Mirror* set, who had died of consumption at twenty-one in 1767.[52]

More than Boswell, more than Burns, Fergusson belonged to
the old subterranean Edinburgh and Canongate rather than to the
tea-tables of George's Square and the New Town. Where English
and Scots poetry of that time tended to the abstract and the timeless, Fergusson had learned from the Latin poets and Allan
Ramsay to deal with the routines and topography of an actual city.
He made permanent the caddies and the law courts and the din of
the Tron Kirk bell, rioting in The Meadows and throwing dead
cats at the City Guard, oysters and gin at Lucky Middlemist's in
the Cowgate or Sunday jaunts to eat sheep's head at Duddingston

or mussel soup at Musselburgh, and fairs in the suburbs, where country John in his 'Sunday claise'

> Rins after Meg wi' *rokelay* new
> An sappy kisses lays on;
> She'll taunting say, Ye silly coof!
> Be o' your gab mair spairing;[53]

and staggering home from Leith Races, and the stolid merchants in their good broadcloth, and the rigmarole of the Council elections, and footmen strutting their livery at the Cross, and the quiet of 'Indian' Peter's coffee-house during the court vacation. And underneath it all, that insistent melancholy nowadays identified as depression.

Fergusson had written a puerile satire on *The Man of Feeling* which he called 'The Sow of Feeling' and it may be that Mackenzie never forgave him. In his memoir *Anecdotes and Egotisms* Mackenzie confused moral and literary values: 'Ferguson, dissipated and drunken, died in early life, after having produced poems faithfully and humourously describing scenes of Edinburgh, of festivity and somewhat of blackguardism … [As for Burns] His great admiration of Ferguson shewed his propensity to coarse dissipation.'[54] Walter Scott objected that Burns, 'having twenty times the abilities of Allan Ramsay and of Ferguson, … talked of them with too much humility as his models.'[55] Stevenson preferred to take on trust Burns's judgement of Fergusson; who would presume to disagree?

~ ⌒ ~

The bookseller, subscription agent, and future Lord Provost William Creech, whose morning levees in the Luckenbooths were as famous for the numbers who attended as for the stinginess of the fare provided, in 1793 took stock of the alterations in his native town in the thirty years since Ossian and the commencement of the North Bridge.

What had once been fields and orchards were now elegant suburbs. Some £3,000,000 sterling, he computed, had been spent on making the New Town, the George's Square developments in the south and other improvements.[56] Rateable rents had doubled.[57] The South Bridge over the Cowgate, of which the foundation stone had been laid in August 1785, was approaching completion, with twenty-two arches over the ruins of the once-fashionable Niddry's, Merlin's (or Marline's) and Pebbles (or Peebles) wynds and the ancient Black Turnpike. Land for shops on each side sold at up to £109,000 per acre, and in Milne's Square at £151,000: prices not seen, Creech adds foolishly, at the height of the Roman Empire.[58] Adam's General Register House was at last nearing completion, at a cost of £36,000, and the foundations had been laid for the new University buildings, which would cost £63,000.[59] The High School, with five hundred boys in new halls built in 1777, was probably the largest in Britain. The spoil of the New Town basements was filling up the North Loch, at a rate of eighteen hundred cartloads a day, to make a causeway from the Old Town known, with admirable simplicity, as The Mound. (Walter Scott hated The Mound, which he used to pass over on his way to Castle Street, and called it 'the most hopeless and irremediable error which has been committed in the course of the improvements of Edinburgh':[60] actually, it provided a fine site for picture galleries.)

Among new industries there were button-makers, shawl-makers, starch-makers, glass-makers, twelve paper mills, sixteen printing-houses, four newspapers, coach-builders. There was a strong market in the south for Scots authors, tapped by the Scots booksellers in London, first Andrew Millar, and then William Strahan and John Murray.[61] By 1783, several ministers and College professors were earning enough from writing to keep carriages.[62]

There were haberdashers and also perfumers, of whom some, said Creech, 'advertised the keeping of bears, to kill occasionally, for greasing ladies' and gentlemen's hair'.[63] By 1783 there were fifteen stage-coaches a week to London, and in 1786 two daily services offered the journey in just sixty hours.[64] When an apprentice named Watt robbed the Forbes bank of £1,200 in 1778, one of the partners, Mr Bartlet, pursued him to London on horseback. The weather was fine, and though Watt had a three-day start, Bartlet took just forty hours in the saddle and was waiting at the correspondent bank in the City, with two Bow Street officers, when Watt came to cash his bill.[65] 'Indian' Peter's caddies delivered letters and bundles every hour from nine in the morning until nine at night within a radius of a mile about the Cross: often enough to take Burns's letters from the Register House to his new love in the Potterrow, sometimes twice or more in a day.[66]

In the larger city, moral restraints had disintegrated. Whisky not ale was now the town drink: in 1790, 1.7 million gallons of spirits were produced from legal stills alone in Scotland, or ten times as much as in 1760.[67] Cock-fighting had caught on and, after 1790, prize-fighting.[68] Kirk collections, said Creech, fell by a third between 1763 and 1783,[69] divorce and separation were more frequent,[70] the number of brothels had increased twenty-fold and ordinary streetwalkers by one hundred-fold.[71] In 1763 the Kirk treasurers collected £154 in fines for bastard children, in 1783 £600. House-breaking, theft and burglary seemed like an epidemic at the peak of Deacon Brodie's night-career in the winter of 1787–8.[72] The country had been used to about three executions a year; but one week in 1783, six persons lay under sentence of death in Edinburgh gaol alone.[73]

Fashionable life altered in the process of migrating first to George's Square and then to St Andrew's Square. Dinner slipped to four or five o'clock, and wine was served after dinner by 'every tradesman in decent circumstances'.[74] Dress improved out of recognition, 'the maid servants dressing almost as fine as their mistresses did in 1763', and umbrellas (pioneered by 'Lang Sandy' Wood in 1780) were coming into use. Dancing was no longer confined to the Old Assembly, starting at six and ending prompt at eleven under the gimlet glare of Miss Nicky. Those rooms passed to John Dho and the City Guard, and there were now three sets of assembly rooms in Edinburgh, as well as another in Leith, and dancers arrived at eight or nine to stay until three or four.[75]

This was the world caught by the pen of John Kay, who quit barbering in 1786 to produce the finest portrait of any city (with the exception of Utamaro's Edo and Hogarth's London) until the age of photography. Here were the balloon mania of 1785 and 1786, the golfers and cockfighters and the fops in The Meadows and Adam Smith walking to the Customs House, and Deacon Brodie at his trial, and the rail-thin Hugo Arnot being pestered by a beggar. Here was the enormous James Bruce, a Stirlingshire laird who had returned from Abyssinia with stories quite as sensational as those of as 'Indian' Peter's tribulations among the Mohawks. His neighbours disbelieved him, though his *Travels to Discover the Sources of the Nile in the years 1768, 1769, 1779, 1771, 1772 and 1773*, published in 1790, is the masterpiece of Scottish travel. (His end was just: on 27 April 1794, while hurrying down to hand a lady to her carriage, the great traveller slipped on his own house-step, fell on his head, and died.)

In this airy and intimate world, where everybody knew far too much about everybody else, one day in late November 1786 a young Ayrshireman appeared, determined to show the drawing-rooms what real genius looked like. For Robert Louis Stevenson, Burns's first visit to Edinburgh was without parallel: 'Such a revolution is not to be found in literary history.'[76] The image of the man in the Nasmyth portrait, in his boots and buckskins, blue coat and waistcoat striped with buff and blue, was as famous as Werther in *his* blue coat was in Germany. Edinburgh was fascinated by Burns's looks and manners, and thrilled and alarmed by his smashing of social form.

Burns set out for Edinburgh in November 1786. His affairs at home were in an inextricable tangle. The Mossgiel farm he had leased with his brother was waterlogged and heavy-going. He was father to an out-of-wedlock daughter by a servant in his father's house, Elizabeth Paton. Early in 1786 it had become clear that Jean Armour, twenty-one-year-old daughter of a respectable master mason, was also pregnant. Robert was ready to marry her but James Armour, Jean's father, was not prepared to accept him. Jean, shipped off to her uncle in Paisley, allowed her father to deface their marriage contract (or could not prevent him). Burns was both distressed and infuriated by what Jean had done, and veered between longing for her and wishing to be rid of her for good. 'I have tryed often to forget her,' he wrote to his friend David Brice, who was now a shoemaker in Glasgow. 'I have run into all kinds of dissipations and riot, Mason-meetings, drinking matches, and other mischief, to drive her out of my head, but all in vain.'[77] (Among these distractions was probably the woman called Mary Campbell, the enigmatic 'Highland Lassie' who has divided Burns scholars and enthusiasts for two centuries.)

Burns had plans to emigrate to Jamaica, as overseer on the plantation of a Mr Charles Douglas, though this was as much a maudlin gesture as a practical step. In this self-pitying mood he was writing poems that he knew were good enough to print, and ''twas a delicious idea that I would be called a clever fellow, even though it should never reach my ears, a poor Negro-driver, or perhaps a victim to that inhospitable clime.'[78] On 25 June he appeared before the Kirk Session and over three Sundays did penance for fornication in Mauchline Kirk. That cleared him of his obligation to marry Jean, but Armour was determined that Burns should put up 'an enormous sum' for the support of the unborn child.[79] On 2 July Burns assigned his share in Mossgiel and any profits from 'the Publication of my Poems presently in the Press' to his brother Gilbert, and thus beyond Armour's power.[80] The copyright was to be held in trust for young Betsey, his daughter by Elizabeth Paton. In the last week of July, to escape being served with Armour's writ, he went into hiding, slipping away from Mossgiel, 'wandering from one friend's house to another'.[81]

Poems, Chiefly in the Scottish Dialect, printed in Kilmarnock on 31 July 1786, was Burns's last throw, and it was a hit. Here, combined at last, were the Fergusson-rackety and the Mackenzie-sentimental, but with effects incomparably better than each manner on its own: as here, from the very first poem in the book, 'The Twa Dogs':

That merry day the year begins,
They bar the door on frosty win's;
The nappy reeks wi' mantling ream,
An' sheds a heart-inspiring steam;
The luntin pipe, an' sneeshin mill,
Are handed round wi' right guid will;
The cantie auld folks crackin crouse,
The young anes rantin thro' the house –
My heart has been sae fain to see them,
That I for joy hae barkit wi' them.

No eighteenth-century poet (except Goethe) had such an effort-less lyrical gift.[82] The book sold just over six hundred copies at three shillings each, brought Burns £20 net of expenses, and found readers all over the Lowlands.[83] Blind Thomas Blacklock (repaying with interest the kindness he had received from Hume) praised the book and urged Burns to bring out a second edition.[84]

Wittingly or not, Burns had set a sentimental trap for Mackenzie and for Edinburgh. In the preface to *Poems, Chiefly in the Scottish Dialect* he described himself as neither a Virgil nor a Theocrites (*sic*) – that is, not a sophisticated urban poet glorify-ing a fantasy world of shepherds and nymphs – but a working farmer: 'Unacquainted with the necessary requisites for com-mencing Poet by rule,' he wrote, 'he sings the sentiments and man-ners, he felt and saw in himself and his rustic compeers around him, in his and their native language.'[85] The mistake of Theocrites for Theocritus might have been designed to show he was no book-scholar.

For the Scottish theorists of the second half of the eighteenth century, genius was a principle of human nature that could be nei-ther taught nor learned. That, at least, was the argument of *The Essay on Genius* by Alexander Gerard, Professor of Divinity at Aberdeen, in 1774. Genius was certainly not confined to the polite strata, and the notion that it might live and die in an obscure and neglected milieu – that 'Some mute inglorious Milton here may rest', in the words of Gray's 'Elegy in a Country Churchyard' –

was a commonplace of sentimental anguish. 'I never look on his dwelling', wrote William Craig and Henry Mackenzie of Michael Bruce in *Mirror* No. 36 in 1779, '– a small thatched house, distinguished from the cottages of the other inhabitants only by a *sashed window* ... fringed with a *honeysuckle* plant, which the poor youth had trained around it ... but I stop my horse involuntarily.'[86]

Pure genius had the power to demolish or suspend social form, and Edinburgh instantly decided that Burns was just such a pure genius. Dugald Stewart invited him to dinner at his country house near Mauchline in October. Like others after him, he found Burns very conscious of his 'genius and worth', but neither forward nor arrogant. Hugh Blair, who was staying with the Stewarts, read the satire on field-preaching, 'The Holy Fair', and called it 'the work of a very great genius'.[87] In November, a highly favourable notice of the Kilmarnock poems appeared in the *Edinburgh Magazine*. Here was 'a striking example of a native genius bursting through the obscurity of poverty ... Who are you, Mr Burns?' asked the author of the review, before comparing him favourably with Ramsay and Fergusson.[88] On 14 November Burns received an express for half a dozen copies of the poems from a local gentlewoman, Mrs Frances Dunlop of Dunlop – a descendant of Sir William Wallace, no less – which inaugurated one of his most rewarding friendships. To cap it all, Jean produced twins.

For the reprint, Burns now set his sights above Kilmarnock. Robert Aiken, a lawyer in Ayr who had drummed up the subscribers for the first edition, gave him introductions to the bookseller William Creech, and the printer William Smellie in Anchor Close.[89] On 27 November Burns set off the sixty miles to the capital on a borrowed pony, spent the night near Biggar in Lanarkshire, and arrived late on the 28th, lodging with a friend in Baxter's Close at the widow Mrs Carfrae – she of the 'Daughters of Belial' – in the Lawnmarket.

Mackenzie's review of the Kilmarnock edition, entitled 'Surprising effects of Original Genius, expemplified in the Poetical Productions of Robert Burns, an Ayrshire Ploughman', appeared in *The Mirror*'s successor, *The Lounger*, in its issue of 9 December. He had swallowed the bait:

> I know not if I shall be accused of such enthusiasm and partiality, when I introduce to the notice of my readers a poet of our own country, with whose writings I have lately become acquainted; but if I am not greatly deceived, I think I may safely pronounce him a genius of no ordinary rank.
>
> The person to whom I allude is ROBERT BURNS, an Ayrshire ploughman, whose poems were some time ago published in a country-town in the west of Scotland, with no other ambition, it would seem, than to circulate among the inhabitants of the country where he was born, to obtain a little fame from those who had heard of his talents. I hope I shall not be thought to assume too much, if I endeavour to place him in a higher point of view, to call for a verdict of his country on the merit of his works, and to claim for him those honours which their excellence appears to deserve.
>
> In mentioning the circumstance of his humble station, I mean not to rest his pretensions solely on that title, or to urge the merits of his poetry when considered in relation to the lowness of his birth, and the little opportunity of improvement which his education could afford. These particulars, indeed, might excite our wonder at his productions; but his poetry, considered abstractly, and without the apologies arising from his situation, seems to me fully entitled to command our feelings, and to obtain our applause.[90]

There was one problem, Mackenzie continued, and it was the usual one of dialect. Mackenzie failed to see that Fergusson had simply abolished dialect as a problem; or, rather, had converted it into opportunity.

> One bar, indeed, his birth and education have opposed to his fame – the language in which most of his poems are written. Even in Scotland, the provincial dialect which Ramsay and he have used is now read with a difficulty which greatly damps the pleasure of the reader; in England it cannot be read at all, without such a constant reference to a glossary, as nearly to destroy the pleasure.[91]

(In reality, Burns's Scots is riddled with literary English, and is easier to understand – even without a glossary – than many passages in old southern writers such as Shakespeare or Marlowe.) Mackenzie then went on to praise some of the poems that were 'almost English' and quoted in full 'To a Mountain Daisy', just the sort of 'pastoral' that Burns was rather bad at. And, as always with Mackenzie, beyond the sentiment there was hard business to be transacted: in this case, subscriptions for the new edition.

> To repair the wrongs of suffering or neglected merit; to call forth genius from the obscurity in which it had pined indignant, and place it where it may profit or delight the world; these are exertions which give to wealth an enviable superiority, to greatness and to patronage a laudable pride.[92]

Burns's chief introduction was to a west-country nobleman of uncertain health, James Cunningham, 14th Earl of Glencairn, a man as tactful as he was generous. By 7 December Burns was writing to his old patron and landlord at the Mossgiel farm, Gavin Hamilton:

> My Lord Glencairn & the Dean of Faculty, Mr H. Erskine, have taken me under their wing; and by all probability I shall soon be the tenth Worthy, and the eighth Wise Man of the world.[93]

Six days later, Burns told the Ayr banker John Ballantine that he had been ill with a headache and stomach complaint, but even so had managed to meet most of Edinburgh society; 'my avowed Patrons & Patronesses are, the Duchess of Gordon – the Countess of Glencairn, with my lord & lady Betty – the Dean of Faculty – Sir John Whitef[o]ord – I have likewise warm friends among the Literati, Professors Stewart, Blair, Greenfield [Blair's successor as Professor of Rhetoric], and Mr McKenzie the Man of Feeling.'[94] Of those, Glencairn and Whitefoord, a Mauchline laird who had lost his estate in the Douglas, Heron bank failure and moved to Edinburgh, arranged in January that the members of the Caledonian Hunt should subscribe for one hundred copies at five shillings each. Creech also agreed to take five hundred copies at the subscription price.

Freemasonry, which ran across the divisions of class in eighteenth-century society, was another *entrée*. On 13 January 1787 Burns attended a meeting of the Grand Lodge of Scotland, where the Grand Master proposed a toast to 'Caledonia, & Caledonia's Bard, brother B——'.[95] He was assumed into the Canongate Kilwinning Lodge at the beginning of February. In addition William Smellie, who was to print the new edition at his works in Anchor Close, introduced him into a club he had founded, the Crochallan Fencibles, which met in the tavern at the opening of the Close. The club officers affected military ranks, in the manner of the militia regiments mustered at the time of the American war.

For most contemporary witnesses, the great literary interest of Burns's visit was the disintegration of hierarchy before the force of Original Genius. Walter Scott, who as a boy of fifteen saw Burns at Adam Ferguson's Kamtschatka in the company of Dugald Stewart, John Home and Dr Hutton and Dr Black, told his biographer Lockhart: 'I never saw a man in company with his superiors in station and information, more perfectly free from either the reality or the affectation of embarrassment.'[96]

Yet Burns was also a man. 'I think I know pretty exactly what ground I occupy, both as a Man & a Poet,' Burns wrote about the middle of December 1786 to William Greenfield, whom he liked and trusted. 'I am willing to believe that my abilities deserved a better fate than the veriest shades of life; but to be dragged forth, with all my imperfections on my head, to the full glare of learned and polite observation, is what, I am afraid, I shall have bitter reason to repent.'[97] Then he strikes a heroic posture: 'You may bear me witness when my buble of fame was at the highest, I stood, unintoxicated, with the inebriating cup in my hand.'[98]

Burns began to record his impressions of Edinburgh society in prose. In the so-called *Second Commonplace Book* – the first had been begun in Ayr in 1783 – a proud and sensitive man gave vent to his resentment at the slights of 'Squire Something' and 'Sir

Somebody' of Edinburgh. As for the *literati*, they were not as clever or as good as they thought they were. Take Hugh Blair:

> When he descends from his pinnacle and meets me on equal ground, my heart overflows with what is called *liking*: when he neglects me for the mere carcase of greatness, or when his eye measures the difference of our points of elevation, I say to myself with scarcely an emotion, what do I care for him or his pomp either? ... In my opinion Dr Blair is merely an astonishing proof of what industry and application can do.[99]

For all the fine Edinburgh women, such as Monboddo's daughter Elizabeth Burnett, Burns missed Jean: 'I don't think I shall ever meet with so delicious an armful again,' he confided to Gavin Hamilton on 7 January 1787.[100] Even in the Crochallan circle, the poet was uncomfortable. At the printing-works in Anchor Close, where the new edition was being set, Smellie's son Alexander later reported that

> He was dressed much in the stile of a plain country man; and walked three or four times from end to end of the composing room, cracking a long hunting whip which he held in his hand, to the no small annoyance of the compositors and press-men; and, although the manuscript of his poems was then lying before every compositor in the house, he never once looked at what they were doing, nor asked a single question. He frequently repeated this odd practice during the course of printing his work, and always in the same strange and inattentive manner, to the great astonishment of the men, who were not accustomed to such whimsical behaviour ... and, though I was at that time very young, the cracking of the whip, and the strangely uncouth, and unconcerned manner of BURNS, always impressed me with the notion that he wished to assume the clownish apearance of a country rustic; and I have never been able to efface the impression that his behaviour proceeded from affectation.[101]

In truth, Burns was ashamed of his own success and of Fergusson's obscurity. On 7 February he wrote to the Canongate magistrates regretting that 'the remains of Robert Fergusson, the so justly celebrated Poet, a man whose talents, for ages to come, will do honor to our Caledonian name, lie in your church-yard, among the ignoble Dead, unnoticed and unknown.'[102] He asked permission to lay 'a simple stone over his revered ashes', which was duly put in place in 1789, and can be seen today.

On 22 March Burns told Mrs Dunlop that he had corrected the last proof sheet and had 'now only the Glossary and sub-scribers names to print'.[103] Those ran to no fewer than fifteen hundred. Two thousand eight hundred copies were printed. Lord Glencairn and his mother took twenty four, and the Duchess of Gordon twenty-one. The book, when it appeared on 17 April with a new dedication to the Caledonian Hunt, contained twenty-two poems not collected in the Kilmarnock edition. Ominously, only two had been written in Edinburgh: the bread-and-butter 'Address to Edinburgh' and the Fergussonian 'Address to a Haggis'. Both had been printed in the *Caledonian Mercury* the preceding December. That same day he assigned to Creech the copyright of the poems – or, rather, his daughter's copyright – for a hundred guineas.

The question now was: What next? For all his social success, Burns had been offered no pension or sinecure. In a gushing letter of 1 February the Earl of Buchan, who had literary propensities, had recommended that he renew his poetic inspiration by visiting Flodden Field and James Thomson's birthplace and other classic scenes of Scottish history and literature.[104] To Mrs Dunlop in the West, Burns wrote: 'The appelation, of a Scotch bard, is by far my highest pride; to continue to deserve it is my most exalted ambi-tion. Scottish scenes, and Scottish story are the themes I could wish to sing. I have no greater, no dearer aim than to have it in my power ... to make leisurely pilgrimages through Caledonia; to sit on the field of her battles; to wander on the romantic banks of her rivers; and to muse by the stately tower or venerable ruins, once the honored abodes of her heroes.' Then, coming down to earth: 'These are all Utopian ideas ... I guess that I shall clear between two and three hundred pounds by my authorship; with that sum I intend, so far as I may be said to have any intention, to return to my old acquaintance, the plough, and, if I can meet with a lease by which I can live, to commence farmer.'[105]

He had an offer from Patrick Miller to view a farm of his near Dumfries. On the way, he wanted to explore some at least of the country of which he was now undisputed laureate. He had also entered with enthusiasm into the scheme of an engraver named James Johnson to collect the words and music of all the traditional Scots songs and arrange them for the pianoforte. The first volume of the *Scots Musical Museum*, including some hundred airs, was already in the press, and Burns was working industriously on the second.[106] Perhaps he already knew that his days as a poet were past and his future was that of a song-writer: he would be not read but heard. 'I have met wt few people whose company & conversation gave me so much pleasure,'[107] he wrote to Johnson, enclosing a song he had from Blacklock, on the eve of his departure.

On 5 May Burns set off for Ayrshire by way of the Border country in the company of a young law student, Bob Ainslie. He now had a mount of his own, christened Jenny Geddes after the heroine of the stool in St Giles. By way of the Border towns of Jedburgh, Kelso, Melrose and Selkirk he travelled as far south as Newcastle in England, returning by Carlisle. He viewed the farm in Dumfries and by 9 June was back at Mossgiel. A celebrity now, he found the Armours delighted to know him. Their 'mean, servile compliance'[108] disgusted him, but Jean did not, and was pregnant again when he left. After a short stay in Edinburgh in early August he was off to Stirlingshire and the West and East Highlands with William Nicol, one of the assistants at the High School, a good scholar but a brutal and difficult man. Distressed by the decay of Stirling Castle, Burns scratched some abusive anti-Hanoverian lines in the pane of his room at the inn. He visited three sets of dukes and duchesses, and his father's relations in Kincardine, before settling back in Edinburgh on 20 October. He moved into rooms in St James's Square, a new development behind the Register House. 'I am determined not to leave Edinr till I wind up my affairs with Mr Creech, which I am afraid will be a tedious business,' he told Patrick Miller.[109] It was.[110]

Burns's social novelty had worn off. He was reluctant to settle to farming and hopeful that the 'whunstane [hard-hearted] gentry' of Edinburgh, who had let poor Fergusson starve, might do something for him.[111] Meanwhile, he toyed again with the idea of Jamaica, or the military. He petitioned his friends to help him to a position in the Excise. Meanwhile, the news from the west was bad. One of the twins born in 1786 had died, evidently in an accident, and when Jean's new pregnancy was discovered, she was turned out of her father's house. Burns's chief pleasure was the *Scots Musical Museum*. From his two attic rooms above the New Town he wrote circular letters appealing for contributions with all the enthusiasm that Walter Scott later brought to his search for Border ballads. 'I have been absolutely crazed about it,' he wrote to the songwriter John Skinner, 'collecting old stanzas, and every information remaining respecting their origin, authors &c &c.'[112]

On 4 December 1787 at a tea-party at a Miss Nimmo's he met Mrs Agnes Craig McLehose, who as 'Nancy' and 'Clarinda' was to transform his last weeks in Edinburgh. Her silhouette by John Miers shows a woman with a generous mouth, a firm bosom, very long eyelashes and a *retroussé* nose. She looks good enough to eat. Married at seventeen and mother of four children by twenty-one, she had separated from her husband in 1780, moved to Edinburgh in August 1782 and lived an existence of precarious respectability in the General's Entry by the Potterrow. Her cousin William Craig, an ambitious lawyer and one of the *Mirror* circle – he praised Michael Bruce – was himself in love with her and watched her with a jealous eye. Her life revolved around the Tolbooth Kirk, where a lecherous prig named John Kemp had succeeded to Webster's pulpit, and bullied her. Nancy McLehose was well-read in the best sentimental authors, and wrote verse. She invited Burns to drink tea with her.

Burns, who was not a man to dawdle where pretty women were concerned, wrote to Mrs McLehose on 6 December, enclosing some verses. Then fate intervened. On 7 December, while leaving a hackney cab when he was not quite sober, Burns injured his knee and was laid up for six weeks. In this period of enforced idleness he wrote daily or even twice daily to Mrs McLchose. Over ninety letters of this correspondence survive, more than fifty of them from Burns. More than any other documents of the eighteenth century they breathe the triumphs and absurdities of the sentimental life. Alarmed by the intensity of this literary emotion, and to create a screen between them, Mrs McLehose contrived for them pastoral personalities, 'Clarinda' and 'Sylvander'. By the end of the month, they had worked one another up to a fine pitch of Arcadian passion.

On 31 December Burns was able to limp round the corner to a Jacobite dinner in Cleland's Gardens with Ruddiman's daughter and son-in law, James Steuart. There he recited his treasonous poem, 'Afar th' illustrious Exile roams', in honour of Prince Charles Edward Stuart, mouldering away the last month of his life in Rome. On 4 January 1788 Sylvander and Clarinda met for the second time.

Burns had no difficulty matching Nancy in the Mackenziean language of virtue and religion: 'Clarinda, my life, you have wounded my soul,' he wrote on 25 January after a passionate visit to the Potterrow. 'If in the moments of fond endearment and tender dalliance, I perhaps trespassed against the *letter* of Decorum's law; I appeal, even to you, whether I ever sinned in the very least degree against the *spirit* of her strictest statute.'[113] Whether that was the case, the affair soon started to attract notice. Colin MacLaurin's son John, newly raised to the bench as Lord Dreghorn and a cousin on the Craig side, cut Nancy in Parliament Square. When she confessed to the Reverend Mr Kemp that she had 'conceived a tender impression', either he or Craig wrote her an admonition which she sent on to Burns. It found him on 13 February in the midst of a dinner party, and he was enraged: 'the half-inch soul of an unfeeling, cold-blooded, pitiful Presbyterian bigot cannot forgive anything above his dungeon-bosom and foggy head.'[114]

Years later, when he had outgrown his juvenile passion for *The Man of Feeling*, Burns wrote to Mrs Dunlop: 'Do not you think,

Madam, there may be a purity, a tenderness, a dignity, an elegance of soul, which are of no use, nay, in some degree, absolutely disqualifying, for the truly important business of making a man's way into life.'[115] As for women, it was a turn of mind 'which may render them eminently happy – or peculiarly miserable'.[116] The Clarinda correspondence might have been designed in sentimental illustration. Early in March Jean gave birth to another pair of twins, but neither lived out the month. That, not Arcadia, was Burns's reality.

Burns left Edinburgh on 18 February. On 13 March he signed a lease with Patrick Miller for the 'farm of Ellisland, on the banks of the Nith, between five and six miles above Dumfries'. He wrote to his old friend Margaret Chalmers: 'I begin at Whitsunday to build a house, drive lime &c.'[117] Back in Edinburgh, he began 'racking shop accounts with Mr Creech' and 'was convulsed with rage a good part of the day'.[118] (The account was not finally settled till a year later, with Burns receiving 'about 440 or 450£'.)[119] Burns left Edinburgh on 24 March, and the following month he wrote to his friend James Smith, who was in the cloth trade, about a printed shawl for 'Mrs Burns'. He had 'lately and privately' given Jean a 'matrimonial title to my corpus'.[120] Both Clarinda and Mrs Dunlop were furious. By the following April, Mackenzie had discovered the German dramatist Heinrich Schiller in French translation and was mad for strong, even grotesque effects.[121] Original Genius was now *passé*.

A harder personality, a careerist such as MacPherson, would have used his Edinburgh celebrity as a spring-board to London, or even Paris. Burns was another type of man. He saw that for a while he had embodied the city's sentimental preoccupations: natural genius, rusticity, good looks, democracy (sort of), picturesque Scottishness, songs, harmless Jacobitism, sex. (The Duchess of Gordon, who entered this story riding down the High Street on a pig, leaves it 'carried ... off her feet' by the conversation of Robert Burns.)[122] He also saw that he was not able to fulfill the town's wishes to its satisfaction, or his own; that he was a town fashion – like those other novelties of 1786 that are now hopelessly entangled with him, haggis and the pianoforte and the portrait silhouette and the hot-air balloon; that the men were snobs and the women teases, and that sentiment and politeness cloaked hard realities of power; and that the place had destroyed his gift. Burns's two winters, though they are the climax of Edinburgh's moment in the eighteenth century, are also its end.

Epilogue

> *'We can never be sure we assume the principles that really*
> *obtain in nature; and that our system, after we have composed it*
> *with great labour, is not mere dream and illusion.'*
> Colin MacLaurin[1]

From his old age in the middle of the nineteenth century, the liberal judge Henry Cockburn remembered seeing the last of the philosophers walking together under the trees of The Meadows.

Cockburn, who was born in 1779 and grew up in a house on the south side of The Meadows, could not have seen David Hume, already three years dead; but looking out from his window or lounging with his friends on the way back from the High School, he had had pointed out to him Principal Robertson and Dr Ferguson, 'the philosopher from Lapland', and Joseph Black and the God-like Dr Alexander Carlyle.

'I knew little then of the grounds of their reputation,' Cockburn wrote in memoirs published two years after his death in 1856. 'We knew enough of them to make us fear that no such other race of men, so tried by time, such friends of each other and of learning, and all of such amiable manners and such spotless characters, could be expected soon to arise, and again ennoble Scotland.'[2]

Thus grew up the notion of a 'Scottish (or 'Edinburgh') Enlightenment'.[3] For liberals such as Cockburn, the dying generation was a race of intellectual heroes who had destroyed the tyranny of dilapidated gods. Their assaults on superstition went right through their target. God himself, though granted a sort of pensionary ontology, was shooed away from the important departments of social life. In demanding that experiment not inherited truth define the business of living, the Edinburgh philosophers stamped the West with its modern scientific and provisional character. They created a world that tended towards the egalitarian and, within reason, the democratic. Their prestige in English-speaking lands was carried on the wave of British and American expansion in to every corner of the world.

It was as if these men knew they were acting the part of philosopher-heroes. 'In judging another man's life,' wrote the Renaissance *savant* Michel de Montaigne, 'I always inquire how he behaved at the last.'[4] David Hume had shown his friends how to die, setting off, as Joseph Black put it, 'in such a happy composure of mind, that nothing could exceed it'.[5] Adam Smith followed in 1790, murmuring to James Hutton and the others in the room: 'I believe we must adjourn this meeting to another place.'[6] On 26 November 1799 Dr Black set his cup of milk and water on his knee, and died: 'as if', Adam Ferguson wrote, 'an *experiment* had been required to shew to his friends the facility with which he departed'.[7]

Yet they left to Scotland a paradox. When, in the case of *Joseph Knight* v. *John Wedderburn*, the Court of Session ruled on 15 January 1778 that a master could not enforce the servitude of a slave in Scotland, the philosopher John Millar felt vindicated: 'This last decision ... is the more worthy of attention, as it condemns the slavery of the negroes in explicit terms, and, being the first opinion of that nature delivered by any court in the island, may be accounted an authentic testimony of the liberal sentiments entertained in the latter part of the eighteenth century.'[8]

If so, that year was the high-water mark of Scots liberalism for some long time. In the course of 1779 there were street disturbances in Edinburgh against an attempt by Lord Advocate Henry Dundas to relieve Roman Catholics of some of the civil disabilities they had laboured under since the Reformation.[9] The Moderates in the General Assembly lost their nerve, Dundas withdrew, and the stage was set for much more serious anti-Catholic disturbances in London the following year, the so-called Gordon Riots. Popular Whiggism died hard.

With the exception of Millar and Boswell, the men of the 'Enlightenment' had been inoculated by the Forty-five against any but the most gradual political change. With their armoury of reason, politeness, patronage and sentiment, the Edinburgh *literati* were as ill-equipped to grapple with the French Revolution as they had been with the American. In the course of the 1790s, under the threat of invasion or radical uprising, intellectual horizons narrowed and liberal sentiment was crushed. The excesses of the Terror caused ranks to close in Scotland, and romantic radicals such as Robert Burns to recant.[10] (In truth, his Jacobinism was no less poetical than his Jacobitism.) There was naked menace in the remarks of a Lord of Session to Dugald Stewart: 'The triumphs of philosophy and reason, daily exhibited in France, ought to have satisfied every thinking and every virtuous man of the dangers of unhinging established institutions.'[11] John Robison took to spinning paranoid conspiracy theories involving Freemasons and *Illuminati*: 'Illumination turns out to be worse than darkness.'[12] Scotswomen were threatened with vague and evil consequences to their position in society if they dared to imitate their sisters in France. Such was the witch-hunt that Eliza Fletcher, wife of the patriot Archibald Fletcher, was rumoured to have set up a guillotine in her yard to practise on small animals against the time of the imposition of 'French principles'.[13] By 1790, the printer William Smellie was writing to London that 'nothing in the way of literature is going on here'.[14]

Edinburgh became notorious for transporting men on the slightest evidence of sedition. At the trials that arose from the

British Convention of the Delegates of the People, convened at Edinburgh in 1793 to press for universal suffrage, Lord Braxfield invented a crime of unconscious sedition: 'endeavoring to create a dissatisfaction in the country ... will very naturally end in overt rebellion: if it has that tendency, though not in the view of the parties at the time ... I apprehend that that will constitute the crime of sedition to all interests and purposes.'[15] At reactionary dinner tables, people quoted with approval his feral growl: 'Let them bring me prisoners, and I'll find them law.'[16] There was a craze for volunteer soldiering.

Scotland was one great big pocket-borough, in which fewer than four thousand voters represented a population of over one million. The pocket was that of Henry Dundas, who became Home Secretary of the United Kingdom in 1791 and delivered to Pitt the Younger's government political control unprecedented even by eighteenth-century standards. To a long-overdue movement to reform the Edinburgh Town Council and the other burghs, Dundas's retort was: 'The fact, indeed, is that the abuses are merely imaginary, and the Scottish nation does not feel them to exist.'[17] At its peak, in the Parliament of 1796–1801, the Dundas 'interest' controlled thirty-six out of forty-five Scottish Members of Parliament. The Scottish campaign for burgh and parliamentary reform never recovered, and when reform came at last, in 1832, it was imposed (as in 1998) from London.

The Edinburgh that emerged at the Peace of Amiens in 1803, the town of Walter Scott, Francis Jeffrey and *The Edinburgh Review*, had lost its innocence. It was clever, caustic, legal, economical, and party-spirited. The broad stream of a humane Scots intellect had branched into the specialised channels of medicine, geology, political economy, finance, military science, engineering. The old Edinburgh had disappeared as comprehensively as the old trees that fell to the feuars of the New Town or the corncrakes that Cockburn used to hear on summer evenings at the mouth of Charlotte Square.[18] For him, the 1780s were the last 'purely Scotch age'.[19]

The new Scotland required a new myth.

In describing the proclamation of King James III and VIII at the Cross in Edinburgh on 17 September 1745, the author of the Woodhouselee manuscript was not sure how to characterise the event. What words would best embrace the Highlanders with their rusty arms and scythes, the heralds in their costumes, the Persian carpet, Mrs Secretary Murray on her prancing horse and the beetling apartment blocks full of people, the grand manifestos of liberty and taxation? 'I saw,' he wrote, 'from a window near the Cross, north syde of the High Streeat, this commick fars or tragic comedy.'[20]

For the new Edinburgh, Scottish history was not quite comedy, nor yet was it tragedy of the sort attempted by John Home. Walter Scott hit on the approach that Cervantes had brought to bear on the superannuated feudalism of seventeenth-century Spain: the past was to be Romantic, twisted a little out of true, sweet, grotesque, no longer shameful or sad. What MacPherson had tried with his Ossian in the Highlands, Scott achieved with his heroes and heroines in the Lowlands and Edinburgh. Starting in 1805, hence the subtitle *'Tis Sixty Years Since*, Scott began work on the romance of the Forty-five that became known as *Waverley*.

Scott was not especially interested in the philosophers, except for their eccentricities. He regaled his respectable generation with tales of Adam Smith chomping lump sugar off his housekeeper's knee, or Dr Black and Dr Hutton recoiling from a dish of snails.[21] What he saw in the streets of Edinburgh was numberless opportunities for myth-making: or, as Stevenson put it with his usual lucidity, for Scott every place 'seemed each to contain its own legend ready made, which it was his to call forth'.[22]

Scott's narrative poem *The Lady of the Lake*, which appeared in May of 1810, brought a deluge of tourists to Loch Katrine.[23] *Waverley*, published in 1814, did the same for Edinburgh, and in recognition the Edinburgh and Glasgow Railway later named its railway station in the capital the Waverley Station. In *The Heart of Midlothian*, his story of the Porteous Riots that appeared in 1818, the town became a character in its own right. 'The reception of this tale in Edinburgh', Lockhart wrote, 'was a scene of all-engrossing enthusiasm, such as I never witnessed there on the appearance of any other literary novelty.'[24]

Scott catapulted Scotland and its ancient capital into the romantic centre of the world, where they have stayed more or less ever since. After the peace that followed the end of the Napoleonic Wars in 1815, English and Continental travellers compulsively visited Edinburgh Castle, Holyroodhouse, Scott country, Burns country, Glencoe, Culloden Moor. King George IV, who went in state to Edinburgh in 1822, left Scotland with the impression that it was inhabited solely by clans and clansmen. On the eve of his departure, Sir Robert Peel wrote to Scott: 'The King wishes to make you the channel of conveying to the Highland chiefs and their followers, who have given to the varied scene which we have witnessed so peculiar and romantic a character, his particular thanks for their attendance, and his warm approbation of their uniform deportment.'[25]

As the architect of modern Edinburgh, Walter Scott outdid even Provost Drummond. Through him it is possible to experience the history of Edinburgh twice: once as disaster, and once as daydream.

Notes

Prologue

1. Adam Fergusson, *An Essay on the History of Civil Society*, ed. Fania Oz-Salzberger (Cambridge, 1995), p. 103.
2. 'The Making of Modern Scotland' in *Some Eighteenth Century Byways and Other Essays* (Edinburgh and London, 1908), p. 113.
3. Tobias Smollett, *The Expedition of Humphry Clinker:* Matt Bramble to Dr Lewis, Manchester, 15 Sept.
4. Patrick Walker, 'Some Remarkable Passages of the Life and Death of Mr Richard Cameron, Late Minister of the Gospel' (1727), in *Biographia Presbyteriana* (Edinburgh, 1827), vol. I, p. 11, and Alexander Carlyle to Sir Gilbert Elliot, 29 July 1761: NLS 110155: 106.
5. Benjamin Franklin to Lord Kames, London, 3 January 1760, in Alexander Tytler, Lord Woodhouselee, *Memoirs of the Life and Writings of the Hon. Henry Home of Kames* (Edinburgh, 1807), vol. I, p. 267. He may have found the evening disputatious, for he wrote in his *Autobiography* (New Haven, 1964): 'Persons of good Sense . . . rarely fall into [disputation] except Lawyers, University Men, and Men of all Sorts that have been bred in Edinborough': p. 60. He made a second visit in 1771.
6. The article, by the Chevalier de Jaucourt, consists of fifty lines of a single column and includes this sentence: 'Elle a été redoutable tant qu'elle n'est pas été incorporée avec l'Angleterre; mais comme dit M. de Voltaire, un état pauvre, voisin d'un riche, devient vénal à la longue, & c'est aussi le malheur que l'Écosse éprouve': *Encyclopédie*, ed. Diderot and d'Alembert (Paris, 1755), p. 351.
7. Woodhouselee, *Life of Kames*, vol. II, Appendix 3.
8. 'Je fis la connaissance de ces hommes respectables par leur génie, lumières et par les moeurs. Également étrangers aux prétentions des petits génies et de l'envie, ils vivaient entre eux en frères qui s'estiment et s'aimaient': *Mémoires de la Princesse Dashkaw, d'après le manuscrit revu et corrigé par l'auteur* (Moscow, 1891), pp. 169–171.
9. In material added to the fifth edition of *England und Italien*, printed in Leipzig in 1787, von Archenholz wrote: 'Die Schottländer haben ihren reichern Nachbarn in den Wissenschaften den Rang abgelaufen. Man findet mehr wahre Gelehrsamkeit und grosse litterarische Talente in

Edinburg, als in Oxford und Cambridge zusammen genommen. Robertson, Hume, Home, Smollet, Monboddo, Macpherson, Ferguson, Watson (?) u.s.w. sind die Ehre Schottlands und der Stolz Britanniens in unsern Tagen; ein Zeitalter, wo der grosse Luxus bey den Südbritten die Cultur der Geistesfähigkeiten einschränkt.' *England und Italien* (Heidelberg, 1993), vol. III, p. 224.

10. *Scots Magazine*, vol. 25 (1763), pp. 362–3.

11. Dugald Stewart, 'Dissertation exhibiting the Progress of Metaphysical, Ethical and Political Philosophy since the Revival of Letters in Europe' in *Works of Dugald Stewart*, ed. Sir William Hamilton (Edinburgh and London, 1854–8), vol. I, p. 551.

Chapter 1: Auld Reekie

1. Alexander Webster in his enumeration of 1755 gives a figure for Edinburgh's ten parishes of 31,112 and a further 4,500 for the Canongate: 'An Account of the Number of People in Scotland in the Year One Thousand Seven Hundred and Fifty Five' in *Scottish Population Statistics*, ed. J.G. Kyd (Edinburgh, 1952), p. 14. William Maitland, a hair merchant turned topographer, in his *History of Edinburgh* (Edinburgh, 1752), p. 220, estimates the population of the city and suburbs in 1747 at 50,120. That is worked back from the median mortality in the parishes multiplied, according to 'the latest discoveries in political Arithmetick', by 28. W.B. Blaikie gives a figure of 40,000 in 'Edinburgh at the Time of the Occupation of Prince Charles' in *The Book of the Old Edinburgh Club*, vol. II (1909).

2. Hugo Arnot, *The History of Edinburgh* (Edinburgh and London, 1779), p. 597.

3. John Ramsay of Ochtertyre reports that very few commoners had incomes over £500–800 at the Union: *Scotland and Scotsmen in the Eighteenth Century* (Edinburgh and London, 1888), vol. II, p. 46.

4. These were the four divisions of St Giles and the two of Greyfriars; Trinity College Church; Christ Church or the Tron; Lady Yester's; and St Cuthbert's or the West Kirk and the Canongate Church outside the walls.

5. Arnot, *History of Edinburgh*, p. 438.

6. Walter Scott, *The Heart of Midlothian* (Edinburgh, 1883), Chapter 28. As late as 1755, the Edinburgh and London mails became mixed up and were sent back to where they had come from: E. Dunbar, *Social Life in Former Days, Chiefly in the Province of Moray* (Edinburgh, 1865), p. 34.

7. Thomas Somerville, *My Own Life and Times, 1741–1814* (Edinburgh, 1861), p. 357. William Creech, in *Letters addressed to Sir John Sinclair, Bart. Respecting the Mode of Living, Arts, Commerce, Literature,*

Manners etc. of Edinburgh in 1763, and since that Period (Edinburgh, 1793), p. 10, says that as late as 1763 the journey took from twelve to sixteen days.

8. The original 'sett' or Decreet Arbitral of 22 April 1583 was revised in 1658 to restrict the magistrates' term of office to two years. Rescinded in 1664, the limit was restored in 1673: *The Sett or Decreet Arbitral of King James the 6th . . . anent the Government of the City of Edinburgh* (Edinburgh, 1683), BL 510 a.40.

9. *A Push for Liberty: in an Address to the Well-Affected Citizens of Edinburgh, concerning the Sett* (Edinburgh, 1746), BL c115 i.3 No. 97.

10. Henry Cockburn, *Memorials of His Time* (Edinburgh, 1910), pp. 87–8.

11. Robert Chambers, *Traditions of Edinburgh* (Edinburgh, c. 1910), p. 1.

12. Daniel Defoe, *A Tour thro' the Whole Island of Great Britain divided into Circuits or Journies*, reprint of 1724–6 edition (London, 1968), vol. II, p. 711.

13. 'Et me jamdadum vexabant *streetia* nostra, ut *coaches* et *chaisi* penderent semijacentes': from William Smellie's macaronic poem of 1786 or 1787, 'Streetum Edinense', in *Memoirs of the Life, Writings, and Correspondence of William Smellie*, ed. Robert Kerr (Edinburgh, 1811), vol. II, p. 232.

14. Boswell says thirteen: F.A. Pottle and C.H. Bennett, eds, *Boswell's Journal of a Tour to the Hebrides with Samuel Johnson, LL.D, now first published from the original manuscript* (London, 1936), p. 25. They replaced buildings of *fifteen* storeys that were destroyed by fire in 1700: Chambers, *Traditions*, pp. 110–11.

15. Oliver Goldsmith to Revd Thomas Contarine, Leyden (n.d. but probably 1754), printed in John Forster, *The Life and Times of Oliver Goldsmith* (London, 1854), vol. I, p. 452.

16. Chambers, *Traditions*, p. 276.

17. W.B. Blaikie gives a list of the occupants of a good land just after 1745. 'First-floor, Mr Stirling, fishmonger; second, Mrs Urquhart, lodging-house keeper; third-floor, the Countess Dowager of Balcarres; fourth, Mrs Buchan of Kelloe; fifth flat, Misses Elliot, milliners and mantua-makers; garrets, a great variety of tailors and other tradesmen': in 'Edinburgh at the time of the Occupation' (see n. 1), p. 2.

18. 'General Account of Edinburgh' in *Provincial Antiquities and Picturesque Scenery of Scotland* (London, 1826), vol. I, p. 73.

19. To John Ballantine, Edinburgh, 14 January 1787, in J. De L. Ferguson, ed., *The Letters of Robert Burns* (Oxford, 1931), vol. I, pp. 77–8.

20. Mr Bruce of Kennet in Chambers, *Traditions*, p. 3. At the end of the seventeenth century, Lord President Dalrymple paid just £100 Scots (£8 6s. 8d. sterling) for his accommodation, and ate roast meat twice a week: Ramsay of Ochtertyre, *Scotland and Scotsmen*, vol. II, pp. 97–8.

21. John Richardson, 'Some Notes on the Early History of the Dean Orphan Hospital' in *Book of the Old Edinburgh Club*, vol. 26 (1949), p. 163.

22. H.D. Erlam, ed., 'Alexander Monro, primus', in *University of Edinburgh Journal*, vol. 17 (1954), p. 103.

23. Letter of 29 March, 1775 in Edward Topham, *Letters from Edinburgh, Written in the Years 1774 and 1775* (Dublin, 1780), vol. II, p. 33.

24. Sir William Forbes of Pitsligo, *Memoirs of a Banking-House* (London and Edinburgh, 1860), p. 4.

25. *Scots Magazine*, vol. 18 (1756), p. 524.

26. Ramsay of Ochtertyre, *Scotland and Scotsmen*, vol. II, p. 81.

27. 6 January, 1738: entry in George Drummond's Diary, EUL DC 1.82, vol. II, fol. 1.

28. J.H. Burton, *Lives of Simon Lord Lovat and Duncan Forbes of Culloden* (London, 1847), p. 303.

29. *Memoirs of the Life of Sir John Clerk of Penicuik, Baronet, 1676–1755*, ed. John M. Gray (Edinburgh, 1892), p. 212.

30. 'My present misfortune is occasioned by drinking. Since my return to Scotland I have given a great deal too much in to that habit, which still prevails in Scotland': Boswell to William Temple, probably 8 March 1767, in *Boswell in Search of a Wife*, ed. F. Brady and F.A. Pottle (New York etc., 1956), p. 38.

31. 'The dress both of men and women alike in the middle and higher ranks exhibited by turns the extremes of gaudy ostentation and disgusting slovenliness': Somerville, *My Own Life* (see n. 7), p. 327.

32. William Creech, *Letters to Sir John Sinclair* (see n. 7), p. 17.

33. 'The Plaid is the Undress of the Ladies, and to a genteel Woman, who adjusts it with a good Air, is a becoming Veil . . . It is made of Silk or fine Worsted, chequered with various lively Colours, two breadths wide, and three Yards in Length; it is brought over the Head, and may hide or discover the Face, according to the Wearer's Fancy or Occasion; it reaches to the Waist behind; one Corner falls as low as the Ancle on one Side; and the other part, in Folds, hangs down from the opposite Arm': Edward Burt, *Letters from a Gentleman in the North of Scotland to his Friend in London* (London, 1754), vol. I, p. 100–1.

34. Chambers reports that a drunken old soldier named Patullo tried to move in to the house in the 1770s, but moved out after a single night: *Traditions*, p. 35.

35. Ibid., pp. 135–6.

36. Ramsay of Ochtertyre, *Scotland and Scotsmen*, vol. I, p. 58. The place was pronounced by Fergusson to rhyme with 'decreed us': 'Hear then, my Bucks! how drunken fate decreed us/For a nocturnal visit to the *Meadows*': 'Epilogue spoken by Mr Wilson, at the Theatre-royal, in the

character of an Edinburgh Buck' in *The Poems of Robert Fergusson*, ed. M.P. McDiarmid (Edinburgh and London, 1956), vol. 2, p. 133. Formerly the site of the Borough Loch, The Meadows were drained in 1722 by Thomas Hope, and hence were sometimes known as Hope Park. According to Alexander Kincaid in his *History of Edinburgh, from the Earliest Accounts to the Present Time* (Edinburgh, 1787), p. 100, Hope made a circular walk of some 2,770 yards in length, 24 feet broad, enclosed on each side by a hedge, a lime avenue and a canal of 9 feet wide. With the residential development of the mid century, such as George's Square, the canal received waste which was 'disagreeable, as well as prejudicial to health, the level from east to west being small'.

37. Scott, *Redgauntlet* (Edinburgh, 1882), p. 338.

38. Even after the Cross had been removed by the Council, to much regret, in 1756 (13 March), it remained the resort for business and society. 'In those days,' wrote the banker Sir William Forbes of Pitsligo of the 1760s and 1770s, 'it was the custom for the merchants and bankers in Edinburgh, to assemble regularly every day at one o'clock at the Cross, where they transacted business with each other, and talked over the news of the day; and as there were among the merchants at that time – I speak of the period before [the slump of] 1772 – several gentlemen of a literary turn, and possessed of considerable powers of conversation, we were joined by many who had no concern in the mercantile world, such as physicians and lawyers who frequented the Cross nearly with as much regularity as the others for the sake of gossiping and amusement merely.' Forbes of Pitsligo, *Memoirs of a Banking-House*, p. 26.

39. James Jamieson, 'The Edinburgh Street Traders and Their Cries' in *The Book of the Old Edinburgh Club*, vol. 2 (1909), pp. 117 222.

40. James Ballantine, *The Gaberlunzie's Wallet* (Edinburgh, 1843), p. 8.

41. 'Behold there is nurro geaks [??] in the whole kingdom, nor any thing for poor sarvants, but a barrel with a pair of tongs thrown a-cross; and all the chairs in the family are emptied into this here barrel once a-day; and at ten o'clock at night the whole cargo is thrown out of a back windore that looks into some street or lane, and the maids call *gardy loo* to the pasengers which signifies *Lord have mercy upon you!*' Smollett, *Humphry Clinker*, 18 July.

42. Burt, *Letters from a Gentleman*, vol. 1, p. 21.

43. Ibid., pp. 22–3.

44. John Buchan, *Sir Walter Scott* (London, 1932), p. 12.

45. 'The constant flow of information and liberality from abroad, which was thus kept up in Scotland in consequence of the ancient habits and manners of the people, may help account for the sudden burst of genius etc.' *Dugald Stewart, Works* (see Prologue, n. 11), vol. 1, p. 551.

46. The Revd John Lee, Principal of Edinburgh University from 1840 to

1859, listed errors such as 'righteousness' for 'unrighteousness' and such examples of bad type as '&adamselcameuntohim': *Memorial for the Bible Societies in Scotland* (Edinburgh, 1824), pp. 152 and 166.

47. *Scots Magazine*, vol. 30 (1768), p. 114.

48. Henry Marchant, quoted in D.B. Horn, *A Short History of the University of Edinburgh* (Edinburgh, 1967), p. 79.

49. Alexander Henderson, *An account of the life and character of Alexander Adam*, (Edinburgh, 1810), pp. 16–17.

50. A Town Council Minute of 1713 complained that 'through the want of professors of physick and chymistry in this Kingdom the youth . . . have been necessitate to travel . . . to the great prejudice of the nation.'

51. In 1713, the Ruddimans' associate James Watson made a premature announcement of a revival in letters in Edinburgh. In 'The Publisher's Preface to the Printers of Scotland' attached to a translation of Jean de la Caille's 1689 *Histoire de l'Imprimerie et de la Librairie*, he wrote: 'Since we are, I trust, all of us honest men, and of better spirits than to propose the Earning our Bread as the chief and only End of our Labour; I entertain a settled well grounded Hope, that [La Caille] will inspire us all with a noble and generous Emulation of equalling, nay, exceeding, if we can, the best Performances of our laudable Ancestors in the Employment. That since our Native Country has at present as many good Spirits, and Abundance of more Authors than in any former Age; we may make it our Ambition, as well as it is our Interest and Honour, to furnish them with Printers that can serve them so well, that they need not . . . go to other Countries to publishe their Writings.' *The History of the Art of Printing &c* (Edinburgh, 1713), pp. 5–6.

52. Richardson, 'Notes on the Orphan Hospital' (see n. 21), pp. 156–7.

53. *The Gentleman's Magazine*, vol. 15 (1745), p. 687.

54. Robert Chambers, *Domestic Annals of Scotland from the Reformation to the Rebellion of 1745* (Edinburgh and London, 1885), p. 392.

55. Arnot, *History of Edinburgh*, pp. 599–600.

56. Adam Petrie, *Rules of Good Deportment, or of Good Breeding for the Use of Youth* (Edinburgh, 1720), p. 86.

57. 'There was then no drawing-room, the lady's bedroom serving the purpose of a drawing-room, to which the company resorted after dinner, if there was a dinner, which was very seldom': *The Anecdotes and Egotisms of Henry Mackenzie*, ed. H.W. Thompson (Edinburgh, 1927), p. 63.

58. Robert Chambers, *The Threiplands of Fingask: A Family Memoir* (London and Edinburgh, 1880) p. 58.

59. Mackenzie, *Anecdotes and Egotisms*, p. 59 and pp. 79–80.

60. Ramsay of Ochtertyre, *Scotland and Scotsmen*, vol. 1, p. 6.

61. Thomas Mathison, 'The Goff', 1743. Both 'Forbes' and 'recreates' are disyllables.

62. Mackenzie, *Anecdotes and Egotisms*, p. 90.

63. Margaret, Countess of Panmure to Earl of Panmure, 24 January 1723, in NAS Dalhousie Muniments, GD 45/14/220.

64. David Hume, *The History of England from the Invasion of Julius Caesar to the Revolution in 1688* (Indianapolis, 1983), vol. 4, p. 123.

65. Alexander Kincaid in his *History of Edinburgh* (see n. 38), p. 63, has this story but gives the honour of throwing the stool to an 'old woman named Hamilton'. He reports her shouting: 'Out, thou false thief, does thou say the Mass at my luggs [ears]?'

66. 'A True Account of the Behaviour and Conduct of Archibald Stewart, Esq., Late Lord Provost of Edinburgh', printed in J.V. Price, *The Ironic Hume* (Austin, Texas, 1965), p. 171.

67. 'The History of the Union of England and Scotland', pp. 20–1, in *The History of the Union of Great Britain* (Edinburgh, 1709).

68. Arnot, *History of Edinburgh*, p. 189.

69. George Lockhart, *Memoirs concerning the Affairs of Scotland from Queen Anne's accession to the throne to the commencement of the Union of the two kingdoms* (3rd edition, London, 1714), p. 224, pp. 228–9.

70. The Borders minister Thomas Somerville recalled the historian William Robertson, Principal of Edinburgh University, telling him of 'the disadvantages our members suffered immediately after the Union. The want of the English language, and their uncouth manners, were much against them. None of them were men of parts, and they never opened their lips but on Scottish business, and then said little. The late Lord Onslow said to him, "Dr Robertson, they were odd-looking dull men. I remember them well."' Somerville, *My Own Life and Times* (see n. 7), p. 271.

71. In his *Autobiography*, ed. J.H. Burton (Edinburgh, 1910), p. 5, Carlyle of Inveresk said the Union had reduced shipping to the Prestonpans district from 20 to 10 vessels in his childhood. Describing a tour of the Borders in 1733, aged 11, he wrote: 'The face of the country was particularly desolate, not having yet reaped any benefit from the union of the Parliaments; nor was it recovered from the effects of that century of wretched government which preceded the Revolution.' Ibid., pp. 28–9.

72. Lockhart, *Memoirs concerning the Affairs of Scotland*, p. 343.

73. That is a sensational allegation, and Lockhart want on to qualify it: 'but this is only a conjecture from some innuendo's I have heard him make': ibid, pp. 71–2.

74. 'The sudden and total dispersion of the rioters, when their vindictive purpose was accomplished, seemed not the least remarkable feature of this singular affair. In general, whatever may be the impelling motive by

which a mob is at first raised, the attainment of their object has usually been only found to lead the way to farther excesses. But not so in the present case. They seemed completely satiated with the vengeance they had prosecuted with such staunch and sagacious activity. When they were fully satisfied that life had abandoned their victim, they dispersed in every direction, throwing down the weapons which they had only assumed to enable them to carry out their purpose.' Scott, *Heart of Midlothian*, p. 59.

75. George Drummond's Diary, EUL DC 1.82, entry for 16 September, vol. I, f. 46.

76. Ibid., f. 169.

77. 10 Geo II c. 35.

78. Scott, *Heart of Midlothian*, p. 160. Robert Wallace, who saw both sides of every question, reported that ministers of the gospel could not reconcile the Government's demands with the Act of Security and did 'not think it becomes them as they are ministers of the gospels of peace . . . to publish severe laws from the pulpit': 'Small Pieces concerning Porteous's Act', EUL La II 620/4, f. 15.

79. Carlyle, *Autobiography* (see n. 71), p. 46.

80. John Home, *History of the Rebellion in the Year 1745*, in *The Works of John Home, Esq.*, ed. H. Mackenzie (Edinburgh, 1822), vol. 3, p. 72.

81. Carlyle, *Autobiography*, p. 242. Of the Jacobite poet Robertson of Strowan, Ramsay of Ochtertyre wrote: 'He was a cavalier in the purest sense of the word, having engaged in every rebellion that took place between 1689 and 1746.' On the last occasion – he was 75 – he was encouraged to go home after Preston with Sir John Cope's chain and furred nightgown: *Scotland and Scotsmen*, vol. 1, pp. 31–4.

82. Carlyle, *Autobiography*, p. 122 and p. 143.

83. 'Correspondance inédite du marquis d'Éguilles, d'après les originaux autographes conservés aux archives des Affairs Etrangères et à la bibliothèque de l'Arsenal' in *Revue Rétrospective*, vol. 3, July–December 1885 (Paris, 1886), p. 146.

Chapter 2: Charlie's Year

1. Robert Louis Stevenson, *Edinburgh: Picturesque Notes*, Edinburgh, 1879, pp. 32 and 23.

2. In James Craig's plan for the New Town, the area lying between Princes Street and the town is shown as an open space. The owners of the houses on Princes Street west of Hanover Street (or, rather, the 'feuars' of the land, under the feudal system of land tenure that existed until recently in Scotland) opposed any building on the south side and, in 1813, per-

suaded the Town Council to abandon any right to develop it. The 1816 Act, which gave statutory effect to the agreement, prohibited any building to the south of the street except on the earth Mound, the St John's Chapel, and a gardener's cottage, greenhouses and conservatories. In return, the proprietors were to enclose and embellish the ground. Gardens were laid out by the proprietors and originally retained for their private use; but it became increasingly hard to exclude the public – or, in the course of the 1840s and after some opposition, the Edinburgh and Glasgow Railway Company. The gardens reverted to the Town Council by Act of Parliament in 1876. David Robertson, *The Princes Street Proprietors and other chapters in the History of the Royal Burgh of Edinburgh* (Edinburgh and London, 1935), pp. 1 61.

3. Princes Street is named not for Charles Edward, of course, but for the Prince of Wales and Regent, the future George IV. Its best time was probably during the Regency in the early years of the nineteenth century, when the New Town style of life still had some novelty for the people of Edinburgh and the 'general promenade' on Princes Street was a sort of cold-weather edition of a Mediterranean *passeggiata*. Thomas Carlyle describes a Saturday in April, 1814: 'There, on a bright afternoon, in its highest bloom probably about 4–5 p.m., all that was brightest in Edinburgh seemed to have stept to enjoy, in the fresh pure air, the finest city-prospect in the world and the sight of one another. From Castle Street or even the extreme west there was a visible increase of bright population, which thickened regularly eastward, and in the sections near the Register Office or extreme east, had become fairly a lively crowd, dense as it could find stepping-ground – never needed to be denser, or to become a crush, so many side-streets offering you free issue all along, and the possibility of returning by a circuit, instead of abruptly on your steps. The crowd was lively enough, brilliant, many coloured, clever-looking (beautiful and graceful young womankind a conspicuous element); crowd altogether elegant, polite, and at its ease though on parade; something as if of unconsciously rhythmic in the movements of it, as if of harmonious in the sound of its cheerful voices, bass and treble, fringed with the light laughters; a quite pretty kind of natural concert and rhythmus of march; into which, if at leisure, and carefully enough dressed (as some of us seldom were) you might introduce yourself, and flow for a turn or two with the general flood. Nothing of the same kind now remains in Edinburgh.' *Reminiscences* (Oxford, 1997), p. 411.

4. Robert Chambers, *Traditions of Edinburgh* (Edinburgh, c. 1910), p. 366.

5. From *The Culloden Papers, Comprising an Extensive and Interesting Correspondence from the Years 1625 to 1748* (London, 1815), pp. 203–4. The Prince disembarked at the farmhouse of Borradale in Arisaig on the mainland on 25 July.

6. Ibid., p. 205.

7. *Memoirs of the Life of Sir John Clerk of Penicuik, Baronet, 1676–1755*, ed. John M. Gray (Edinburgh, 1892), p. 180.

8. The French envoy, Boyer d'Éguilles, wrote to d'Argenson on 6 April 1746: 'Forbes de Culloden . . . est l'homme le plus considérable et le plus considéré du pais, par son crédit à la cour, par sa place et par son mérite. Ses ennemis mêmes le donnent pour un homme doux et modéré, plein d'esprit, de savoir et de probité . . . C'est luy qui, par son éloquence, son activité et ses intrigues, a retenu un tiers de nos amis, a fait agir bien des indifférens et a rassemblé tous nos ennemis. D'ailleurs, il n'a jamais parlé du Prince qu'avec respect et estime.' In 'Correspondance inédite du marquis d'Éguilles, d'après les originaux autographes conservés aux archives des Affairs Etrangères et à la bibliothèque de l'Arsenal' in *Revue Rétrospective*, vol. 3, July–December 1885 (Paris, 1886), p. 143.

9. 'Je passai dans la batterie entre Leith et Kinghorn avec le président de Forbes, et il me dit pour nouvelle que le prince était debarqué en Ecosse et qu'il allait dans le Nord, pour empêcher autant qu'il pouvait les chefs de clans de se joindre à lui, qu'il y en avait qu'il ne pouvait pas empêcher, connaissant leur zèle pour cette cause, dont il etait bien fâché, parce que le prince ne pouvait illuminer qu'un feu de paille qui serait bientôt éteint par les soins du général Cope, et qui finirait par la ruine de plusieurs très honnêtes gens dont il plaignait de sort. Et entre autres il plaignait le duc de Perth and M. Cameron de Lochyell . . .' From 'Extraits de journal de Lord Elcho' in P. Chamley, 'Documents relatifs à sir James Steuart', *Annales de la Faculté de Droit et des Sciences Politiques et Economiques de Strasbourg*, vol. 14 (Paris, 1965), p. 102.

10. *The History of the Rebellion in the Year 1745*, in *The Works of John Home, Esq.*, ed. H. Mackenzie (Edinburgh, 1822), vol. 3, pp. 30–1.

11. *Caledonian Mercury*, 16 September 1745.

12. *The Diary of John Campbell: A Scottish Banker and the Forty-Five* (Edinburgh, 1995), p. 31.

13. Alan Cameron, *Bank of Scotland, 1605–1995: A Very Singular Institution* (Edinburgh, 1995), p. 46.

14. Lord President Forbes to Lord Lovat, 19 September 1745, in *The Culloden Papers* (see n. 5), p. 221.

15. 'From the time that the ships sailed, the people of Edinburgh were continually looking up to the vanes and weather-cocks, to see from what point the wind blew, and computing how soon they might expect the general and his army': Home, *History of the Rebellion*, p. 40.

16. James Hogg, *The Jacobite Relics of Scotland; Being the Songs, Airs and Legends of the Adherents of the House of Stuart*, 2nd Series, vol. 2 (1821), p. 113.

17. 'It is easy to conclude, that such a Situation as this could never be pick'd

out for a City or Town, upon any other Consideration than that of Strength to defend themselves from the suddain Surprizes and Assaults of Enemies': Daniel Defoe, *A Tour thro' the Whole Island of Great Britain divided into Circuits or Journies*, reprint of 1724–26 edition (London, 1968), vol. 2, p. 709.

18. Home, *History of the Rebellion*, pp. 31–2.

19. Memorandum to Tweeddale, possibly by Lord Stair, in *Culloden Papers*, pp. 217–18.

20. *Memoirs of Sir John Clerk* (see n. 7), p. 181.

21. *Memorials of John Murray of Broughton, Sometime Secretary to Prince Charles Edward, 1740–1747*, ed. R.F. Bell (Edinburgh, 1898), p. 195.

22. 'The Citty is surrounded from the Castle upon the South Side till the Cowgate Port, with a pretty high wall flank att particular distance though not att proper ones, the wall is thine and in very bad repair from the Cowgate Port to the nether bow Port, along St Marys wind the houses compose part of the Town wall.' Ibid., p. 196. I have assumed that Murray intended to punctuate after 'repair'.

23. Ibid., pp. 196–7.

24. *Memoirs of Sir John Clerk*, p. 181.

25. The election was due to take place on the first Tuesday after Michaelmas, that is, on Tuesday, 1 October 1745. On Friday, 13 September the trade incorporations – surgeons, goldsmiths, skinners, furriers, hammerers, wrights, masons, tailors, bakers, fleshers, shoemakers, weavers, waukers, and bonnetmakers – each submitted 'leets' of six candidates for office to the outgoing Council. The Council struck out half of them and, on the 14th, selected from the remaining three a deacon, who was to be presented on the following Wednesday, 18 September. As it turned out, no magistrates were elected, and the Provost and Baillies resigned on 1 October, leaving Edinburgh without a city government other than proclamation from Holyrood. The Council's constitution or 'sett' is described in the *Scots Magazine*, vol. 25 (1763), pp. 378ff.

26. So a defence of Provost Stewart attributed to Hume, 'A True Account of the Behaviour and Conduct of Archibald Stewart, Esq., Late Lord Provost of Edinburgh', printed in J.V. Price, *The Ironic Hume* (Austin, Texas, 1965), p. 158. The attribution is by Sir Walter Scott.

27. *Memoirs of Sir John Clerk*, p. 182.

28. 'Mr McLourin's Journall of What Passed relating to the Defence of Edinburgh from Monday, September 2d till Monday September 16, 1745': typed transcription by Walter Blaikie, NLS MS 299, f. 1.

29. The Latin is from the epitaph placed by his son John on the south outside wall of Greyfriars Church.

30. 'It having been found in Tryall that all in the shops had been bought up of late by ladies who had been sent for them.' Op. cit., f. 2.

31. Ibid., f. 3.

32. Ibid.

33. Ibid., ff. 7–9. 'There was a plain collusion,' he wrote to Lord President Forbes, on 9 December 1745: *The Collected Letters of Colin MacLaurin*, ed. Stella Mills (Nantwich, 1982), p. 132. Of the memoirists, Murray of Broughton and Carlyle say Stewart was just a politician, Elcho and Home that he was plotting for the Stuarts. According to Elcho, Provost Stewart was 'a zealous supporter of the Prince, formed the design of causing the arms to fall into the hands of Prince's army, which stood in great need of them, since it was unarmed': *A Jacobite Miscellany: Eight Original Papers on the Rising of 1745–1746*, ed. Henrietta Tayler (Oxford, 1948), p. 144. Murray of Broughton says that half the muskets were useless, and there were not two hundred men who needed them (*Memorials*, p. 198).

34. Ibid., f. 7.

35. W.B. Blaikie, 'Edinburgh at the Time of the Occupation of Prince Charles', in *The Book of the Old Edinburgh Club*, vol. 2 (1909), p. 18.

36. Hogg, *Jacobite Relics* (see n. 16), vol. 2, p. 113.

37. 'Mr McLourin's Journall', f. 8.

38. Alexander Carlyle, *Autobiography*, ed. J.H. Burton (Edinburgh, 1910), p. 125.

39. 'A most unlucky Signall was pitched on to Call them to their Arms, the Ringing of the Fire Bell which never fails to raise a Pannick in Edinburgh. This happened in time of Divine Service, the Churches Dismissed in Confusion and Terror and this was the first appearance of Fear in the place and this signal ought not to have been proposed or allowed by the Magistrats in such a time the Rebells not being far from us': 'Mr McLourin's Journall', f. 10.

40. Carlyle, *Autobiography*, p. 126.

41. *The Woodhouselee MS: A Narrative of Events in Edinburgh and District during the Jacobite Occupation, September to November, 1745, printed from original papers ,in the Possession of C.E.S. Chambers* (Edinburgh, 1907), p. 15.

42. It was not the Gauls but the Veientes: Livy, *Hist.* II, 48–50.

43. Carlyle, *Autobiography*, pp. 126–7. Sir Walter Scott records that one of these civilians, 'a very worthy man, a writing master by occupation, . . . had ensconced his bosom beneath a professional cuirass, consisting of two quires of long foolscap writing paper; and doubtful that even this defence might be unable to protect his valiant heart from the claymores, amongst which its impulses might carry him, had written on the outside, in his best flourish, "This is the body of J— M—; pray give it Christian

burial"': 'Life and Works of John Home', in *Miscellaneous Prose Works* (Edinburgh, 1835), vol. 19, p. 293.

44. *Scots Magazine*, vol. 64 (1802), p. 382.
45. Carlyle, *Autobiography*, p. 128.
46. *Woodhouselee MS*, p. 17.
47. Carlyle, *Autobiography*, p. 129.
48. *Woodhouselee MS*, p. 17.
49. *Scots Magazine*, vol. 64, p. 382.
50. In the end, some 90 of the Town Guard and 363 men of the Edinburgh Regiment did march out to Coltbrigg, but did not engage the enemy there: ibid.
51. Carlyle, *Autobiography*, p. 129–130.
52. 'More men were imployed this day than ever before and everybody seemed to Exert themselves': 'Mr McLourin's Journall', ff. 10–11.
53. According to Murray of Broughton, here was precisely where an attack would be made: 'On the north Side again a strong diversion would have been made by the Phisick garden to render themselves a master of the Sluice upon the north Loch, whille a party attacked the hospital att the foot of Leith wind, and these carried on whille the principal attempt would have been made in St Marys wind by taking possession of the houses on the east side, and setting fire to those on the west.' *Memorials* (see n. 21), p. 197. The town walls had been in a poor state here two centuries earlier: on the night of Darnley's murder, 9 February 1567, Bothwell had tried to slip out of the town over a ruined section in Leith Wynd, but was forced to go back to the Nether Bow and bribe the porter to let him out.
54. Carlyle, *Autobiography*, p. 132.
55. It is attributed to Patrick Crichton, a Canongate saddler and ironmonger, who bought the Woodhouselee estate in 1734 and still owned it at the time of the Forty-five.
56. *Woodhouselee MS*, p. 20. From the Jacobite side, Lord Elcho recorded: 'On the 16 the Princes army march'd on the high road to Edinburgh with a designe to attack the Dragoons but they, whenever they perceived the Highlanders, were struck with such a pannick, that they wheel'd about and galloped away in the greatest confusion, pass'd by the town of Ednr, droping their Bagadge and arms upon the road; and a great many of them never stopped until they gott to Haddington, which is fourteen miles off': David, Lord Elcho, *A Short Account of the Affairs of Scotland in the years 1744, 1745, 1746*, ed. The Hon. E. Charteris (Edinburgh, 1907), p. 255.
57. Home, *History of the Rebellion* (see n. 10), p. 57.
58. Carlyle, *Autobiography*, p. 133.
59. 'Narrative of Sundry Services performed together with an account of

money disposed in the service of Government during the late rebellion' in W.B. Blaikie, *Origins of the Forty-Five* (Edinburgh, 1916), p. 341. Grossett gave evidence for the prosecution of the rebel lords and was promoted to Inspector-General of Customs on the recommendation of the Duke of Cumberland, but was brutally hounded out of Scotland and lost his fortune in mining speculations in Italy. He died in poverty in London in 1760.

60. Carlyle, *Autobiography*, p. 133.

61. 'Two Letters from Magdalen Pringle' in *A Jacobite Miscellany* (see n. 33), p. 38. These letters are so charming, well-written, verifiable and to the historical point that one must question their authenticity; in addition, the first is unsigned.

62. *Scots Magazine*, vol. 7 (1745), p. 436.

63. Home, *History of the Rebellion*, p. 62.

64. Ibid., p. 63.

65. 'Mr McLourin's Journall', f. 12.

66. Carlyle, *Autobiography*, p. 134.

67. In their flight, the dragoons scattered the road with 'pistols, swords, scullcaps etc.' enough to fill a close cart and two creels on horseback the next morning: Carlyle, *Autobiography*, p. 137.

68. Elcho spent much of his life trying to recover it.

69. 'Mr McLourin's Journall', f. 12.

70. This is from the exordium of Fergusson's 'Auld Reikie, A Poem' of 1773: 'On stair wi' TUB or PAT in hand,/The barefoot HOUSEMAIDS looe to stand,/That antrin fock [strangers] may ken how SNELL [sharp]/Auld Reikie will at MORNING SMELL' (ll.33–6), in *The Poems of Robert Fergusson*, ed. M.P. McDiarmid (Edinburgh and London, 1956), vol. 2, p. 110.

71. Elcho, *A Short Account* (see n. 56), pp. 257–61. In contradiction, Murray of Broughton reported that 'a good many' Edinburgh men volunteered for the Duke of Perth's regiment that day: *Memorials*, p. 198.

72. Home, *History of the Rebellion*, pp. 70–1. The description of the Prince compares well with the little portrait Charles Edward himself gave to Cameron of Lochiel which is still in that family's possession.

73. Carlyle, *Autobiography*, p. 142.

74. *A Jacobite Miscellany*, p. 39.

75. *Woodhouselee MS*, p. 26.

76. Margaret Murray was rebuked by the Whigs and anti-feminists, which infuriated her husband. 'The same day the Chevalier was proclaimd over the Market Cross, and the two following declarations read by the heralds in there robs when was present a great Concourse of people of the best fashion in the place, not a few women only, as some of the Grub-street writers on this affair would make believe': Murray of Broughton, *Memorials*, p. 198.

77. *Woodhouselee MS*, pp. 26–7.

78. *Scots Magazine*, vol. 9 (1747), p. 618.

79. Ibid.

80. Ibid., p. 624.

81. *Woodhouselee MS*, pp. 28–9.

82. *A Jacobite Miscellany*, p. 39.

83. Ibid.

84. *The Caledonian Mercury*, 18 September 1745. Ruddiman, born in 1674, began printing the revived *Mercury* in 1724. He had earlier edited Gavin Douglas, Drummond of Hawthornden and George Buchanan, wherein his Jacobite aspersions caused a sensation in 1715. However, *DNB* reports that he spent the Forty-five in retirement. His *Rudiments of the Latin Tongue* (1714) was the standard Latin primer for two generations. He was succeeded by Hume as Keeper of the Advocates' Library in 1752.

85. A Jacobite pamphlet of 20 September 1745 (BL c115 i.3 No. 71) has: 'The Prince clothed in a plain Highland Habit, cocked his Blue Bonet, Drew his Sword, Threw away the Scabard and Said, GENTLEMEN, Follow me, By the Assistance of God, I will, this Day, make you a Free and Happy People.'

86. Carlyle, *Autobiography*, p. 141.

87. Ibid., pp. 149–50.

88. Ibid., p. 151.

89. Ibid., pp. 151–2.

90. Laurence Oliphant the younger, aide-de-camp to the Chevalier, in T.L. Kington Oliphant, *The Jacobite Lairds of Gask* (London, 1870), p. 110. *The Scots Magazine*, vol. 7 (1745), p. 439, gives seven or eight minutes; Carlyle, *Autobiography*, p. 151, 'ten or fifteen minutes'; Lady George Murray, in a letter to her brother-in-law, 'about half an hour', in *Jacobite Correspondence of the Atholl Family, During the Rebellion 1745–1746* (Edinburgh, 1840), p. 22; Hamilton of Bangour, in his poem 'Gladsmuir', one hour . . . et cetera.

91. Dougal Graham, *An Impartial History of the Rise, Progress and Extinction of the Late Rebellion in Britain, in the Years 1745 and 1746* (Glasgow, 1774), pp. 24–6. Graham was by his own account 'an Eye witness to most of the Movements of the Armies from the Rebels first crossing of the Ford of Frew to their final Defeat at Culloden.'

92. Gen. Wightman to President Forbes, 26 September 1745, in *Culloden Papers* (see n. 5), p. 225.

93. Ibid. Reaching the English border, some Dragoon officers quartered themselves for the night in the house where Sir John and Lady Clerk had taken refuge. Woken up, 'we thought Hell had broken louse, for I never heard such Oaths and imprecations branding one another with

Cowardice and neglect of duty': *Memoirs of Sir John Clerk* (see n. 7), p. 187.

94. Hogg, *Jacobite Relics* (see n. 16), vol. 2, p. 114.

95. Lord George Murray to the Duke of Atholl, Edinburgh, 24 September 1745: *Atholl Correspondence* (see n. 90), p. 24. *The Scots Magazine*, vol. 7 (1745), p. 440, gives 500 killed on the government side, 900 wounded and 1,400 other prisoners.

96. Carlyle, *Autobiography*, p. 152.

97. Proclamation of 23 September 1745: BL c115 1.3.

98. The notorious exception was the West Kirk under the Castle guns on the morrow of the battle, where the Revd Neil McVicar prayed: 'Bless the King; Thou knows what King I mean; may the crown sit long easy on his head &c. And for this man that is come amongst us to seek an earthly crown, we beseech Thee, in mercy, to take him to Thyself and give him a crown of glory.'

99. Lord President Forbes to Mr Mitchell, 13 November 1745, in *The Culloden Papers*, p. 250. The bankrupts were presumably led by Murray of Broughton, who was now lending out money at 5 per cent interest, which 'was very Suspicious as everyone knew he was not worth 100pd before the Affair': Elcho, *Short Account* (see n. 56), p. 281.

100. *Memoirs of Sir John Clerk*, p. 187.

101. 'A Portion of the Diary of David, Lord Elcho, 1721–87' in *A Jacobite Miscellany* (see n. 33), p. 148.

102. BL C115 i.3/64.

103. Magdalen Pringle in *A Jacobite Miscellany*, p. 40.

104. *Woodhouselee MS* (see n. 41), p. 55.

105. To hold off a run by the Edinburgh banks the teller of the Arms Bank, founded in Glasgow in 1750, once took 34 working days to count out £2,893 in coin: Alan Cameron, *Bank of Scotland* (see n. 13), p. 55.

106. *Diary of John Campbell* (see n. 12), p. 38.

107. Ibid., p. 39.

108. *Woodhouselee MS*, p. 59. I have repunctuated the passage.

109. Ibid., pp. 58–9.

110. BL c115 i.3/65.

111. Letter of 13 October, *A Jacobite Miscellany*, pp. 40–1.

112. Lord George Murray to the Duke of Atholl, Edinburgh, 11 October 1745: *Atholl Correspondence* (see n. 90), p. 80.

113. Éguilles to the Duke of Atholl, Edinburgh, 15 October 1745: ibid., p. 94.

114. 'Correspondance inédite' in *Revue Rétrospective* (see n. 8), p. 104.

115. 'In his conversations with us, however, he gave us to understand that it was all one to France whether George or James was King of England but that, if the Scotch wished to have a King for themselves, then the King

of France would help them to the utmost of his power. This proposal had many supporters, but the Prince declared himself strongly against it, and said often he would have the three kingdoms or nothing at all': 'Portion of Elcho's Diary' in *A Jacobite Miscellany* (see nn. 33 and 101), p. 148.

116. *Scots Magazine*, vol. 7 (1745), p. 493.

117. J.H. Burton, *Lives of Simon Lord Lovat and Duncan Forbes of Culloden* (London, 1847), p. 382.

118. *A Push for Liberty; in an Address to the Well-Affected Citizens of Edinburgh, concerning the Sett* (Edinburgh, 1746), BL c115 i.3/97.

119. Hugo Arnot, *The History of Edinburgh* (Edinburgh and London, 1779), pp. 444ff.

120. Janet Adam Smith, 'Some Eighteenth-Century Ideas of Scotland' in N.T. Phillipson and Rosalind Mitchison, *Scotland in the Age of Improvement* (Edinburgh, 1970).

121. *A Sermon Preached in the Ersh Language, to His Majesty's First Highland Regiment of Foot, commanded by Lord John Murray, at their Cantonment at Camberwell, on the 18th day of December 1745, being appointed as a Solemn Fast* (London, 1746), p. 9.

122. *The Wrath of Man Praising God: A Sermon Preached in the High Church of Edinburgh, May 18, 1746, before his Grace the Lord High Commissioner* (Edinburgh, 1746), p. 33.

123. 'Congratulatory Letter to HRH William, Duke of Cumberland, May 20, 1746', in *Acts of the General Assembly of the Church of Scotland 1638–1842* (Edinburgh, 1843), p. 686.

124. John Ramsay of Ochtertyre, *Scotland and Scotsmen in the Eighteenth Century* (Edinburgh and London, 1888), vol. 2, p. 87.

125. Lord Justice-Clerk Milton, the deputy governor of the Royal Bank, 'found fault with almost every part of the directors' conduct': entry for Monday, 28 October 1745 in *Diary of John Campbell* (see n. 12), p. 51.

126. 'A True Account' in *The Ironic Hume* (see n. 26), pp. 156. Adam Smith in his Glasgow lectures used the example of the Forty-five in the Lowlands to show the 'bad effect of commerce' on 'the courage of mankind': 'Report dated 1766', in *Lectures on Jurisprudence* (Oxford, 1978), pp. 540–1.

Chapter 3: The Disease of the Learned

1. 'Proceedings against THOMAS AIKENHEAD, for Blasphemy. 8 William III. A.D. 1696' in T.B. Howell, ed., *A Complete Collection of State Trials* (London, 1812), vol. 13, No. 401, col. 919.

2. Ibid., col. 918.

3. Ibid., col. 920.

4. Ibid., cols 925–6.

5. Ibid., col. 925.

6. John Buchan and George Adam Smith, *The Kirk in Scotland 1560–1929* (Edinburgh, 1930), p. 53.

7. Ibid., p. 59.

8. 'His Majesty's gracious Letter to the Assembly, October 17, 1690' in *Acts of the General Assembly of the Church of Scotland 1638–1842* (Edinburgh, 1843), p. 222.

9. Robert Wodrow, *Analecta, or Materials for a History of Remarkable Providences* (Edinburgh, 1843), vol. 3, p. 303.

10. Hugh Arnot, *The History of Edinburgh* (Edinburgh and London, 1779), p. 167.

11. Alison Hay Dunlop, *Anent Old Edinburgh* (Edinburgh and Glasgow, 1890), p. 38.

12. Elizabeth Mure of Caldwell, 'Some remarks on the change of manners in my own time 1700–1790' in *Selections from the Family Papers preserved at Caldwell*, ed. William Mure (Glasgow, 1854), part 1, pp. 265–6.

13. John Ramsay of Ochtertyre, *Scotland and Scotsmen in the Eighteenth Century* (Edinburgh and London, 1888), p. 235.

14. Edward Topham, *Letters from Edinburgh, Written in the Years 1774 and 1775* (Dublin, 1780), pp. 235–6.

15. John Jackson, *The History of the Scottish Stage* (Edinburgh, 1793), p. 418.

16. 'All the villainous profane and obscene books and playes printed at London by Curle and others, are gote doun from London by Allan Ramsey, and lent out, for an easy price to young boyes, servant weemen of the better sort, and gentlemen': Wodrow, *Analecta*, vol. 3, p. 515. Lord Grange, a sanctimonious drunkard, denounced Ramsay to the Town Council but the poet had, according to Wodrow, 'nottice an hour before' and the magistrates found nothing offensive: ibid., p. 516.

17. George Anderson, *The Use and Abuse of Diversions: A Sermon on Luke XIX. 13* (Edinburgh, 1733), p. 22.

18. The four women declared 'that the dread of the pillory was the cause of their murders': Arnot, *History of Edinburgh*, p. 193. Robert Wallace, a leading moderate minister at mid-century, learned the facts of life at the pulpit's foot: 'I remember well that when I was very young I could see a reason why an unmarried woman who had born a child ought to do pennance because she was not married, but I saw no reason why any man should make a publick appearance; att that time I knew nothing of the congressus Venereus. I asked my father's servants on a Sunday's evening after a fornicator had done pennance why he appeared & what he had done, whether he had bid the woman bear the child.' EUL La II 620, 'Of

Venery, or of the Commerce of the 2 Sexes', printed in Norah Smith, ed., *Texas Studies in Literature and Language*, vol. 15 (1973), p. 436.

19. Patrick Walker, 'The Life and Death of Mr Alexander Peden, Late Minister of the Gospel', in *Biographia Presbyteriana* (Edinburgh, 1827), vol. 1, p. 88.

20. Council Register, 24 May 1721, in Arnot, *History of Edinburgh*, p. 204.

21. 'The Young Laird and Edinburgh Katy' by Allan Ramsay senior.

22. 'It was alleged that nothing had incensed [the Presbytery of Glasgow] so much as his inculcating, in the strongest terms, the necessity of moral virtue, without which there could be no real Christianity': Ramsay of Ochtertyre, *Scotland and Scotsmen*, vol. 1, p. 273.

23. 'Mr Simson's Speech to the Presbytery of Glasgow, March 29, 1715' in *The Case of Mr John Simson* (Glasgow, 1715), p. 63.

24. He was allowed to retain his professorship, and a cow he kept on College land.

25. James McCosh, *The Scottish Philosophy* (London, 1875), p. 198. Thomas Blacklock, the blind poet, presented to a country living in the 1760s, was so unpopular that he lost heart, resigned his living and set up as a crammer in Edinburgh.

26. *Acts of the General Assembly* (see n. 8), pp. 634, 640.

27. '. . . whereas by this act [of Patronage], the man with the gold ring and gay clothing is preferred unto the man with the vile raiment and poor attire': J. Fisher, ed., *The Whole Works of the Late Rev. Mr Ebenezer Erskine, Minister of the Gospel at Stirling* (London, 1810), vol. 1, p. 508.

28. George Drummond's Diary, EUL DC 1.82, f. 84.

29. Alexander Carlyle of Inveresk, *Autobiography*, ed. J.H. Burton (Edinburgh, 1910), p. 58.

30. John Thomson, *Life of Cullen* (1859), vol. 2, p. 507.

31. John Cunningham, *The Church History of Scotland, from the Commencement of the Christian Era to the Present Century* (Edinburgh, 1882), vol. 2, p. 368.

32. 'The Kirk,' James Rows preached in St Giles in 1746, 'was a bonny trotting Naig . . . but the Bishops after they gat on her back, Cross-langed her and Hapshackled her': *A Sermon Preached by Mr James Rows in St Geil's Kirk at Edinburgh* (Edinburgh, 1746) (BL 4255 a.60).

33. Ramsay of Ochtertyre, *Scotland and Scotsmen*, vol. 1, pp. 241–2.

34. Wodrow, *Analecta* (see n. 9), vol. 4, p. 129.

35. *The Regard due to Divine Revelation, and the Pretences to it, Considered: A Sermon preached before the Provincial Synod of Dumfries, at their meeting in October, 1729* (London, 1731), pp. 37 and 44 (BL 693 e. 16 (6)).

36. 'Fund for a Provision for the Widows and Children of Ministers of the Church of Scotland', 17 Geo. II.

37. The mathematics is set out in 'Biographical Sketch of the Late Alex. Webster, D.D.', in *The Scots Magazine*, vol. 64 (1802), p. 281.

38. In letters to Wallace of 23 and 24 May 1743: *The Collected Letters of Colin MacLaurin*, ed. Stella Mills (Nantwich, 1982), pp. 105–8.

39. George Drummond's Diary, 5 June 1737: EUL DC 1.82, vol. 1, f. 219.

40. 'Biographical Sketch', in *Scots Magazine* (see n. 37), p. 284.

41. C. Ryskamp and F.A. Pottle, eds, *Boswell: The Ominous Years, 1774–1776* (London etc., 1963), p. 35.

42. D. Butler, *John Wesley and George Whitefield in Scotland; or the Influence of the Oxford Methodists on Scottish Religion* (Edinburgh and London, 1898), pp. 57–8.

43. 'Claudero to Whitefield' in *Miscellanies in Prose and Verse on Several Occasions, by Claudero* (Edinburgh, 1766), p. 13.

44. Butler, *Wesley and Whitefield*, p. 32.

45. Ibid., p. 28.

46. Ibid., p. 38.

47. Ibid., p. 39. The revival began in February when the hell-fire preaching of the local minister, William M'Culloch, began to attract crowds from Glasgow and its surrounding district. A minister visiting from Dundee stayed at M'Culloch's house in the spring and reported many people 'in Darkness and great Distress about their Souls condition, and with many Tears bewailing their Sins and original Corruption, and especially the Sin of Unbelief, and Slighting of precious Christ.' Others described themselves as 'sick of Love' for the Lord: James Robe, M.A. (attr), *A Short Narrative of the Extraordinary Work of Cambuslang; in a Leter to a Friend, with proper Attestations by Ministers, Preachers and Others* (Glasgow, May 1742). Whitefield arrived on 18 June, and on 11 July preached to more than 20,000.

48. Butler, *Wesley and Whitefield*, p. 35.

49. There were censuses in Sweden in 1749 and in Austria in 1754.

50. Adam Smith to George Chalmers, 10 November 1785, in *Correspondence of Adam Smith*, ed. E.C. Mossner and I.S. Ross (Oxford, 1987), p. 288. Adam Smith, *Wealth of Nations*, IV.v.b.30.

51. Hugh Blair in the first issue of the *Edinburgh Review* said of Hutcheson: 'He has observed, that it was the glory of the present age, to have thrown aside the method of forming hypotheses in natural philosophy, and to set themselves to make observations and experiments on the material world itself. He was convinced, that, in like manner, a true scheme of morals could not be the product of genius and invention, or of the greatest precision in metaphysical reasoning; but must be drawn from proper observations on the several powers and principles which we are conscious of in our bosoms': No. 1, January–July 1755, p. 13.

52. Carlyle, *Autobiography* (see n. 29), p. 78.

53. *Correspondence of Adam Smith*, p. 309.

54. The elders of Armagh are alleged to have said to Hutcheson's father: 'We a' feel a muckle wae for your mishap; but it cannot be concealed. Your silly son Frank has fashed a' the congregation wi' his idle cackle, for he has been babbling this 'oor aboot a guid and benevolent God, and that the souls of the heathen will themsels gang to heaven if they follow the licht o' their ain consciences. Not a word does the daft boy ken, speer, nor say aboot the gude auld comfortable doctrines of election, reprobation, original sin, and faith. Hoot, man, awa' wi' sic a fellow.' J.S. Reid, *The History of the Presbyterian Church in Ireland* (Edinburgh, 1834), vol. 3, pp. 405–6.

55. Bernard de Mandeville, *The Fable of the Bees: or, Private vices, public benefits* (London, 1723), p. 10.

56. Francis Hutcheson, *Inquiry into the Original of Our Ideas of Beauty and Virtue, in Two Treatises . . .* (London and Dublin, 1725), Treatise 2, Section 2, p. 7.

57. *The Spectator*, vol. 1, No. 10.

58. Hutcheson, *Inquiry into the Original*, p. ix.

59. Ibid., p. vi.

60. 'Formam quidem ipsam, Marce fili, et tamquam faciem honesti vides, "quae si oculis cernereter, mirabiles amores", ut ait Plato, "excitaret sapientiae"': *De Officiis*, 1.V.15; *Phaedr.* 250B.

61. Hutcheson, *Inquiry into the Original*, p. 164.

62. 'The Moment of Evil produced by any Agent, is as the Product of his Hatred into his Ability, or $v = H \times A$': Hutcheson, *Inquiry into the Original*, p. 173.

63. Ibid., p. 275.

64. Carlyle, *Autobiography*, p. 78.

65. Ibid., pp. 93–4. The point is made more forcefully by Dugald Stewart who wrote, in a note to his Life of Adam Smith, that Hutcheson's lectures 'appear to have contributed very powerfully to diffuse, in Scotland, that taste for analytical discussion, and that spirit of liberal inquiry, to which the world is indebted for some of the most valuable productions of the eighteenth century': 'Account of the Life and Writings of Adam Smith, LL.D.', in Adam Smith, *Essays on Philosophical Subjects*, ed. W.P.D. Wightman and J.C. Bryce (Oxford, 1980), p. 334.

66. Ramsay of Ochtertyre, *Scotland and Scotsmen* (see n. 13), vol. 1, p. 275.

67. Ibid., p. 60.

68. Carlyle, *Autobiography*, p. 217.

69. Hugh Blair, *Sermons* (Edinburgh and London, 1790), vol. 3, Sermon xvii, pp. 351ff.

70. Carlyle, *Autobiography*, pp. 257–8.

71. N. Morren, *Annals of the General Assembly of the Church of Scotland* (Edinburgh, 1838), vol. 1, p. 233.

72. Ibid., pp. 245–6.

73. *Scots Magazine*, vol. 14 (1752), p. 264.

74. Carlyle, *Autobiography*, p. 268.

75. David Hume, *A Treatise of Human Nature* [1739], ed. E.C. Mossner (London, 1985), p. 311.

76. David Hume, 'My Own Life', printed with *An Enquiry concerning Human Understanding*, ed. Anthony Flew (Chicago and La Salle, 1988), p. 10.

77. G. F. Hegel, 'Vorlesungen über die Geschichte der Philosophie', in *Werke* (Hamburg, 1978), vol. 20.

78. According to Alexander Carlyle, Dr John Gregory ceased to attend the Poker Club, 'afraid, I suppose, of disgusting some of the ladies he paid court to by falling sometimes there with David Hume': Carlyle, *Autobiography*, pp. 483–4.

79. *Caldwell Papers* (see n. 12), Pt 1, pp. 39–40.

80. *Memoirs of the Life, Writings and Correspondence of William Smellie*, ed. Robert Kerr (Edinburgh, 1811), vol. 1, p. 354.

81. James Boswell, *The Life of Samuel Johnson, LL.D.* (London, 1953), pp. 1214–15 and pp. 838–9.

82. The estate of Ninewells recorded rents in 1713 of £192 sterling in money and £35 sterling in victuals: E.C. Mossner, *The Life of David Hume* (Oxford, 1980), p. 24.

83. 'My Own Life' in *Enquiry concerning Human Understanding* (see n. 76), p. 4.

84. To Gilbert Elliot of Minto, 10 March 1751: *The Letters of Hume*, ed. J.Y.T. Greig (Oxford, 1932), vol. 1, p. 154.

85. 'He never had entertained any belief in Religion since he began to read [John] Locke and [Samuel] Clarke.' That was presumably at College or soon after. 'An Account of My Last Interview with David Hume, Esq.', 7 July 1776, in *Boswell in Extremes, 1776–1778*, ed. C.McC. Weis and F.A. Pottle (London, 1971), p. 11.

86. Recipient unknown, March or April 1734, in Greig, *Letters of Hume*, vol. 1, p. 13.

87. Ibid.

88. Ibid.

89. Ibid., p. 14.

90. Ibid.

91. Ibid., p. 17.

92. Andrew Michael Ramsay, known as the Chevalier Ramsay, a Roman Catholic convert and sometime tutor to the young Prince Charles Edward and his brother. Ramsay to Stevenson, 24 August 1742, EUL La II 301.

93. François de Paris, a follower of Cornelius Jansen, had died in 1727. There were reports of the sick being healed and the blind made to see at his tomb, and Hume attacked these in the section of *Enquiry concerning Human Understanding* called 'Of Miracles'.

94. Hume, *Treatise of Human Nature* (see n. 75), p. 318.

95. 'My Own Life' in *Enquiry concerning Human Understanding*, p. 4. 'All, all but Truth, drops dead-born from the Press,/ Like the last Gazette, or the last Address': Alexander Pope, *Epilogue to the Satires, Dialogue II*, ll. 226–7.

96. *The History of the Works of the Learned*, Nov.–Dec. 1739, vol. 2, pp. 353–404.

97. *Letters of Hume*, vol. 1, p. 158.

98. *An Account of Sir Isaac Newton's Philosophical Discoveries*, 2nd edn (London, 1750), p. 9.

99. *Opticks: or, a Treatise on the Reflections, Refractions, Inflections and Colours of Light* (London, 1718), p. 381.

100. To Hutcheson, 17 September 1739, in *Letters of Hume*, vol. 1, p. 34.

101. 'In the production and conduct of the passions, there is a certain regular mechanism, which is susceptible of as accurate a disquisiton, as the laws of motion, optics, hydrostatics, or any part of natural philosophy' in 'Of the Passions', *Four Dissertations* (London, 1757).

102. Hume, *Treatise of Human Nature*, p. 43.

103. 'An Abstract of *A Treatise of Human Nature*', in *Enquiry concerning Human Understanding*, p. 29. Hume was indebted to Newton, not just for his ambitions but also for his style. The *Treatise* is full of Newtonian metaphors, as in this beautiful passage on money which might have come from the *Opticks*. 'Thus the pleasure, which a rich man receives from his possessions, being thrown upon the beholder, causes a pleasure and esteem; which sentiments again, being perceiv'd and sympathiz'd with, encrease the pleasure of the possessor; and being once more reflected, become a new foundation for pleasure and esteem in the beholder ... [and so on to] a third rebound of the original pleasure; after which 'tis difficult to distinguish the images and reflections, by reason of their faintness and confusion.' *Treatise of Human Nature*, p. 414.

104. Ibid., p. 415.

105. Epigraph to the *Treatise*.

106. To Henry Home, 2 December 1737 in *Letters of Hume*, vol. 1, p. 25.

107. *Treatise of Human Nature*, p. 153.

108. 'If there be any suspicion that ... the past may be no rule for the future, all experience becomes useless': *Enquiry concerning Human Understanding* (see n. 76), p. 81.

109. *The Works of Thomas Reid, DD* (Edinburgh, 1812), vol. 1, p. v.

110. Ibid., p. 11.

111. Reid had used 'hinder to do' where Hume preferred 'hinder from doing': to Reid, 25 February 1763, in *Letters of Hume*, vol. 1, p. 201.
112. *Treatise of Human Nature*, p. 316.
113. Ibid.
114. 'I cannot but consider myself a kind of resident or ambassador from the dominions of learning to those of conversation; and shall think it my constant duty to promote a good correspondence betwixt these two states, which have so great a dependence on each other. I shall give intelligence to the learned of whatever passes in company, and shall endeavour to import into company whatever commodities I find in my native country proper for their use and entertainment. The balance of trade we need not be jealous of, nor will there be any difficulty to preserve it on both sides. The materials of this commerce must chiefly be furnish'd by conversation and common life: the manufacturing of them alone belongs to learning.' From 'Of Essay Writing' in *Essays, Moral and Political* (Edinburgh, 1742), vol. II, p. 4.
115. 'Of Interest' in *Political Essays*, ed. K. Haakonssen (Cambridge, 1994), pp. 138–9.
116. Paul Hazard, *European Thought in the Eighteenth Century* (Harmondsworth, 1965), p. 320. Reid was spiteful: 'It is probable that the *Treatise of Human Nature* was not written in company; yet it contains manifest indications, that the author every now and then relapsed into the faith of the vulgar, and could hardly for half a dozen pages, keep up the sceptical character.' *Inquiry into the Human Mind* in *The Works of Thomas Reid* (see n. 109), vol. 1, p. 24.
117. *Treatise of Human Nature*, p. 316.
118. 1 April 1767, in *Letters of Hume*, vol. 2, p. 134.
119. *Enquiry concerning Human Understanding*, p. 56.

Chapter 4: The Philosopher's Opera

1. David Hume, 'My Own Life', printed with *Enquiry concerning Human Understanding*, ed. Anthony Flew (Chicago and La Salle, 1988), p. 5.
2. Ibid.
3. 'Of the Original Contract' (1748), in *Political Essays*, ed. K. Haakonssen (Cambridge, 1994), p. 191.
4. 'Of Refinement in the Arts' (1752) in *Political Essays*, p. 112.
5. David Hume, *A Letter from a Gentleman to his Friend in Edinburgh* (Edinburgh, 1745), p. 33.
6. 'My Own Life', loc. cit., p. 5.
7. Ibid., p. 6.
8. 'And would they but allow themselves to think, with any degree of cool-

ness, they would soon be convinced, that the peace of society is an object of greater importance than the right of any particular man can be, supposing him to be descended from a thousand kings': Henry Home, *Essays upon Several Subjects Concerning British Antiquities* (Edinburgh, 1747), p. 217.

9. 'I should prefer David Hume to any man for a colleague; but I am afraid the public would not be of my opinion; and the interest of the society [College] will oblige us to have some regard to the opinion of the public': Adam Smith to Cullen, November 1751, in *Correspondence of Adam Smith*, ed. E.C. Mossner and I.S. Ross (Oxford, 1987), p. 5.

10. The offending titles were the *Contes* of La Fontaine, *L'Écumoire* of Crébillon, and *L'histoire amoureuse des Gaules* of the comte de Bussy.

11. Carlyle said he gave the salary to 'families in distress': Alexander Carlyle of Inveresk, *Autobiography*, ed. J.H. Burton (Edinburgh, 1910), pp. 287–8.

12. *Correspondence of Adam Smith*, p. 9.

13. Robert Chambers, *Traditions of Edinburgh* (Edinburgh, *c.* 1910), p. 56.

14. Richard Hurd, *Moral and Political Dialogues* (London, 1759), p. 304.

15. 'My Own Life', loc. cit., n. 1, p. 7.

16. David Hume, *The History of Great Britain: The Reigns of James I and Charles I* (Edinburgh, 1754), p. 62.

17. Hume, *Enquiry concerning Human Understanding* (see n. 1), p. 166.

18. 'As we arise in philosophy [science] towards the first cause, we obtain more extensive views of the constitution of things, and see his [the great Being's] influences more plainly': Colin MacLaurin, *Account of Newton's Philosophical Discoveries*, 2nd edn (London, 1750), pp. 23–4.

19. David Hume, 'The Natural History of Religion', Section 14 in *Four Dissertations* (London, 1757).

20. Ibid., Section 15.

21. Adam Smith to Alexander Wedderburn, 14 August 1776, in *Correspondence of Adam Smith*, p. 204.

22. David Hume, *Dialogues Concerning Natural Religion* (Edinburgh, 1779).

23. Hume, 'Natural History of Religion', Section 12.

24. *The Letters of David Hume*, ed. J.Y.T. Greig (Oxford, 1932), vol. 1, p. 106.

25. Carlyle, *Autobiography* (see n. 11), p. 288.

26. 'Anecdotes of Hume' in E.C. Mossner, *The Life of David Hume* (Oxford, 1980), p. 214.

27. David Hume, *Treatise of Human Nature*, ed. E.C. Mossner (London, 1985), p. 537.

28. John Witherspoon, *Ecclesiastical Characteristics: or the Arcana of Church Policy, being an Humble Attempt to open up the Mystery of Moderation* (Glasgow, 1753), p. 15.

29. Ibid., pp. 27–8.

30. 'Have you seen our Friend Harrys Essays? They are well wrote; and are an unusual instance of an obliging method of answering a Book. Philosophers may judge of the question; but the Clergy have already decided it, & say he is as bad as me. Nay some affirm him to be worse, as much as a treacherous friend is worse than an open Enemy.' *Letters of Hume*, vol. 1, p. 162.

31. George Anderson, *Estimate of the Profit and Loss . . .* (Edinburgh, 1753), pp. 77–8 and 389–92.

32. 'Life of Lord Loughborough' in J.L. Campbell, *The Lives of the Lord Chancellors* (London, 1847), vol. 6, pp. 17–18.

33. '7th August, 1754. The Committee having refused the following question, "Whether the law of Queen Joan of Naples, allowing licensed stews, would be of advantage to a nation", Mr Wedderburn, who proposed it, appealed to the Society, and the determination of the appeal was delayed till next session.' Quoted by Campbell in *Lives of the Chancellors*, vol. 6, p. 32.

34. 'Rules and Orders of the Select Society', NLS Adv. Ms 23.1.1.

35. *Edinburgh Review*, No. 1, January–July 1755, pp. ii–iii.

36. *Edinburgh Review*, No. 2, July 1755–January 1756, pp. 26–7.

37. William Robertson, *The Situation of the World at the Time of Christ's Appearance* (Edinburgh, 1755), p. 41.

38. *An Analysis of the Moral and Religious Sentiments Contained in the Writings of Sopho, and David Hume, Esq; Addressed to the consideration of the Reverend and Honourable Members of the General Assembly of the Church of Scotland* (Edinburgh, 1755), pp. 26–49.

39. Ibid., p. 42.

40. 'The General Assembly, being filled with the deepest concern on account of the prevalence of infidelity and immorality, the principles whereof have been, to the disgrace of our age and nation, so openly avowed in several books published of late in this country, and which are but too well known among us; do, therefore, judge it proper and necessary for them at this time to express the utmost abhorrence of these impious and infidel principles, which are subversive of all religion, natural and revealed, and have such pernicious influence on life and morals. And they do earnestly recommend it to all the ministers of this Church to exercise the vigilance, and exert the zeal, which becomes their character, to preserve those under their charge from the contagion of these abominable tenets.' From 'Act against Infidelity and Immorality' in *Acts of the General Assembly of the Church of Scotland 1638–1842* (Edinburgh, 1843), p. 721.

41. *Letters of Hume*, vol. 1, p. 224.

42. *Scots Magazine*, vol. 17 (1755), p. 241.

43. 'Observations upon an Analysis of Sopho and Hume', *Scots Magazine*, vol. 17 (1755), pp. 233ff.

44. *Scots Magazine*, vol. 18 (1756), p. 147.

45. Carlyle, *Autobiography* (see n. 11), p. 324.

46. N. Morren, *Annals of the General Assembly of the Church of Scotland* (Edinburgh, 1838), vol. 2, pp. 86–7.

47. Campbell, *Lives of the Chancellors* (see n. 32), vol. 6, pp. 21–5.

48. Robert Wallace, 'Necessity or Expediency of the Churches inquiring into the Writings of David Hume Esq.', EUL La II/97, ff. 21–3.

49. Hume to Allan Ramsay, *Letters of Hume*, vol. 1, p. 224.

50. John Ramsay of Ochtertyre, *Scotland and Scotsmen in the Eighteenth Century* (Edinburgh and London, 1888), vol. I, p. 315–17.

51. E.C. Mossner, *The Life of David Hume* (Oxford, 1980), p. 355.

52. *Correspondence of Adam Smith* (see n. 9), p. 44.

53. 'Were I to change my habitation, I would retire to some provincial town in France, to trifle out my old age, near a warm sun in a good climate, and amidst a sociable people.' David Hume to Dr Clephane, 4 April 1756, in *Letters of Hume*, vol. 1, p. 232.

54. Hugo Arnot, *The History of Edinburgh* (Edinburgh and London, 1779), p. 366. John Jackson, *The History of the Scottish Stage* (Edinburgh, 1793), p. 22.

55. 'Admonition and Exhortation of November 30, 1727' printed as appendix to George Anderson, *The Use and Abuse of Diversions: A sermon on Luke XIX.13* (Edinburgh, 1733), p. 145.

56. George Anderson, *A Reinforcement of the Reasons proving that the Stage is an Unchristian Diversion* (Edinburgh, 1733), pp. 41ff.

57. 'To the Honourable Duncan Forbes of Culloden, etc. The Address of Allan Ramsay' in *The Poems of Allan Ramsay* (London, 1800), vol. 1, p. xliii, n.

58. The Revd Thomas Somerville, who admits to being fond of the theatre, 'perhaps to a culpable degree', includes a playbill of 22 February 1762 in which the supposed musical pieces are not even listed: *My Own Life and Times, 1741–1814* (Edinburgh, 1861), pp. 113–14.

59. John Jackson, *The History of the Scottish Stage* (Edinburgh, 1793), pp. 23–4.

60. 'Life and Works of John Home' in Walter Scott, *Essays on Chivalry, Romance and the Drama* (London, 1888), p. 376.

61. *Boswell's London Journal, 1762–1763*, ed. F.A. Pottle (London, 1950), p. 137.

62. Carlyle, *Autobiography* (see n. 11), p. 325.

63. 'The Canongate Play-House in Ruins', in *The Poems of Robert Fergusson*, ed. M.P. McDiarmid (Edinburgh and London, 1956), vol. 2, p. 60.

64. *Scots Magazine*, vol. 18 (1756), p. 623.

65. Carlyle, *Autobiography*, p. 329.

66. *Correspondence of Adam Smith* (see n. 9), p. 20.

67. *Boswell's Life of Johnson* (London, 1953), p. 595.

68. 'In his compositions, we regret, that many irregularities, and even absurdities, should so frequently disfigure the animated and passionate scenes intermixed with them': 'Appendix to the Reign of James I', in Hume, *History of England* (Indianapolis, 1983), vol. 5.

69. To Joseph Spence, 15 October 1754 in *Letters of Hume* (see n. 24), vol. 1, p. 204.

70. John Home, *Douglas: A Tragedy* (London, 1757), p. 34.

71. Carlyle, *Autobiography*, p. 327.

72. Carlyle, *An Argument to Prove that the Tragedy of Douglas ought to be Publickly burnt* (Edinburgh, 1757), p. 17.

73. Ibid., p. 18.

74. Ibid., p. 24.

75. Ibid., p. 12.

76. John Hill Burton, *The Life and Correspondence of David Hume, Esq.* (Edinburgh, 1846), vol. 1, p. 420.

77. *Douglas, A Tragedy, weighed in the Balances and found wanting* (Edinburgh, 1757), p. 36.

78. Ibid., p. 20.

79. Ibid., p. 23.

80. *Scots Magazine*, vol. 19 (1757), p. 18.

81. From Carlyle's own account in NLS 17601: 70.

82. Campbell, *Lives of the Chancellors* (see n. 32), vol. 6, p. 28.

83. 'Recommendation to Presbyteries to take care that none of the Ministers of this Church attend the Theatre', in *Acts of the General Assembly* (see n. 40), p. 729.

84. John MacLaurin, *The Philosopher's Opera* (Edinburgh, 1757), p. 15.

85. John Murray, *A Letter from a Gentleman in Edinburgh to his Friend in the Country: Occasioned by the Late Theatrical Disturbance* (Edinburgh, 1766).

86. Jackson, *History of the Scottish Stage* (see n. 59), p. 64.

87. Ibid., p. 78.

88. Carlyle, *Autobiography*, pp. 338–9. 'The clergy of Scotland', the Englishman Thomas Pennant wrote in 1769, 'are at present much changed from the furious, illiterate and enthusiastic teachers of the old times': *A Tour in Scotland MDCCLXIX*, 3rd edn (Warrington, 1774), p. 155.

89. 'The giddy parson is one of the most contemptible of types, and the ministers who flocked after Mrs Siddons and made a parade of their little liberties have an indescribable air of naughty urchins': *Some Eighteenth Century Byways and other Essays* (Edinburgh and London, 1908), p. 187.

90. Campbell, *Lives of the Chancellors*, vol. 6, pp. 47–8.

91. Carlyle, *Autobiography*, p. 491.

92. Ibid., p. 249.

93. *The Epigoniad: A Poem in Nine Books* (Edinburgh, 1757), pp. xxvi–xxviii.

94. *The Monthly Review; or Literary Journal*, September 1757, vol. 17, p. 228.

95. Ibid., p. 229. The July number of the *Critical Review*, edited by the Scot Tobias Smollett, lampooned the poem as strained, anachronistic and dull, 'resembling an epic poem in very little else but the outward form, and extent of it, dragging its slow length along through nine tedious books.' *The Critical Review: or Annals of Literature*, July 1757, vol. 4, p. 28.

96. *A Critical Essay on the Epigoniad; wherein the Author's horrid Abuse of Milton is examined* (Edinburgh, 1757): BL 8403.i.I (3).

97. *The Yale Edition of Horace Walpole's Correspondence*, ed. W.S. Lewis (London and New Haven, 1952), vol. 15, pp. 42–3.

98. Robert Wallace, *Characteristics of the Present Political State of Great Britain* (London, 1758), p. 132.

99. Ibid., p. 127.

100. Ibid., p. 136–7.

101. *Memorials of John Murray of Broughton, Sometime Secretary to Prince Charles Edward, 1740–1747*, ed. R.F. Bell (Edinburgh, 1898), pp. 257–8.

102. Hugh Paton, ed., *A Series of Original Portraits and Caricature Etchings by the late John Kay, Miniature Painter, Edinburgh* (Edinburgh, 1877), vol. I, p. 4.

Chapter 5: Smaller Joys from Less Important Causes

1. '... but I forgot that you have seen the plan – Ld Kaims has a book upon Law ready to appear – Davd Hume is well advanced in the reign of Henry ye 7th – You see we endeavour to fill up some part of the System here. Yours is the first. Pulchrum benefacere reipublicae, etiam benedicere haud absurdum [It is a fine thing to do well by the commonwealth, but to speak well of it is not contemptible].' Wedderburn to Sir Gilbert Elliot of Minto, 2 July 1757 in NLS MS. 11008, f. 58–9.

2. See particularly the chapter added to Part 1, 'Of the corruption of our moral sentiments', in Adam Smith, *The Theory of Moral Sentiments*, ed. D.D. Raphael and A.L. MacFie, I. iii. 3 (Oxford, 1979), pp. 61–6.

3. Dugald Stewart's 'Account of the Life and Writings of Adam Smith, LL.D.' in Adam Smith, *Essays on Philosophical Subjects*, ed. W.P.D. Wightman and J.C. Bryce (Oxford, 1980), p. 314.

4. H.T. Buckle, *History of Civilization in England* (London, 1857–61), vol. 1, p. 194 and vol. 2, p. 437.

5. *Correspondence of Adam Smith*, ed. E.C. Mossner and I.S. Ross (Oxford, 1987), p. 168.

6. To William Strahan, Glasgow, 4 April 1760, in *Correspondence of Adam Smith*, p. 68.

7. Daniel Defoe, *Tour thro' the Whole Island of Great Britain* (see Ch. 2, n. 17) vol. 2, pp. 780–1.

8. Stewart's 'Life of Smith' in *Essays on Philosophical Subjects*, p. 269.

9. John Rae, *Life of Adam Smith* (London, 1895), pp. 4–5.

10. Stewart's 'Life of Smith' in *Essays on Philosophical Subjects*, p. 270.

11. H.W. Davis, *A History of Balliol College* (Oxford, 1963), p. 150.

12. Adam Smith, *Wealth of Nations*, V.i.f.34 (Oxford, 1979), p. 772.

13. Stewart's 'Life of Smith' in *Essays on Philosophical Subjects*, p. 271.

14. Ibid.

15. John Ramsay of Ochtertyre, in *Scotland and Scotsmen of the Eighteenth Century* (Edinburgh and London, 1888), says Smith wrote and spoke English 'with great purity': vol. 1, p. 461.

16. *Correspondence of Adam Smith*, p. 3.

17. Ibid.

18. In his introduction to *Essays on Philosophical Subjects* (see n. 3), pp. 7–9, W.P.D. Wightman lays out the arguments as to dating.

19. To le duc de La Rochefoucauld, Edinburgh, November 1, 1785, *Correspondence of Adam Smith*, pp. 286–7.

20. Ibid.

21. 'History of Astronomy' in *Essays on Philosophical Subjects*, p. 46.

22. Ibid., pp. 45–6.

23. Adam Smith, *Lectures on Rhetoric and Belles Lettres*, ed. J.C. Bryce (Oxford, 1983), p. 146. In *The Wealth of Nations*, Smith speaks of Greek science and ethics showing 'the beauty of a systematical arrangement of different observations connected by a few common principles': V.I.f.25.

24. *Essays on Philosophical Subjects*, p. 105.

25. Einstein said of his own work that 'there could be no sweeter destiny for any physical theory than that it should point the way to a more comprehensive theory, and live on as a limiting case': *Über die spezielle und die allgemeine Relativitätstheorie* (Braunschweig, 1920), Sect. 22.

26. *Essays on Philosophical Subjects*, p. 49.

27. Stewart's 'Life of Smith' in *Essays on Philosophical Subjects*, p. 272.

28. Rae, *Life of Smith* (see n. 9), p. 30.

29. Ramsay of Ochtertyre, *Scotland and Scotsmen*, vol. 1, p. 28.

30. For example, the goddess Scotia declaims: ''Tis done, my sons! 'Tis nobly done!/Victorious over tyrant power:/How quick the race of fame

was won!/The work of ages in one hour': James Hogg, *The Jacobite Relics of Scotland* . . . (see Ch. 2, n. 16), 2nd series, p. 119.

31. In the preface, Smith praises Hamilton's translations or adaptations of the ancient poets. Those included the famous barter scene of Glaucus and Diomed from the sixth book of the *Iliad*, which has survived to baffle modern economists in *The Wealth of Nations*.

32. Hume to Smith, 8 June 1758, in *Correspondence of Adam Smith*, p. 24.

33. Thomas Ruddiman, *A Dissertation upon the Way of Teaching the Latin Tongue* (Edinburgh, 1733), p. 32.

34. Thomas Ruddiman, *Audi Alteram Partem* (Edinburgh, 1756), p. 56.

35. Ramsay of Ochtertyre, *Scotland and Scotsmen*, vol. 1, p. 9.

36. Ibid., pp. 211–12. Alexander Carlyle was always grateful to a widowed aunt who taught him to read English 'with just pronunciation and a very tolerable accent – an accomplishment which in those days was very rare': Alexander Carlyle of Inveresk, *Autobiography*, ed. J.H. Burton (Edinburgh, 1910), p. 4.

37. Some of his despised Scotticisms, such as 'in the long run', have triumphed: *Scots Magazine*, vol. 22 (1760), pp. 686–7.

38. Here is a specimen of Sheridan's teaching: 'The next progression of number is when the same note is repeated, but in such a way that one makes a more sensible impression on the ear than the other, by being more forcibly struck, and therefore having a greater degree of loudness. As ti-tum or tum-ti-tum-ti or when two weak notes precede a more forcible one, as ta-ta-tum; or when they follow one, as tum-ti-ti, tum-ti-ti.' J.L. Campbell, *The Lives of the Lord Chancellors* (London, 1847), vol. 6, pp. 36–7.

39. Ibid.

40. Lecture Three, 'Of the origin and progress of language', appeared in expanded form as 'Considerations concerning the first formation of Languages' in *The Philological Miscellany* (1761).

41. Hugh Blair, *Lectures on Rhetoric and Belles Lettres* (see Ch. 6, n. 74), vol. 1, p. 381.

42. Adam Smith, *Lectures on Rhetoric and Belles Lettres* (see n. 23), p. 42.

43. Ibid., p. 137.

44. William Wordsworth, 'Essay Supplementary' to the *Preface to the Lyrical Ballads*. Rae (see n. 9) quotes a couplet from Caleb Coton's 'Hypocrisy' which calls Smith, for all his eminence in commerce and banking, a bankrupt in poetic matters: 'But when on Helicon he dar'd to draw/His draft return'd and unaccepted saw' – in other words, the Muses bounced his cheque.

45. Stewart's 'Life of Smith' in *Essays on Philosophical Subjects*, p. 274.

46. Ibid., p. 293.

47. 'He who thus considers things in their first growth and origin, whether

a state or anything else, will obtain the clearest view of them': *Politica*, 1252a. Tacitus, *Ann.* 3, 26 'Vetustissimi mortalium . . .' and *Germania*, throughout.

48. Smith, *Essays on Philosophical Subjects*, p. 293. The problem, as Dugald Stewart was not the first to recognise, is that facts can be inconvenient. 'In most cases, it is of more importance to ascertain the progress that is most simple, than the progress that is most agreeable to fact; for, paradoxical as the proposition may appear, it is certainly true, that the real progress is not always the most natural.' Ibid., p. 296. Adam Ferguson had already made the point: 'Our method, notwithstanding, too frequently, is to rest the whole on conjecture; to impute every advantage of our nature to those arts which we ourselves possess; and to imagine, that a mere negation of all our virtues is a sufficient description of man in his original state.' In *An Essay on the History of Civil Society*, ed. Fania Oz-Salzberger, (Cambridge, 1995), p. 75. Ferguson's passage is presumably the source of Stewart's word 'conjectural'.

49. Dugald Stewart, *Works* (see Prologue, n. 11), vol. 10, p. 35.

50. To Smith, 12 April 1759, in *Correspondence of Adam Smith*, p. 34.

51. Smith, *Lectures on Rhetoric and Belles Lettres* (see n. 23), p. 9.

52. 'The Indians themselves allowed that Murphy died with great heroism, singing, as his death song, the Drimmendoo': *Humphry Clinker*, July 13.

53. 'Auld Reekie', ll. 227–30, in *The Poems of Robert Fergusson*, ed. M.P. McDiarmid (Edinburgh and London, 1954), vol. 2, p. 116.

54. Stewart's 'Life of Smith' in *Essays on Philosophical Subjects*, p. 274.

55. Adam Smith, *The Theory of Moral Sentiments*, VII, iv.37 (Oxford, 1979), p. 342.

56. Stewart's 'Life of Smith' in *Essays on Philosophical Subjects*, p. 275.

57. Or earlier. For the journey in the other direction in 1767, Boswell told his friend William Temple to 'set out in the fly on Monday morning, and reach Glasgow by noon': *Boswell in Search of a Wife* (see Ch. 1, n. 30), p. 78.

58. 'A Dictionary of the English Language by Samuel Johnson' in *Essays on Philosophical Subjects*, p. 232.

59. Hume to Allan Ramsay, April or May 1755, in *Letters of Hume*, ed. J.Y.T. Greig (Oxford, 1932), vol. 1, p. 220–1.

60. David Hume, *Political Essays*, ed. K. Haakonssen (Cambridge, 1994), p. 107.

61. Carlyle, *Autobiography*, p. 293.

62. Ramsay of Ochtertyre, *Scotland and Scotsmen*, vol. 1, p. 468.

63. Sir Walter Scott, 'The Life and Works of John Home' in *Essays on Chivalry, Romance and the Drama* (London, 1888), p. 388.

64. Smith, *Lectures on Rhetoric and Belles Lettres*, p. 112.

65. Smith, *Theory of Moral Sentiments*, p. 80.

66. Ibid., p. 181.
67. Ibid., p. 182.
68. Ibid., p. 57.
69. *Correspondence of Adam Smith*, pp. 34–5.
70. David Hume, *Treatise of Human Nature*, p. 627.
71. Smith, *Theory of Moral Sentiments*, p. 9.
72. Ibid., p. 26.
73. Ibid., p. 112.
74. Ibid., p. 130.
75. Ibid., p. 181.
76. Ibid., pp. 182–3.
77. Ibid., p. 183.
78. Ibid., p. 86.
79. Ibid., pp. 184–5.
80. Ibid., p. 36, p. 289.

Chapter 6: The Thermometer of the Heart

1. The month is conjecture. We know Home spent two to three weeks each summer in Moffat, and Carlyle arrived on 2 October.
2. Henry Mackenzie, ed., *Report of the Committee of the Highland Society of Scotland, appointed to inquire into the nature and authenticity of the Poems of Ossian* (Edinburgh, 1805), Appendix, p. 68.
3. Ibid., Appendix, p. 69.
4. Alexander Carlyle of Inveresk, *Autobiography*, ed. J.H. Burton (Edinburgh, 1910), p. 417.
5. Samuel Johnson, *A Journey to the Western Islands of Scotland* (London, 1984), p. 119. Boswell, *Life of Johnson* (London, 1953), p. 579.
6. *Report*, p. 152.
7. 'With a genius truly poetical, [MacPherson] was one of the first literary impostors in modern times': Malcolm Laing, ed., *The Poems of Ossian* (Edinburgh, 1805), vol. 1, p. liv.
8. *The Edinburgh Review, or Critical Journal*, vol. 6, No. 12 (April–July 1805), p. 458.
9. D.S. Thomson, *The Gaelic Sources of MacPherson's 'Ossian'* (Edinburgh and London, 1952).
10. 'If we know little of the ancient highlanders, let us not fill the vacuity with Ossian. If we have not searched the Magellanick regions, let us however forbear to people them with "patagons".' Johnson, *Journey to the Western Islands*, p. 119.
11. 'A Dissertation concerning the Antiquity etc. of the Poems of Ossian, the Son of Fingal' in *Fingal, an ancient epic poem in six books, together*

with several other poems, composed by Ossian the son of Fingal, trans-lated from the Galic Language by James MacPherson (London, 1762), p. xiv.

12. To Sir David Dalrymple, 23 June 1760, printed in Laing, ed., *Poems of Ossian*, vol. 1, p. xvi.

13. J.H. Burton, *The Life and Correspondence of David Hume, Esq.* (Edinburgh, 1846), vol. 1, p. 148.

14. Margaret M. McKay, ed., *The Rev. Dr John Walker's Report on the Hebrides of 1764 and 1771* (Edinburgh, 1980), p. 4.

15. In the end, he adapted *Fragments* IX, the story of Rivine, which was per-formed as *The Fatal Discovery* at Drury Lane in February, 1769. Home travelled north with MacPherson on his second manuscript hunt in the summer of 1761. He wrote to Bute in June to say he was setting off with 'the highland bard' to gather 'some of nature's gems to adorn Rivine the daughter of Kew': Home to Bute, Edinburgh, 12 June 1761 in R.G. Thomas, 'Lord Bute, John Home and Ossian: Two Letters', *Modern Language Review*, vol. 51 (1956), pp. 73–5.

16. Hume to Smith, 27 February 1754: 'I am writing kind of circular Letters, recommending Mr Blacklock's Poems [to] all my Acquaintance, but especially to those whose Approbation would contribute most to recommend them [to] the World ... What a Prodigy are they, when con-sidered as the Production of a man, so cruelly dealt with, both by Nature [and] Fortune.' *Correspondence of Adam Smith* (see Ch. 5, n. 5), p. 10.

17. 'Remarks on the Poems of Ossian', *The Mirror*, 13 (9 March 1779).

18. See especially Edmund Burke, *A Philosophical Enquiry into the Origin of our Ideas of the Sublime and Beautiful*, 2nd edn (London, 1759).

19. Shenstone to M'Gouan, 24 September 1761: EUL, La III/423/203.

20. R.B. Sher, *Church and University in the Scottish Enlightenment* (Edinburgh, 1985), p. 258.

21. *Fragments of Ancient Poetry Collected in the Highlands of Scotland and Translated from the Galic or Erse Language* (Edinburgh, 1760), p. 9.

22. *Life of Johnson*, p. 581.

23. Gray was not at all impressed by bragging letters from MacPherson. 'I am gone mad about them. They are said to be translations (literal & in prose) from the *Erse*-tongue, done by one Macpherson, a young clergy-man [*sic*] in the Highlands. He means to publish a Collection he has of these Specimens of antiquity, if it be antiquity: but what plagues me is, I can not come at any certainty in that head. I was so struck, so *extasié* with their infinite beauty, that I writ into Scotland to make a thousand enquiries. The letters I have in return are ill-wrote, ill-reason'd, unsatis-factory, calculated (one would imagine) to deceive one, & yet not cun-ning enough to do it cleverly. In short, the whole external evidence would make one believe these fragments (for so he calls them, tho' nothing

could be more entire) counterfeit: but the internal is so strong on the other side, that I am resolved to believe them genuine, spite of Devil & the Kirk.' To Thomas Wharton, *c.* 20 June 1760, in *Correspondence of Thomas Gray*, ed. P. Toynbee and L. Whibley (Oxford, 1935), vol. 2, p. 680.

24. D.S. Thomson (see n. 9) identifies just one authentic ballad source and that only in Book One of *Temora*.

25. *Boswell's London Journal 1762–1763*, ed. F.A. Pottle (London, 1950), 23 May 1763, p. 265.

26. John Ramsay of Ochtertyre, *Scotland and Scotsmen in the Eighteenth Century* (Edinburgh and London, 1888), vol. 1, p. 545.

27. Ibid., p. 544.

28. The preface to the first number promised to encourage new writing so that 'the Caledonian muse might not be restrain'd by want of a Publick Echo to her Song'.

29. 'I am entitled to conclude, that those first essays of his muse in English, if fairly produced, would afford the same convincing detections [unmasking] of Ossian': Malcolm Laing to Lord Banatyne, 30 March 1802, in *Report* (see n. 2), p. 86.

30. *The Hunter*, Canto 1, ll. 139–43.

31. 'I married in summer 1759 [?illegible] and I mind perfectly well that a year or two before then upon a Sunday . . .': Donald MacPherson of Laggan to the Revd John Adamson, October 1797: NLS MS 73.2.13, No. 23, f. 47.

32. 'In the same humour I proceeded to inforce the thought by showing Greek, Latin, English & French Poets brought themselves forward to esteem by such means that if he did not take up some or other at that new period, it would sleep for ever. In my memory, the generality & people of taste [–] themselves upon hearing these poems with full applaus from the auditors of all degrees, & so went on he being mostly muit all the time & I followed this to push strong as I coud then the Bell rang when we went in to preaching.' Ibid.

33. *Report*, Appendix, p. 68.

34. *Scots Magazine*, vol. 18 (1756), p. 15.

35. *Report*, Apendix, pp. 68–9.

36. Laing, ed., *Poems of Ossian*, vol. 1, p. xix.

37. Ramsay of Ochtertyre, *Scotland and Scotsmen*, vol. 1, p. 546.

38. 'I was inclined to be a little incredulous on that head, but Mr Home removed my scruples, by informing me of the manner in which he procured them from Mr Mcpherson, the translator': Hume to an unknown recipient, Edinburgh, 16 August 1760, in *The Letters of David Hume*, ed. J.Y.T. Greig (Oxford, 1932), vol. 1, p. 328. 'My Scepticism extends no farther, nor ever did, than with regard to the extreme Antiquity of

those Poems; and it is no more than Scepticism': Hume to Blair, Paris, 23 August 1765, in ibid., vol. 1, p. 516.

39. To William Mason, *c.* 31 August 1760, in *Correspondence of Thomas Gray* (see n. 23), vol. 2, p. 695.

40. 'Burns's Unpublished Common-place Book', pp. 5–6, in *Macmillan's Magazine*, vol. 39 (1878–9), pp. 455–6.

41. *Fragments* (see n. 21), p. iii.

42. Ibid.

43. Ibid., p. vii.

44. 'On my own Bottom' is a term taken from nautical commerce, indicating that the same merchant owns both the vessel and the cargo.

45. 'An Account of the Number of People in Scotland in 1755', in *Scottish Population Statistics*, ed. J.G. Kyd (Edinburgh, 1952).

46. Edward Burt, *Letters from a Gentleman in the North of Scotland to his friend in London* (London, 1754), vol. 2, p. 161.

47. Dr John MacPherson, minister at Sleat, to Dr Blair, 27 November 1763, in *Report* (see n. 2), Appendix, p. 10.

48. 'Dissertation on Ossian's Antiquity', in *Fingal* (see n. 11), p. xv.

49. To Henry Mackenzie, 4 March 1801, in *Report*, p. 40.

50. To Henry Mackenzie, 26 March 1798, in *Report*, Appendix, p. 65.

51. Donald Macleod to Dr Blair, Glenelg, 26 March 1764, in *Report*, Appendix, p. 29.

52. Lachlan M'Pherson to Dr Blair, Strathmashie, 22 October 1763, in *Report*, Appendix, p. 8.

53. Malcolm MacPherson's affidavit, Scalpa, 5 September 1800, in *Report*, Appendix, pp. 92–3.

54. Ibid.

55. Ewan MacPherson's Declaration, Knock in Sleat, 11 September 1800, in *Report*, Appendix, p. 95.

56. Ibid.

57. Ibid., p. 96.

58. Angus MacNeill to Dr Blair, Hovemore in South Uist, 23 December 1763, in *Report*, Appendix, p. 20.

59. Declaration of Lachlan Mac Vuirich, made at Torlum in Barra, 9 August 1800, in *Report*, Appendix, p. 278.

60. Ibid., p. 279. Ewan MacPherson remembered it differently. He said that Niel gave him 'a book the size of a New Testament, and of the nature of a common-place book', which contained – amid family exploits and a history of the great Montrose – 'some of the poems of Ossian': *Report*, Appendix, p. 96. But the *Leabhar Dearg*, a Gaelic folio manuscript apparently containing some Ossian poems, was actually with Lt. Donald MacDonald in Edinburgh. James MacPherson obtained 'an order' from Clanranald senior on this volume: ibid., p. 97. Angus

MacNeill, the minister of South Uist, confused matters further by telling Blair that a large ancient manuscript, 'treating of the wars of Fingal and Comhal his father', had been spirited off some time before to Ireland and lost. Fortunately, Clanranald himself had copied more than one hundred pages and directed MacPherson to pick up the copy from one Donald MacDonald, a 'merchant in the Luckenbooths', who was contemplating publishing it: MacNeill to Blair, in *Report*, Appendix, p. 19.

61. *Report*, Appendix, p. 278–9 (also NLS ADV MS. 73.2.12.66).
62. Andrew Gallie to Charles MacIntosh, Writer to the Signet, Kincardine, 12 March 1799, in *Report*, p. 36.
63. 'Traditions are uncertain; poetry delivered down from memory, must lose considerably; and it is a matter of surprise to me, how we have now any of the beauties of our ancient Gaelic poetry remaining. Your collection, I am informed, is pure, as you have taken pains to restore the style.' *Report*, Appendix, p. 154.
64. 'The Ossian Manuscripts: A Note by Gordon Gallie MacDonald', in *Bulletin of the New York Public Library*, vol. 34 (1930), p. 80.
65. Ibid.
66. To Revd Dr Kemp, Kincardine, 4 March 1801, in *Report*, p. 44.
67. *Report*, p. 34.
68. To Henry Mackenzie, Peebles, 26 March 1798, in *Report*, Appendix, p. 63.
69. Ramsay of Ochtertyre, *Scotland and Scotsmen*, vol. 1, pp. 548–9.
70. To Revd Mr MacLagan, Edinburgh, 16 January 1761, in *Report*, Appendix, p. 155.
71. Hume to Strahan, Edinburgh, 9 February 1761, in *Letters of Hume*, vol. 1, p. 343.
72. Sir Hew Dalrymple, an abusive man who quarrelled with Sir James Steuart: NAS GD 110/1171.
73. David Hume, *Treatise of Human Nature*, p. 414.
74. Hugh Blair, *Lectures on Rhetoric and Belles Lettres* (London and Edinburgh, 1783), vol. 2, p. 406. In the printed lectures, Blair lists Ossian with Wilkie as a great epic poet, but refrains from discussing either.
75. Hugh Blair, *A Critical Dissertation on the Works of Ossian, the son of Fingal* (London, 1763), p. 24.
76. Ibid.
77. 'Semper ad eventum festinat, et in medias res . . . [The poet always hurries on to the business, and plunges into the midst of the action]', *De Arte Poetica*, 1.148. Blair, *Critical Dissertation*, p. 26.
78. Blair, *Critical Dissertation*, p. 28.
79. *Fingal* (see n. 11), Bk iii, p. 37.
80. Boswell, *Life of Johnson* (see n. 5), p. 1207.
81. Hugh Blair, *Critical Dissertation*, p. 17.

82. *Fingal*, Bk vi, p. 77.

83. Hugh Blair, *Critical Dissertation*, p. 3.

84. Ibid., p. 11.

85. Ibid., p. 20.

86. Ibid., p. 15.

87. Ibid., p. 21.

88. Ibid., p. 13.

89. Testimony or Declaration of Hugh M'Donald of Killepheder, South Uist, 12 August 1800, in *Report* (see n. 2), Appendix, p. 47.

90. 'On Sensibility', in Hugh Blair, *Sermons* (London and Edinburgh, 1790), vol. 3, p. 33. Edmund Burke had written: 'In grief, the *pleasure* is still uppermost; and the affliction we suffer has no resemblance to absolute pain, which is always odious; and which we endeavour to shake off as soon as possible': *A Philosophical Enquiry into the Origin of our Ideas of the Sublime and Beautiful*, 2nd edn (London, 1759), p. 55.

91. 'On the Benefits to be derived from the House of Mourning', in Blair, *Sermons*, vol. 2, p. 373. 'Let not a false idea of fortitude, or mistaken conceptions of religious duty, be employed to restrain the bursting emotion. Let the heart seek its relief, in the free effusion of just and natural sorrow': 'On Death', in *Sermons*, vol. 3, p. 99. 'There is *a time to mourn*, as well as *a time to rejoice*. There is a virtuous *sorrow, which is better than laughter*': ibid., p. 107.

92. 'On Death', *Sermons*, vol. 3, p. 92.

93. The translation is by Catullus, but is still a translation.

94. *Dionysii Longini de Sublimitate*, Sect. 44, 6.

95. *Critical Dissertation* (see n. 75), p. 16.

96. Burke, *Philosophical Enquiry*, p. 149.

97. Ibid.

98. Burt, *Letters from a Gentleman* (see n. 46), vol. 2, p. 10. In the early 1750s Oliver Goldsmith, who studied medicine at the College in Edinburgh then, was still describing the mountains about Drumlanrig as having 'the wildest and most hideous aspect of any in all the south part of Scotland': *Tour in Great Britain* (London, 1753), vol. 4, p. 124.

99. Blair, *Lectures on Rhetoric and Belles Lettres* (see n. 74), vol. 1, p. 48.

100. Blair, *Critical Dissertation* (see n. 75), p. 68.

101. Blair, *Lectures on Rhetoric and Belles Lettres*, vol. 1, p. 45.

102. Ibid., vol. 1, p. 52.

103. *Boswell's London Journal 1762–1763*, ed. F.A. Pottle (London, 1950), p. 182.

104. Ibid.

105. *Temora, an ancient epic Poem in eight Books: together with several other poems composed by Ossian the son of Fingal. Translated from the Galic Language by James MacPherson* (London, 1763), p. xxii.

106. Ibid., p. xxxiv.

107. Blair's 'Appendix' to vol. 2 of *The Works of Ossian* (London, 1765), p. 448.

108. Ibid., p. 457.

109. Captain A. Morison's answers to queries, *Report* (see n. 2), Appendix, p. 177.

110. Johnson to Baretti, London, 10 June 1761, in *Life of Johnson*, p. 257.

111. The Treaty of Paris preserved Britain's gains in Canada, its supremacy in India, certain Caribbean islands, Senegal and Minorca, but returned to France its West Indian possessions and its fishing rights off Newfoundland, and restored Cuba and the Philippines to Spain.

112. *Boswell's London Journal*, pp. 71–2.

113. Ibid., p. 72.

114. *The Prophecy of Famine: A Scots Pastoral, Inscribed to John Wilkes, Esq.* (London, 1763), p. 16.

115. Ibid., p. 7.

116. Ibid., p. 13.

117. *Boswell's London Journal*, p. 234.

118. 'Dr Blair, relying on the internal evidence of their antiquity, asked Dr Johnson whether he thought of any man of a modern age could have written such poems? Johnson replied, "Yes, Sir, many men, many women, and many children."' *Life of Johnson*, p. 280.

119. *Boswell's London Journal*, pp. 258 9.

120. Ibid., p. 255.

121. *Fingal* (see n. 11), Bk ii, p. 34.

122. Churchill, *The Prophecy of Famine*, p. 11.

123. Hume to Blair, London, 19 September 1763, in *Letters of Hume*, vol. 1, p. 399.

124. Ibid.

125. Ibid., p. 400.

126. Ibid.

127. Ibid.

128. Ibid., pp. 400–1.

129. Hume to Blair, London, 6 October 1763, in *Letters of Hume*, vol. 1, p. 403.

130. Ibid.

131. Hume to Sir Gilbert Elliot of Minto, Paris, 22 September 1764, in *Letters of Hume*, vol. 1, p. 470.

132. Ibid.

133. Hume to Blair, Paris, possibly December 1763, in *Letters of Hume*, vol. 1, pp. 418–19.

134. Hume to Ferguson, Fontainebleau, 9 November 1763, in *Letters of Hume*, vol. 1, p. 413.

135. Hume to Blair, Paris, 6 April 1765, in *Letters of Hume*, vol. 1, p. 497.

136. *The Works of Ossian* (London, 1765), vol. 1.

137. Hume to Blair, Compiègne, 20 July 1765, in *Letters of Hume*, pp. 513–14.

138. Hume to Strahan, Edinburgh, 30 January 1773, in *Letters of Hume*, vol. 2, p. 269.

139. 'It is hard to tell whether the Attempt or the Execution be worse': Hume to Adam Smith, Edinburgh, 10 April 1773, in *Correspondence of Adam Smith*, p. 167.

140. *A Journey to the Western Islands*, p. 118.

141. Ibid., p. 116.

142. To Boswell, 7 February 1775, in *Life of Johnson*, p. 578.

143. Hume was equally suspicious of oral traditions. In Section One of 'The Natural History of Religion', he wrote: 'An historical fact, while it passes by oral tradition from eye-witnesses and contemporaries, is disguised in every successive narration, and may at last retain but very small, if any, resemblance of the original truth, on which it was founded. The frail memories of men, their love of exaggeration, their supine carelessness; these principles, if not corrected by books and writing, soon pervert the account of historical events; where argument or reasoning has little or no place, nor can ever recal the truth, which has once escaped those narrations.' *Four Dissertations* (London, 1757), p. 8.

144. *A Journey to the Western Islands*, p. 119.

145. This is the version in Boswell's *Life of Johnson*, p. 579.

146. 'If we could, with safety, indulge the pleasing supposition, that Fingal lived, and that Ossian sung, the striking contrast of the situation and manners of the contending nations might amuse a philosophic mind': Edward Gibbon, *History of the Decline and Fall of the Roman Empire* (London, 1776), Ch. 6.

147. Hume to Gibbon, Edinburgh, 18 March 1776, in *Letters of Hume*, vol. 2, p. 310.

148. *Boswell: The Ominous Years 1774–1776*, ed. C. Ryskamp and F.A. Pottle (London, 1963), p. 73.

149. 'Of the Poems of Ossian' in J.H. Burton, *Life and Correspondence of David Hume* (Edinburgh, 1846).

150. Creech, *Letters to Sir John Sinclair* (see Ch. 1, n. 7), p. 12.

151. 'The Songs of Selma' in *Fingal*, p. 218.

Chapter 7: Torrents of Wind

1. Defoe, *Tour thro' Great Britain* (see Ch. 1, n. 12), vol. 2, p. 710.

2. 'Of the External Senses' in Adam Smith, *Essays on Philosophical Subjects*, ed. W.P.D. Wightman and J.C. Bryce (Oxford, 1980), p. 152.

3. Drummond said at the foundation-laying ceremony for the North

Bridge that he was 'only beginning to execute what the Duke of York . . . had suggested so far back as 1681 when residing at Holyrood': William Baird, 'George Drummond: An Eighteenth-Century Lord Provost' in *The Book of the Old Edinburgh Club*, vol. 4 (1911), p. 50.

4. Andrew Fletcher of Saltoun, 'Second Discourse concerning the Affairs of Scotland', in *Political Works* (see Ch. 8, n. 4), p. 172.

5. For his work in achieving the passage of this Act, Drummond was paid expenses of £1,420 sterling and £500 as a 'compliment': A.J. Youngson, *The Making of Classical Edinburgh, 1750–1840* (Edinburgh, 1966), p. 23.

6. 'The access to them would be easy on all hands, and the situation would be agreeable and convenient, having a noble prospect of all the fine ground towards the sea, the frith of Forth, and coast of Fife. One large and long street, in a straight line, where the Long Gate [or Lang Dykes] is now; on one side of it would be a fine opportunity for gardens down to the North Loch and one, on the other side, towards Broughton. No houses to be on the bridge . . . Another bridge might also be made on the other side of the Town, and almost as useful and commodious as that on the north. The place where it could be most easily made is St Mary's Wynd and the Pleasants. The hollow there is not so deep as where the other bridge is proposed; so that it is thought two stories of arches might raise it near upon a level with the street at the head of St Mary's Wynd. Betwixt the south end of the Pleasants and the Potter-Row, and from thence to Bristo Street, and by the back of the wall at Herriot's Hospital, are fine situations for houses and gardens.' Quoted by Sir John Sinclair in *The Statistical Account of Scotland, Drawn up from the Communications of the Ministers of the Different Parishes* (Edinburgh, 1791–), vol. 8, pp. 648–9.

7. *Guy Mannering*, Ch. 36.

8. Hugo Arnot, *The History of Edinburgh* (Edinburgh and London, 1779), p. 382.

9. [Sir Gilbert Elliot of Minto], *Proposals for carrying on certain Public Works in Edinburgh* (Edinburgh, n.d.), advertisement.

10. Ibid.

11. Ibid., p. 1.

12. Ibid., p. 5.

13. Ibid., p. 8.

14. Ibid., pp. 20–1.

15. The quantities of linen stamped for excise purposes in Scotland were 2.2 million yards in 1727, 7.4 million in 1748, 11.8 million in 1768, and 25.5 million in 1788.

16. *Proposals*, pp. 25–6.

17. Ibid., p. 30.

18. Ibid., p. 31.

19. Ibid., p. 31.

20. Ibid., p. 32.

21. 'Quelques gens ont pensé qu'en assemblant tant de peuple dans une capitale, on diminuait le commerce, parce que les hommes ne sont plus à une certaine distance les uns des autres. Je ne le crois pas; on a plus de désirs, plus de besoins, plus de fantaisies, quand on est ensemble.' 'Du Luxe', in Montesquieu, *De l'Esprit des Lois* (1748), VII, 1.

22. *Proposals*, p. 32.

23. Ibid.

24. In the pamphlet, Dalrymple made play with the hackneyed English complaint about Edinburgh's sanitary arrangements: 'In a word, London is opulent and beautiful, because it abounds in necessary-houses; Edinburgh, from being deprived of these conveniences, is poor, and wallowing in mire.' Sir David Dalrymple (Lord Haites), *Proposals for carrying on a certain Public Work in the City of Edinburgh* (n.p., n.d.), p. 6.

25. Robert Fergusson, 'Auld Reikie' ll. 313–14, in *The Poems of Robert Fergusson*, ed. Matthew McDiarmid (Edinburgh, 1954), vol. 2, p. 118.

26. Adam Smith, *Wealth of Nations*, V.iii.89 (Oxford, 1976), p. 944.

27. 23 August 1736: George Drummond's Diary, EUL DC 1.82, vol. 1, f. 43.

28. Ibid., f. 46.

29. Ibid., f. 48.

30. Ibid., f. 96.

31. Ibid., f. 296.

32. Ibid., f. 301.

33. Ibid., f. 298.

34. George Drummond's Diary, EUL DC 1.83, vol. 2, f. 7.

35. Ibid., f. 12.

36. Ibid., f. 164.

37. Ibid., ff. 164–5.

38. Ibid., f. 167.

39. Ibid., f. 165.

40. Ibid., f. 169.

41. Ibid.

42. Alexander Carlyle of Inveresk, *Autobiography* ed. J.H. Burton (Edinburgh, 1910), p. 491.

43. Ibid., p. 130.

44. Drummond's Town Council created five professorships: Chemistry, the Theory of Physic and the Practice of Physic, Midwifery, and *Belles-lettres* and Rhetoric.

45. *My Own Life and Times* (see Ch. 1, n. 7), p. 47. The date is my conjecture from Somerville's recollection on p. 48: 'The Town Council were

then taking in plans and estimates for building the North Bridge, and the North Loch was in progress of being drained.'

46. Ibid.

47. Town Council Minutes, Wednesday 6 May 1752: Edinburgh City Archives SL1/1/70, f. 138–9.

48. Ibid.

49. Ibid., 1 July 1752, SL1/1/70, f. 186.

50. Ibid., f. 187.

51. Arnot, *History of Edinburgh* (see n. 8), p. 521.

52. Town Council Minutes, 13 September 1769, SL 1/1/85, f. 278.

53. Ibid., 1 July 1752, SL1/1/70, f. 188.

54. Forbes of Pitsligo, *Memoirs of a Banking-House* (London and Edinburgh, 1860), p. 26.

55. Arnot, *History of Edinburgh*, p. 312.

56. *Scots Magazine*, vol. 21 (1759), pp. 383 and 442.

57. Alastair Smart, *The Life and Art of Allan Ramsay* (London, 1952), pp. 125–6.

58. Arnot, *History of Edinburgh*, p. 318.

59. Ibid., p. 324.

60. Thomas Pennant, *A Tour in Scotland, MDCCLXIX* (Warrington, 1774), p. 57.

61. Robert Chambers, *Traditions of Edinburgh* (Edinburgh, c. 1910), p. 5.

62. Scott, *Provincial Antiquities and Picturesque Scenery of Scotland* (London, 1826), p. 75.

63. 'What business has he to upbraid us,' I said, 'with the change of our dwelling from a more inconvenient to a better quarter of the town? What was it to him if we chose to imitate some of the conveniences or luxuries of an English dwelling-house, instead of living piled up above each other in flats? Have his patrician birth and aristocratic fortunes given him any right to censure those who dispose of the fruits of their own industry, according to their own pleasure.' *Redgauntlet*, Letter 5.

64. Arnot, *History of Edinburgh*, pp. 528–9.

65. *Redgauntlet*, Ch. 11.

66. John Ramsay of Ochtertyre, *Scotland and Scotsmen in the Eighteenth Century* (Edinburgh and London, 1888), vol. 1, p. 186.

67. Ibid., p. 187.

68. 'Before his promotion the dinner-hour of people of fashion was three o'clock, and that of writers [attorneys], shopkeepers, etc., two, when the bell rung; but his late and irregular hours made the ladies agree to postpone their meal till four': Ramsay of Ochtertyre, *Scotland and Scotsmen*, vol. 1, p. 337. The bookseller William Creech, reminiscing in 1793, put the hours at two or 'a little after' for the quality, and for shopkeepers at one: *Letters addressed to Sir John Sinclair, Bart* (see

Ch. 1, n. 7), p. 32. Scott, who is accurate in these matters, had Fairford senior invite Herries of Birrenswerk to dinner thus: 'We will expect the honour of seeing you in Brown's Square at three o'clock'. *Redgauntlet*, Letter V.

69. 'At Edinburgh, the old judges had a practice at which even their barbaric age used to shake its head. They had always wine and biscuits *on the bench*, when the business was clearly to be protracted beyond the usual dinner hour. The modern judges – those I mean who were made after 1800, never gave in to this; but with those of the preceding generation, some of whom lasted several years after 1800, it was quite common. Black bottles of strong port were set down beside them on the bench, with glasses, caraffes of water, tumblers and biscuits; and this without the slightest attempt at concealment. The refreshment was generally allowed to stand untouched, and as if despised, for a short time, during which their Lordships seemed to be intent only on their notes. But in a little, some water was poured into the tumbler, and sipped quietly, as if to sustain nature. Then a few drops of wine were ventured upon, but only with the water: till at last patience could endure no longer, and a full bumper of the pure black element was tossed over; after which the thing went on regularly, and there was a comfortable munching and quaffing, to the great envy of the parched throats in the gallery. The strong-headed stood it tolerably well, but it told plainly enough, upon the feeble.' Henry Cockburn, *Memorials of His Time* (Edinburgh, 1910), pp. 326–7.

70. *Scots Magazine*, vol. 25 (1763), p. 177.

71. Ibid., p. 362.

72. Ibid., pp. 362–4.

73. Ibid., p. 363.

74. Ibid., p. 364.

75. Ibid., p. 582.

76. Ibid.

77. Town Council Minutes, 7 November 1764, SL 1/1/80, f. 230.

78. Ibid., 16 January 1765, SL 1/1/80, ff. 279–80.

79. Ibid., 20 February 1765, SL 1/1/80, f. 335.

80. Ibid., 16 July 1765, SL 1/1/81, ff. 113–17.

81. 'Auld Reikie', ll. 344–50, McDiarmid, *Poems of Fergusson* (see n. 25) p. 19.

82. Arnot, *History of Edinburgh* (see n. 8). p. 315.

83. Robert Chambers, *Traditions of Edinburgh* (Edinburgh, *c.* 1910), p. 6.

84. Hugh Paton, ed., *A Series of Original Portraits and Caricature Etchings by the late John Kay, Miniature Painter, Edinburgh* (Edinburgh, 1877), vol. 1, p. 11.

85. Forbes of Pitsligo, *Memoirs of a Banking-House* (see n. 54), p. 39.

86. David Hume to Adam Smith, Edinburgh, 27 June 1772, in *Correspondence of Adam Smith*, ed. E.C. Mossner and I.S. Ross (Oxford, 1987), p. 162.

87. Youngson, *Making of Classical Edinburgh* (see n. 5), p. 71.

88. Ibid.

89. T.C. Smout, *A History of the Scottish People*, 1560–1830 (London, 1972), p. 347.

90. D. Daiches, *The Paradox of Scottish Culture* (London, 1964), p. 68.

91. *Plan of the new streets and squares intended for the City of Edinburgh*, 1767.

92. Town Council Minutes, 14 October 1767, SL 1/1/83, f. 227.

93. Ibid.

94. Ibid., 13 September 1769, SL 1/1/85, f. 278.

95. Pennant, *Tour of Scotland* (see n. 60), p. 58.

96. J. Farington, Notebook 3, Edinburgh Room, Edinburgh Central Library.

97. F.A. Pottle and C.H. Bennett, eds, *Boswell's Journal of a Tour to the Hebrides with Samuel Johnson, LL.D.* (London, 1936), p. 24.

98. Town Council Minutes, 13 September 1769, SL 1/1/85, f. 278.

99. Alexander Kincaid, *The History of Edinburgh; from the Earliest Accounts to the Present Time* (Edinburgh, 1787), pp. 125–6.

100. Town Council Minutes, 31 January 1781, SL 1/1/100 f. 313–17.

101. 'For Beef and Cabbage (a charming Dish), and old Mutton and old Claret, no body excels me': to Sir Gilbert Elliot of Minto, Edinburgh, 16 October 1769, in *The Letters of David Hume*, ed. J.Y.T. Greig (Oxford, 1932), vol. 2, p. 208.

102. *Caldwell Papers* (see Ch. 3, n. 12), part 2, vol. 2, pp. 177–8. The name St David's Street was supposedly chalked on the wall of Hume's house by his great favourite Nancy Ord as a joke; but it stuck. Mossner, *Life of Hume* (see Ch. 4, n. 26), p. 621, weighs the evidence for this story. Hume is listed in *Williamson's Directory* (see n. 106 below) at St Andrew's Street.

103. 'I have formerly noticed the extensive speculations which were entered into by some Scotchmen for the purchase and cultivation of lands in the newly acquired West India Islands; as well as the spirit which took place about the time for improvements in agriculture at home. Some of the houses which carried on the banking business in Edinburgh, having embarked in these speculations, required a larger capital than their own resources could command. To this must be added, the rage which then began to take place for building larger and more expensive houses, than had been customary in Edinburgh before the plan of the New Town was set on foot; and large houses necessarily led to more extensive establishments, as to furniture, servants, and equipages. At the same time those projectors and improvers, flattering themselves with the prospect of the immense advantage to be derived from their

speculations, launched out into a style of living up to their expected profits, as if they had already realised them. Such causes combined had induced those gentlemen to have recourse to the ruinous mode of raising money by a chain of bills on London.' Forbes of Pitsligo, *Memoirs of a Banking-house*, p. 39.

104. *The precipitation and fall of Mess. Douglas, Heron, and Company, late Bankers in Air, with the causes of their . . . ruin, investigated . . . by a Committee of Inquiry, appointed by the proprietors* (Edinburgh, 1778).

105. To James Burness, Lochlee, 21 June 1783, in *The Letters of Robert Burns*, ed. J. de L. Ferguson (Oxford, 1931), vol. 1, p. 16.

106. *Williamson's Directory for the City of Edinburgh, Canongate, Leith, and Suburbs from June, 1775 to June 1776* (Edinburgh, 1776).

107. Arnot gives a survey figure of 13,806 families, assessed at six members each, for a total of 80,836. To that figure, he added a further 1,400 single persons in the Castle garrison, the Charity Workhouse and the hospitals. *History of Edinburgh* (see n. 8), pp. 339–40.

108. Ibid., p. 353.

109. Ibid., p. 318.

110. Cockburn, *Memorials of His Time*, p. 27.

111. M. Tait and W.F. Gray, 'George Square, Annals of an Edinburgh Locality, 1766–1926' in *The Book of the Old Edinburgh Club*, vol. 26 (1948).

112. Scott, *Provincial Antiquities* (see n.62), p. 75.

113. Town Council Minutes, 3 March 1784, SL 1/1/105, f. 69.

114. Ibid., 17 September 1788, SL 1/1/1/112, ff. 188–9.

115. Ibid., 26 October 1791, SL 1/1/118, ff. 308–9.

116. Youngson, *Making of Classical Edinburgh* (see n. 5), p. 90.

117. *Edinburgh: Picturesque Notes*, pp. 99–100.

118. Ibid., p. 16.

119. Creech, *Letters Addressed to Sir John Sinclair, Bart* (see Ch. 1, n. 7), p. 7.

120. At least two Nories, John and Robert, were active as decorative painters in Edinburgh in the 1740s.

121. Chambers, *Traditions of Edinburgh*, pp. 9–10.

122. [James Wilson], 'A Sermon preached by Claudero on the Condemnation of the Netherbow Porch of Edinburgh, 9th July 1764, before a crowded Audience' in *Miscellanies in Prose and Verse on Several Occasions, by Claudero, Son of Nimrod the Mighty Hunter* (Edinburgh, 1766), p. 57.

123. Ibid.

124. 'The Echo of the Royal Porch of the Palace of Holy-Rood-House which fell under Military Execution Anno 1753' in ibid., p. 2.

125. David Robertson, 'The Surplus Fire Fund' in *The Princes Street Proprietors* (Edinburgh, 1935), pp. 216–17.

126. *Edinburgh: Picturesque Notes*, p. 29.

127. Ibid., p. 33.

128. Chambers, *Traditions of Edinburgh*, p. 8.

Chapter 8: The Savage and the Shopkeeper

1. 'They would sometimes present their piece; and upon being asked what they wanted, answer *A Penny*; with which they would rest satisfied': *The Scots Magazine*, vol. 7 (1745), p. 442.

2. There are several versions of Balmerino's instructions to the headman. This is from Robert Chambers's *History of the Rebellion of 1745–6* (Edinburgh, 1869), p. 456.

3. Adam Smith, *Wealth of Nations*, III.iv.8 (Oxford, 1976), p. 416.

4. 'A Discourse of Government with Relation to Militias' (Edinburgh, 1698), in *The Political Works of Andrew Fletcher, Esq.* (London, 1732), p. 54ff; and 'The Second Discourse Concerning the Affairs of Scotland; Written in the Year 1698' in *Political Works*, p. 121ff.

5. 'There seems not to be the smallest chance, that there shall be any sudden increase of mankind, equal to what appeared in ancient times': Robert Wallace, *A Dissertation on the Numbers of Mankind in antient and modern Times* (Edinburgh, 1753), p. 147. Wallace confesses, in the little manuscript essay entitled 'On Venery' in the EUL (see Ch. 9, n. 92), that for a long time he was uncertain about the facts of life.

6. David Hume, *Political Essays*, ed. K. Haakonssen (Cambridge, 1994), p. 98. Hume thought the Spartan constitution was 'a philosophical whim or fiction' that just happened to be supported by positive and circumstantial report: ibid., p. 97.

7. *Political Essays*, p. 107.

8. 'An Address to the Reverend the Clergy of the Church of Scotland by a Layman of their Communion on occasion of composing, acting & publishing the Tragedy called DOUGLAS', EUL La II 620/2 f. 22.

9. 'Prisca juvet alius; ego me nunc denique natum/Gratulor   '

10. Robert Wallace, *Characteristics of the Present Political State of Great Britain* (London, 1758), pp. 156–7. Even Lord Elibank, who had answered the *Political Discourses* with a backwoods and anti-semitic attack on the public debtholders, had by 1758 awoken to the virtues of paper currency: *An Inquiry into the Original and Consequences of the Public Debt* (Edinburgh, 1753) and *Thoughts on Money, Circulation and Paper Currency* (Edinburgh, 1758).

11. Robert Wallace, *Various Prospects of Mankind, Nature and Providence* (London, 1761), p. 58 and p. 67.

12. Ibid.

13. 'The republic of Lycurgus represents the most perfect plan of political

oeconomy, in my humble opinion, anywhere to be met with, either in ancient or modern times. That it existed cannot be called into question, any more than that it proved the most durable of all those established among the Greeks; and if at last it came to fail, it was more from the abuses which gradually were introduced into it, than from any vice in the form.' Sir James Steuart, *An Inquiry into the Principles of Political Oeconomy*, ed. A. Skinner (Edinburgh, 1966), p. 218.

14. For example: '"Mercantilism" was the economic outlook which dominated the Stuart century and beyond, aptly reaching its apogee in the writings of Sir James Steuart, before being shot down by Adam Smith's *Wealth of Nations*': Roy Porter, *Enlightenment Britain and the Creation of the Modern World* (London, 2000), p. 385.

15. His great-grandfather, Sir James Stewart of Kirkfield and Coltness (1608–81), was Lord Provost of Edinburgh under both King Charles I and Cromwell, and presided over the execution of Montrose. He was notorious for having a bible in every room of his house: 'that still [permanently] lay there, as part of the furnitour': Robert Wodrow, *Analecta, or Materials for a History of Remarkable Providences* (Edinburgh, 1843), vol. 1, p. 71.

16. He courted Henrietta Baillie, who married Robert Dundas the younger. At the Midlothian parliamentary election of 1743, Steuart found his name omitted on a technicality from the electoral roll of freeholders. Calling over the roll was Robert Dundas the elder, Lord Arniston, a justice of the Court of Session. Steuart sued but, on 9 February, Arniston's fellow judges sided with their colleague: *Elchies Decisions of the Court of Session from the Year 1733 to the year 1754, Collected and Digested in the Form of a Dictionary*, ed. W.M. Morison (Edinburgh, 1813), vol. 1, 9 February 1744.

17. Elcho in 'Documents relatifs' (see Ch. 2, n. 9), p. 101.

18. 'La plupart, comme le comte de Buchan, les chevaliers Macdonald de Slate et Stuart de Coltness [Sir James Steuart], Mylord Lovat, furent d'avis de'envoyer un express au prince pour lui dire qu'il pouvait venu dans son royaume avec 6,000 hommes de troupes réglées, des armes pour 10,000 et 30,000 louis d'or argent comptant, qu'il pouvait venir et que tous ses partisans se joindiraient à lui, mais en cas qu'il ne pût venir avec tous ces secours, on lui conseilla très fort de ne pas venir, car sa présence ne pouvait causer que sa propre ruine, la ruine de sa cause, et de tous ceux qui s'embarqueraient avec lui.' Elcho in 'Documents relatifs', p. 102.

19. 'It's ane insult upon common sence, Sir James his practice, he has been ane oppen tool and now that the Prince is come he, Sir James of Gutters and Lord Cardros [i.e., his cousin the Earl of Buchan] whom he has seduced, ar under arrest by the Prince as is pretended and owt upon

paroll.' *The Woodhouselee MS* (see Ch. 2, n. 41), p. 49. Robert Chambers, *History of the Rebellion* (see n. 2), pp. 164–5, describes their arrest at the Cross. John Ramsay of Ochtertyre: 'When the Rebellion broke out, he did not act with the same generosity and manliness that were displayed by the other partisans of the Stuart family. Instead of openly venturing his life and fortune to procure its restoration, he was contented with being the Prince's confidential man, and drawing up his manifestoes and other public papers.' *Scotland and Scotsmen in the Eighteenth Century* (Edinburgh and London, 1888), vol. 1, p. 362.

20. Murray of Broughton, in his interrogation, said that its authors were the Prince's old tutor, Sir Thomas Sheridan, and Steuart: Chambers, *History of the Rebellion*, p. 149. Elizabeth Mure, in a letter to Mrs Calderwood Durham of 1787, said that Steuart 'wrote the Manifesto': 'Documents relatifs', p. 116.

21. 'With respect to the pretended Union of the two Nations, the King cannot possibly ratify it, since he has had repeated Remonstrances against it from each Kingdom; and since it is incontestable that the principal point then in View, was the exclusion of the Royal Family from their undoubted Right to the Crown, for which Purpose the grossest Corruptions were openly used to bring it about: But whatever may be hereafter devised for the joint Benefit of both Nations, the King will most readily comply with the Request of his Parliaments to establish.' BL C115 i.3 67.

22. According to Lord Elcho, the prince said it was 'below his dignity to Enter into such a reasoning with Subjects, and order'd the paper to be laid aside': Elcho, *Short Account* (see Ch. 2, n. 56), p. 290.

23. *Scots Magazine*, vol. 9 (1747), p. 259.

24. Ibid., vol. 10 (1748), p. 507.

25. Ramsay of Ochtertyre, *Scotland and Scotsmen*, vol. 1. p. 362.

26. Henry Mackenzie reported that when Steuart returned from exile, he invited some friends to dine in celebration, and wishing for sober conversation, proposed the wine and glasses be cleared after dinner. When he looked up, many of his guests were asleep. 'I see, Gentlemen, this won't do in Scotland,' he is said to have said, and called for wine: *Anecdotes and Egotisms* (see Ch. 1, n. 57), p. 2.

27. In documents turned up by the author in the National Archives of Scotland (NAS CD 110/1053 Nos 2–4), Steuart's son James, a Captain of Horse, is accused of conducting a love affair with his married aunt, Lady Ann Hamilton. Among the documents is a contract, signed by Sir James Steuart, that the young man will leave Scotland for seven years and not return without written permission from Lady Ann's husband, John Hamilton of Bargany.

28. Steuart, *Inquiry* (see n. 13), p. 28.

29. Ibid., p. 154.

30. Ibid., p. 45.

31. Ibid., p. 44. He reasoned too coolly over the balance of trade in Corsica for Boswell's taste, and civilly dropped the passage in the corrected edition: *Inquiry*, p. 360, and F. Brady and F.A. Pottle (eds), *Boswell in Search of a Wife, 1766–1769* (London, 1956), p. 92.

32. Steuart, *Inquiry*, p. 281.

33. Ibid., p. 209.

34. Ibid., p. 218.

35. *Wealth of Nations*, IV.i.19.

36. Steuart, *Inquiry*, pp. 315–16. Likewise Elibank in his *Thoughts on Money* (see n. 10), p. 18: 'The word *Money*, in its original and proper sense, is only a relative term to express the value of Commodities, as much as a Tun, a Pound, or a Yard, are made of, to denote their quantity; but, like a statue in a Popish Church, it is constantly mistaken by the Vulgar, and has that worship bestowed on it, which is only due to the Saint it was meant to represent.'

37. Steuart, *Inquiry*, p. 217.

38. Ibid., p. 279.

39. Ibid., p. 16.

40. Ibid., p. 296.

41. Ibid., p. 278.

42. Ibid., p. 217.

43. *The Critical Review: or, Annals of Literature*, London, June 1767, vol. 23, p. 412.

44. Ibid.

45. *The Monthly Review, or Literary Journal*, London, June 1767, vol. 26, p. 464.

46. *Georg Wilhelm Friedrich Hegels Leben* (Berlin, 1844), p. 86.

47. The passage is this: 'Diese mannigfachen Arbeiten der Bedürfnisse als Dinge müssen ebenso ihren Begriff, ihre Abstraktion realisieren; ihr allgemeiner Begriff muss ebenso ein Ding sein, wie sie, das aber als Allgemeines alle vorstellt. Das Geld ist dieser materielle, existierende Begriff, die Form der Einheit, oder der Möglichkeit aller Dinge des Bedürfnisses. Das Bedürfnis und die Arbeit in diese Allgemeinheit erhoben, bildet so für sich in einem grossen Volk ein ungeheureres System von Gemeinschaftlichkeit und gegenseitiger Abhängigkeit, ein sich bewegendes Leben des Toten, das in seiner Bewegung blind und elementarish sich hin und her bewegt, und als ein wildes Tier einer beständigen strengen Beherrschung und Bezähmung bedarf.' J. Hoffmeister, ed., *Jenenser Realphilosophie* I (Leipzig, 1932), pp. 239–40.

48. 'Wealth, commerce, extent of territory and knowledge of arts ... are of little significance, when only employed to maintain a timid, dejected,

and servile people': *An Essay on the History of Civil Society*, ed. Fania Oz-Salzberger (Cambridge, 1995), p. 60.

49. Most notable is Part IV, Section IV in which Ferguson has an imaginary contemporary traveller visit ancient Greece: Adam Ferguson, *Essay on Civil Society*, pp. 185–8.

50. Scott, *Miscellaneous Prose Works* (Edinburgh, 1835), pp. 331–2. John Small, *Biographical Sketch of Adam Ferguson, LL.D, FRSE* (Edinburgh, 1864), p. 3.

51. *A Sermon Preached in the Ersh Language* (see Ch. 2, n. 121), p. 3.

52. In a letter to Lord Milton at that time, Drummond wrote: 'You know I devote Saturday and Monday to our friends of the town [the trades] to strengthen our harmony in the Councill. On the Saturday I mix some other folks with them . . . so tomorrow I am to move the Councill to call our Clergy together for their *avisamentum* on Friday next, and the Wednesday following we will elect Fergusone unanimously by the merchants, and I imagine by the trades too.' Quoted in T.C. Smout, *Provost Drummond* (Edinburgh, 1978), p. 16.

53. Alexander Carlyle of Inveresk, *Autobiography*, ed. J.H. Burton (Edinburgh, 1910), pp. 297–8.

54. Adam Ferguson, *The Morality of Stage Plays Seriously Considered* (Edinburgh, 1757).

55. Carlyle, *Autobiography*, p. 297.

56. Fletcher of Saltoun, *Political Works* (see n. 4), p. 48. Fletcher objected to standing or mercenary armies not merely because they increased the power of the Crown and were a threat to the liberty of the freeholder but also because of the licentiousness of both officers and men. His own millitia was to be a 'school of virtue', in which young men in their early twenties would submit to religious as well as military exercises, women were proscribed, and 'the crime of abusing their own bodies any manner of way, punished with death.' Ibid., pp. 54–64.

57. Adam Ferguson, *Reflections Previous to the Establishment of a Militia* (London, 1756), pp. 24–5. At the same time, Wallace sent a typically able paper to the Lord Advocate that argued 'Brittain can not be secure against being conquered by France without a powerfull militia or a powerfull standing army' and, since the last was a danger to liberty, 'Upon the whole the best method is to make a whole people capable of fighting . . . The Highlands may be excepted or any Jacobite country, but indeed it would be bad policy to except any part of the nation whatsoever.' Robert Wallace, 'Copy of a Scheme for a Militia in Brittain', 27 March, 1756, EUL La ll. 620/6, f. 19.

58. 'Ferguson has very much polished and improved his Treatise on Refinement, and with some Amendments it will make an admirable Book': Hume to Smith, London, 12 April 1759, in *Correspondence of Adam Smith*, ed. E.C. Mossner and I.S. Ross (Oxford, 1987), p. 34.

59. [Alexander Carlyle], *The Question Relating to a Scots Militia, Considered in a Letter to the Lords and Gentlemen who have concerted the form of a law for that establishment, by a Freeholder* (Edinburgh, 1760), p. 32.

60. *The History of the Proceedings in the Case of Margaret, Commonly called Peg, only lawful Sister to John Bull, Esq.* (London, 1761), p. 37. In his *Autobiography*, Carlyle wrote: 'Adam Ferguson fell luckily on the name of "Poker", which we perfectly understood, and was at the same time an enigma to the public.' J.H. Burton, editor of the 1910 edition, glossed 'poker' as 'An instrument for stirring up the militia question': *Autobiography* (see n. 53), p. 439.

61. Members reconstituted the old Diversorium at Nicolson's new premises near the site of the Cross, until a quarrel caused them to move to the more fashionable, and expensive, Fortune's tavern in Stamp-Office Close on the north side of the High Street: *Autobiography*, pp. 439–44.

62. 'I do not think them fit to be given to the Public, neither on account of the Style nor the Reasoning; the Form nor the Matter': David Hume to Blair, London, 11 February 1766, in *The Letters of David Hume*, ed. J.Y.T. Greig (Oxford, 1932), vol. 2, p. 12.

63. Carlyle, *Autobiography*, p. 299.

64. *Boswell in Search of a Wife* (see n. 31), pp. 37–8.

65. Dugald Stewart no doubt derived his phrase 'conjectural history' from Ferguson's critical use of the word 'conjectures' in Part One of the *Essay*, p. 9.

66. Adam Ferguson, *Essay on Civil Society*, p. 14.

67. Ibid., p. 10.

68. Ibid., pp. 119–20.

69. Ibid., p. 36.

70. Ibid., p. 140.

71. 'Were Lycurgus employed anew to operate on the materials we have described, he would find them, in many important particulars, prepared by nature herself for his use. His equality in matters of property being already established, he would have no faction', etc.: ibid., p. 93.

72. Ibid., p. 177.

73. Ibid., p. 200.

74. Ibid., p. 174.

75. Ibid., p. 177.

76. Ibid., p. 218.

77. Ibid., p. 239.

78. Matt Bramble to Dr Lewis, 2 September.

79. Matt Bramble to Dr Lewis, 15 July.

80. Ferguson to Smith, Edinburgh, 23 January 1774, in *Correspondence of Adam Smith*, p. 170.

81. Adam Ferguson, *Remarks on a Pamphlet lately Published by Dr Price,*

intitled, Observations on the Nature of Civil Liberty, the Principles of Government, and the Justice and Policy of the War with America &c (London, 1776), p. 59. Blair 'prayed against' the American colonists from the High Kirk in 1776: C. McC. Weis and F.A. Pottle, eds, *Boswell in Extremes, 1776–1778* (London 1971), p. 56 and pp. 357–8.

82. Ferguson to Carlyle, 2 October 1797: EUL DC 4.41.57.

83. Ibid.

84. In *Medico-Chirurgical Transactions*, vol. 7, p. 230.

85. Henry Cockburn, *Memorials of His Time* (Edinburgh, 1910), p. 45.

86. Ibid., p. 44.

87. 'Early Draft of Part of *The Wealth of Nations*/ Chap. 2: Of the nature and causes of public opulence' in *Lectures on Jurisprudence*, ed. R.L. Meek, D.D. Raphael and P.G. Stein (Oxford, 1978), p. 562.

88. *Theory of Moral Sentiments*, ed. D.D. Raphael and A.L. MacFie (Oxford, 1976), VII.iv. 37, p. 342.

89. *Lectures on Jurisprudence*, p. 561.

90. 'Early Draft' in *Lectures on Jurisprudence*, p. 562.

91. Ibid., p. 574.

92. 'Report dated 1766' in *Lectures on Jurisprudence*, p. 541.

93. 'Tho I have had no concern myself in the Public calamities, some of the friends for whom I interest myself the most have been deeply concerned in them; and my attention has been a good deal occupied about the most proper method of extricating them': to William Pulteney, Kirkcaldy, 3 September 1772, in *Correspondence of Adam Smith* (see n. 58), p. 163.

94. With Steuart, at least, it was deliberate neglect. As he wrote to William Pulteney in September 1772, 'I have the same opinion of Sir James Stewarts Book that you have. Without once mentioning it, I flatter myself, that every false principle in it, will meet with a clear and distinct confutation in mine': *Correspondence of Adam Smith*, p. 164.

95. Much, not all: 'The man whose whole life is spent in performing a few simple operations, of which the effects, too, are perhaps, always the same, or very nearly the same, has no occasion to exert his understanding, or to exercise his invention in finding out expedients for removing difficulties which never occur. He naturally loses, therefore, the habit of such exertion, and generally becomes as stupid and ignorant as it is possible for a human creature to become.' *An Inquiry into the Nature and Causes of the Wealth of Nations*, ed. R.H. Campbell, A.S. Skinner and W.B. Todd (Oxford, 1976), v.i.f.50, p. 782. Marx, who knew nothing of Adam Smith's lectures or unpublished works, thought that the author of *The Wealth of Nations* had been influenced to denounce the division of labour by his 'teacher, A. Ferguson': *Capital*, vol. 1, p. 124, n. 1. Carlyle, more plausibly, reports that Smith accused Ferguson of plagiarism (or, rather, 'of having borrowed some of his inventions'): *Autobiography*, p. 299.

96. *Wealth of Nations*, I.ii. 1, p. 25.

97. Ibid., IV.ix.51, p. 687.

98. 'Report dated 1766', in *Lectures on Jurisprudence*, pp. 543–5.

99. *Wealth of Nations*, V.i.a.41, p. 707.

100. Ibid., V.i.a.27, p. 701.

101. Ibid., IV.vii.c.79, pp. 625–6.

102. 'In the present imperfect condition of society, luxury, though it may proceed from vice or folly, seems to be the only means that can correct the unequal distribution of property. The diligent mechanic, and the skilful artist [artisan], who have obtained no share in the division of the earth, reaceive a voluntary tax from the possessors of land; and the latter are prompted, by a sense of interest, to improve those estates, with whose produce they may purchase additional pleasures.' *The Decline and Fall of the Roman Empire in the West* (London, 1776), vol. 1, p. 54.

103. 'Dissertation Exhibiting the Progress of Philosophy' in Dugald Stewart, *Works* (see Prologue, n. 11), vol. 1, p. 504.

104. Samuel Johnson and James Boswell, *A Journey to the Western Islands of Scotland*, ed. P. Levi (London, 1984), p. 35.

105. Ibid., p. 151.

106. '*Homines mullius originis* . . . is, I am afraid, barbarous – Ruddiman is dead.' To James Boswell, 21 August 1766, in *Life of Johnson*, p. 367.

107. *Journey to the Western Islands*, p. 151.

108. *Boswell for the Defence*, ed. W.K. Wimsatt, Jr, and F.A. Pottle (New York and London, 1960), p. 202. Boswell had moved from Hume's rooms in James's Court to a bigger apartment on the ground floor on the Lawnmarket side. It is unthinkable that Johnson would have agreed to stay on infidel property. Robert Chambers, *Traditions of Edinburgh* (Edinburgh, c. 1910), p. 60.

109. Hugo Arnot, *The History of Edinburgh* (Edinburgh and London, 1779), p. 352.

110. F.A. Pottle and C.H. Bennett, eds, *Boswell's Journal of a Tour to the Hebrides with Samuel Johnson, LL.D, now first published from the original manuscript* (London, 1936), p. 12.

111. Ibid., p. 24.

112. The information comes from a hand-written note on the fly-leaf of the British Library copy of *An Account of a Savage Girl, Caught Wild in the Woods of Champagne* (Edinburgh, 1768): BL 957 d. 22.

113. Ibid., pp. xvii–xviii.

114. 'Went with Dr Solander and breakfasted with Monboddo, who listened with avidity to the Doctor's description of the New Hollanders, almost brutes – but added with eagerness, "Have they tails, Dr Solander?" "No, my Lord, they have not tails."' *Boswell for the Defence*, p. 146.

115. *Of the Origin and Progress of Language* (Edinburgh, 1773), pp. 174–5.

116. *Tour to the Hebrides*, p. 27.

117. Ibid., p. 53.

118. Ibid., p. 56.

119. 'It would certainly be very unbecoming in me to exhibit my honoured father, and my respected friend, as intellectual gladiators, for the entertainment of the publick': ibid., p. 375.

120. Ibid., p. 386.

121. Ibid.

122. Ibid., p. 115.

123. *Life of Johnson*, p. 1129.

124. *Tour to the Hebrides*, p. 116.

125. Ibid.

126. *Tour to the Hebrides*, p. 117.

127. Ibid.

128. *Journey to the Western Islands*, p. 97.

129. Ibid., p. 100.

130. *Journey to the Western Islands*, p. 99.

131. Ibid., p. 254.

132. *Tour to the Hebrides*, p. 159.

133. Ibid.

134. Ibid., p. 161. If this is a quotation, I cannot find its source. It scans (more or less) as the back end of a hexameter. Johnson may have had it in mind to compose a Latin couplet of this nature: 'How money must give way before virtue! How avarice must surrender to fidelity!'

135. *Journey to the Western Islands*, p. 73.

136. *Wealth of Nations*, III.iv.10.

137. John Walker, *An Economical History of the Hebrides and Highlands of Scotland* (Edinburgh, 1808), vol. II, pp. 404–14.

138. *Capital* (New York, 1967), vol. 1, p. 681.

139. Letter of 24 January 1775, in Edward Topham, *Letters from Edinburgh, Written in the Years 1774 and 1775* (Dublin, 1780), vol. 1, p. 159.

140. *Journey to the Western Islands*, p. 151.

141. Ibid., p. 152.

Chapter 9: The Art of Dancing

1. William Alexander, *The History of Women from the Earliest Antiquity to the Present Time* (London, 1779), vol. 2, p. 313.

2. 'Their condition is naturally improved by every circumstance which tends to create more attention to the pleasures of sex, and to increase the value of those occupations that are suited to the female character; by the cultivation of the arts of life; by the advancement of opulence;

and by the refinement of taste and manners.' John Millar, *The Origin of the Distinction of Ranks* (London, 1779), p. 69.

3. 'Their drawing-rooms are deserted; and after dinner and supper, the gentlemen are impatient till they retire': John Gregory, *A Father's Legacy to his Daughters* (London and Edinburgh, 1774), p. 40. 'In 1783 – The drawing-rooms were totally deserted . . .': William Creech, *Letters to Sir John Sinclair* (see Ch. 1, n. 7), p. 33.

4. Lord Chesterfield to his son, London, 5 September 1748: Charles Strachey, ed., *The Letters of the Earl of Chesterfield to His Son* (London, 1932), vol. 1, p. 261.

5. 'Nature has given *man* the superiority above *woman* by endowing him with greater strength both of mind and body': 'Of the Rise and Progress of the Arts and Sciences' in Hume, *Political Essays*, ed. K. Haakonssen (Cambridge, 1994), p. 74.

6. 'Of Refinement in the Arts' in Hume, *Political Essays*, p. 107.

7. Lt-Col J. Forrester, *The Polite Philosopher* (London, 1736), p. 47.

8. Smellie to ——, 1763 in *Memoirs of Smellie* (see Ch. 1, n. 13), vol. 1, pp. 165–6.

9. 'I have often thought that if a well-grounded affection be not really a part of virtue, 'tis something extremely akin to it. Whenever the thought of my E. warms my heart, every feeling of humanity, every principle of generosity kindles in my breast.' J. De L. Ferguson, ed., *The Letters of Robert Burns* (Oxford, 1931), vol. 1, p. 6.

10. Henry Mackenzie, *Julia de Roubigné* (Edinburgh, 1777), vol. 2, pp. 73–4.

11. Ibid., vol. 2, p. 75.

12. Kames to Miss Gordon, Dumfries, 12 May 1764, in Stewart of Afton Papers, BL Add. 40635, vol. 3, f. 18.

13. James Fordyce, 'On Female Virtue, with Intellectual Accomplishments' in *Sermons to Young Women* (London, 1767), vol. 1, p. 273.

14. James Hutton, *An Investigation of the Principles of Knowledge and of the Progress of Reason from Sense to Science and Philosophy* (Edinburgh and London, 1794), vol. 3, pp. 587–9.

15. 'Of the Rise and Progress of the Arts and Sciences' in Hume, *Political Essays*, p. 74.

16. 'We wish to feel the inferiority of the other sex, as one that does not debase but endear it': *Julia de Roubigné*, vol. 2, pp. 73–4.

17. William Robertson, *History of Scotland during the Reigns of Queen Mary and of King James VI* (London, 1759), vol. 2, p. 243.

18. William Hunter, 'On the Uncertainty of the Signs of Murder in the Case of Bastard Children' in *Medical Observations and Inquiries*, vol. 6 (1784), pp. 272–3.

19. Hugo Arnot, *The History of Edinburgh* (Edinburgh and London, 1779), p. 193.

NOTES   397

20. Ibid., p. 194.
21. The four women declared 'that the dread of the pillory was the cause of their murders'. Ibid., p. 193.
22. 'What was most extraordinary was, that, except in conversation for a few weeks only, this enormous act, committed in the midst of the metropolis of Scotland . . . was not taken the least notice of by any of her own family, or by the King's Advocate or Solicitor, or any of the guardians of the laws.' Alexander Carlyle of Inveresk, *Autobiography*, ed. J.H. Burton (Edinburgh, 1910), p. 13.
23. 'Upright, firm, yet easy Manner of the ladies Walking in *Edinburgh*': Edward Burt, *Letters from a Gentleman* (see Ch. 1, n. 33), vol. 1, p. 102.
24. Ramsay of Ochtertyre, *Scotland and Scotsmen* (see Ch. 1, n. 3), vol. 2, p. 61.
25. 'Some remarks on the change of manners' in *Caldwell Papers* (see Ch. 3, n. 12), part 1, pp. 262–3.
26. W.C. Dickinson and G. Donaldson, *A Source Book of Scottish History* (Edinburgh, 1954), vol. 3, p. 407.
27. To Susan, Marchioness of Tweeddale, 14 February 1716, NLS, Yester Papers, MS 14419, f. 136.
28. *Edinburgh Evening Courant*, 2 July 1752.
29. Robert Chambers, *Traditions of Edinburgh* (Edinburgh, c. 1910). p. 231.
30. *Heart of Midlothian*, Chapter 28.
31. *Atholl Correspondence* (see Ch. 2, n. 90), p. 95.
32. To Archibald Campbell of Knockbuy, 21 February 1745, in *Letters of MacLaurin* (see Ch. 2, n. 33), p. 124.
33. Mrs Robertson of Myreton, who died an old woman in 1756, remembered as a child coming in to the room at Kincardine where her grandfather and his friends were drinking. All the gentlemen had dirks stuck into the table except one, Graham of Gartur, who had a pistol before him. Her grandfather, by the way, was the minister at Kincardine. Ramsay of Ochtertyre, *Scotland and Scotsmen*, vol. 2, p. 48.
34. Ibid., vol. 2, p. 77.
35. 'In 1763 – It was the fashion for gentlemen to attend the drawing-rooms of the ladies in the afternoons, to drink tea, and to mix in the society and conversation of the women.' William Creech, *Letters Addressed to Sir John Sinclair, Bart., respecting the Mode of Living, Arts, Commerce, Literature, Manners, etc. of Edinburgh* (Edinburgh, 1793), p. 33.
36. C. McC. Weis and F.A. Pottle, eds, *Boswell in Extremes, 1776–1778* (London, 1971), p. 77.
37. 'An Eclogue' in *Poems of Fergusson*, vol. 2, p. 88.
38. Ramsay of Ochtertyre, *Scotland and Scotsmen*, vol. 2, p. 72.
39. Ibid., p. 73.
40. Arnot, *History of Edinburgh* (see n. 19), p. 167.

41. Burt, *Letters from a Gentlemen* (see Ch. 1, n. 33), vol. 1, pp. 236–8.

42. John Jackson, *The History of the Scottish Stage* (Edinburgh, 1793), p. 418. The house is illustrated in Robert Chambers, *Traditions of Edinburgh* (Edinburgh, *c.* 1910), p. 44.

43. 'Missive from the Countess of Panmure to ye Earle' [in France], 24 January 1723: NAS Dalhousie Muniments GD 45/14/220.

44. Anne Stuart to Elizabeth Dunbar, Edinburgh, 28 January, 1723, in E. Dunbar, ed., *Social Life in Former Days, Chiefly in the Province of Moray, Illustrated by Letters and Family Papers* (Edinburgh, 1865), p. 118.

45. Chambers, *Traditions*, p. 157.

46. *Biographia Presbyteriana* (Edinburgh, 1827), vol. 1, p. 138.

47. Burt, *Letters from a Gentleman*, vol. 1, p. 234.

48. Ibid., p. 235.

49. Ibid., p. 235.

50. Wodrow, *Analecta* (see Ch. 3, n. 9), vol. 3, p. 514.

51. Youngson, in *Classical Edinburgh* (see Ch. 7, n. 5), pp. 251–2, says that between 1746 and 1776 the Assembly donated £2,495 each to the Royal Infirmary and Charity Workhouse, and £1,439 to the Directresses' charities.

52. NAS GD 45/24/79.

53. Oliver Goldsmith to Robert Bryanton, Edinburgh, 26 September 1753, printed in John Forster, *The Life and Times of Oliver Goldsmith* (London, 1854), vol. 1, pp. 446–7.

54. Ibid., p. 447.

55. Burt, *Letters from a Gentleman*, vol. 1, p. 236.

56. Chapter 11.

57. Robert Chambers, *History of Edinburgh*, p. 382.

58. Ibid., p. 265.

59. Boswell to George Dempster, Edinburgh, 23 February 1769, in *Boswell in Search of a Wife*, ed. F. Brady and F.A. Pottle (London, 1957), p. 187.

60. Alexander Kincaid, in his *History of Edinburgh*, described the George Street rooms as they were about to open in January 1787. He says £6,300 had been raised in subscription for a Grand Ballroom of 92 ft by 42 ft and 40 ft high, with seven crystal lustres, including a central chandelier of forty lights measuring 10 ft 8 in. in height; a tea room (52 ft × 35 ft), two card rooms (each 33 ft × 18 ft) and a grand salon (24 ft square): *History of Edinburgh, from the Earliest Accounts to the Present Time* (Edinburgh, 1787), p. 125.

61. *The Woodhouseless MS* (see Ch. 2, n. 41), p. 83.

62. *The Culloden Papers* (see Ch. 2, n. 5), p. 250.

63. 'Après avoir parlé des hommes nos ennemis, il faut parler de *leurs femmes nos amies*. La plus belle et la plus singulière est Anne Fackirson,

fille du chef de ce nom et femme du chef des Mackintosh, dont on a parlé cy-dessus et qui vient d'etre fait prisonier. Elle aimoit éperdûment son mari qu'elle espéra longtems de gagner au Prince; mais, ayant appris qu'il s'étoit enfin engagé, avec le Président [i.e., Forbes], à servir la maison d'Hanovre, elle ne voulut plus le voir. Elle ne s'en tint pas là: elle souleva une partie de ses vassaux, à la teste desquels elle mit un très-beau cousin qui, jusques-là, l'avait aimée inutilement. Mackintosh fut obligé de quitter son lit, sa maison et ses terres. L'intrépide ladi, un pistolet d'une main et de l'argent de l'autre, parcourt le pais, menace, donne, promet, et, en moins de quinze jours, ramasse 600 hommes. Elle en avoit envoyé la moitié à Fakirt [i.e., Falkirk], qui y arriva la veille de la bataille. Elle avoit retenu l'autre moitié *pour se garder de son mari* et de Loudoun qui, à Inverness, n'étoient qu'à trois lieues de son château. Le prince logea chez elle, à son passage. Elle s'offrit à luy avec la grace et la noblesse d'une divinité, car rien n'est ce beau que cette femme. Elle luy présenta toute sa petite armée qu'elle avoit rassemblé, et, après avoit parlé aux soldats de ce qu'ils devoient à la situation, aux droits et aux vertus de leur Prince, elle jura très-catégoriquement de casser la tête au premier qui s'en tourneroit, après avoir, à ses yeux, brulé sa maison et chassé sa famille.' Boyer d'Éguilles to d'Argenson, Findhorn, 6 April 1746, in 'Correspondance inédite', *Revue Rétrospective* (Paris, 1886), pp. 145 6.

64. Dougal Graham, *Impartial History* (see Ch. 2, n. 91), p. 94.
65. 'About twenty years after the Fire; and in even that time . . . as the gay humour came on'. *The Complete English Tradesman* (London, 1738), vol. 2, p. 332.
66. David, Lord Elcho, *A Short Account of the Affairs of Scotland in The Years 1744, 1745, 1746,* ed. Hon. E. Charteris (Edinburgh, 1907), p. 261.
67. Carlyle, *Autobiography* (see n. 22), p. 163.
68. 'Ce n'est pas qu'il coquet ou galant, c'est peut-être, au contraire, parce qu'il ne l'est pas, et que les Ecossoises, naturellement sérieuses et passionnées, en concluent qu'il est véritablement tendre et qu'il seroit constant.' 'Correspondance inédite' in *Revue Rétrospective*, p. 147.
69. That is, Col. Gardiner's, who fell at Prestonpans.
70. Letter of 13 October in Tayler, ed., *A Jacobite Miscellany* (see Ch. 2, n. 34), pp. 40–2.
71. *Atholl Correspondence* (see Ch. 2, n. 90), p. 30.
72. Dougal Graham, *Impartial History*, p. 120.
73. *Memoirs of Sir John Clerk* (see Ch. 1, n. 29), p. 197.
74. Hume to Sir James Johnstone, Portsmouth, June 6, 1746, *Letters of Hume* (see Ch. 4, n. 24), vol. 1, p. 91.
75. *Tales and Essays by Sir Walter Scott*, ed. A. and W. Galignani, Paris, 1829, p. 36.

76. *Scotland and Scotsmen* (see Ch. 1, n. 3), vol. 2, pp. 87–8.

77. Ibid., p. 86.

78. William Creech, *Letters Addressed to Sir John Sinclair, Bart.* (see n.35), p. 17.

79. Ibid.

80. W.F. Gray, 'Edinburgh in Lord Provost Drummond's Time', in *The Book of the Old Edinburgh Club*, vol. 27 (1949), p. 18.

81. Creech, *Letters Addressed to Sir John Sinclair, Bart.*, p. 36.

82. *Boswell in Search of a Wife*, p. 76 and p. 81.

83. 'But don't for Heavn's sake meddle with her as a Piece.' To Robert Ainslie, Dumfries, June 1787 or 1788: *Letters of Burns*, vol. 1, p. 226.

84. *Boswell: The Applause of the Jury, 1782–1785*, ed. I.S. Lustig and F.A. Pottle (London, 1981), p. 27.

85. M.J.C. Meiklejon, 'Eighteenth Century Scotland: Some Gleanings from Family Papers', *Glasgow Herald*, 21 September 1929. The papers themselves have not been found. 'The most usual wages of common labour are now [1776] . . . ten-pence, sometimes a shilling about Edinburgh': *Wealth of Nations* (Oxford, 1976), I.viii.34, p. 94.

86. *Wealth of Nations*, V.i.f. 47, p. 781.

87. Lord Kames to Miss Gordon, Edinburgh, 17 June 1765, in Stewart of Afton Papers, BL Add.40635, f. 61.

88. John Gregory, *A Father's Legacy to his Daughters* (London and Edinburgh, 1774), p. 32. Alexander Monro *secundus* also wrote an essay for his daughter Margaret.

89. William Creech, *Edinburgh: Fugitive Pieces* (Edinburgh, 1791), p. 298.

90. Smellie to Mrs Riddell, Edinburgh, 27 March 1792, in *Memoirs of Smellie* (see Ch. 1, n. 13), vol. 2, p. 362.

91. 28 March 1768, *Boswell in Search of a Wife*, p. 156.

92. Robert Wallace, 'Of Venery – of the Commerce of the 2 Sexes', EUL La II 620/12, printed in Norah Smith, ed., *Texas Studies in Literature and Language*, vol. 15 (1973), 3, p. 434.

93. Ibid., p. 439.

94. EUL La II 620/12.

95. David Hume, *Essays Moral and Political* (Edinburgh, 1742), vol. II, p. 180 and p. 191.

96. Adam Smith, *Theory of Moral Sentiments* (Oxford, 1976), IV.2.10, pp. 190–1.

97. Ibid., V.2.9, p. 205.

98. Ibid., 1.ii.2.2, p. 32.

99. Ibid., 1.ii.2.1, p. 31.

100. *Anecdotes and Egotisms of Henry Mackenzie*, ed. H.W. Thompson (Edinburgh, 1927), p. 176.

101. Frances Hutcheson, *Inquiry concerning Virtue* (see Ch. 3, n. 56), VI, 5.
102. Ibid.
103. John Millar, *The Origin of the Distinction of Ranks* (London, 1779), p. 14.
104. Ibid., p. 17.
105. Ibid., pp. 75–6.
106. Ibid., p. 79.
107. 'Warriors went forth to realise the legends they had studied; princes and leader of armies dedicated their most serious exploits to a real or to a fancied mistress. But whatever was the origin of notions, often so lofty and so ridiculous, we cannot doubt of their lasting effects on our manners. The point of honour, the prevalence of gallantry in our conversations, and on our theatres . . .' Adam Fergusson, *An Essay on the History of Civil Society*, ed. Fania Oz-Salzburger (Cambridge, 1995), Part 4, Section 4, p. 193.
108. *Origin of Ranks*, p. 121.
109. Ibid., p. 108.
110. Ibid., pp. 108–9. Gregory, in his admonition to his daughters published a couple of years later, used very similar language: modern men treat women 'not as domestic drudges, or the slaves of our pleasures, but as our companions and equals': *A Father's Legacy*, p. 6.
111. *Origin of Ranks*, p. 109.
112. Ibid., p. 123.
113. Ibid., p. 123.
114. James Hutton, *Investigation of the Principles of Knowledge* (Edinburgh, 1794), vol. 3, p. 587.
115. The passage in Boswell's journal covering their journey together from Paris to London in early February 1766 was destroyed just before the Boswell papers from Malahide Castle were sold to Col. Isham. He had, however, read the passage and left an account of its contents. *Boswell on the Grand Tour: Italy, Corsica and France 1765–1766*, ed. F. Brady and F.A. Pottle (London, 1955), pp. 265–6.
116. Kames to Miss Gordon, Edinburgh, August 19–25 1764, in Stewart of Afton Papers, BL Add. 40635, vol. 3, f. 36.
117. Kames to Miss Gordon, Edinburgh, 27 June 1764, in ibid., f. 27.
118. 'Lord Abercromby told me, that one night, after supper, in his own house, he [Kames] spake in rapturous terms of a young lady's *legs*. In a vein of dignified irony Mrs Drummond said to him: "I thought, my lord, you had never gone *so low* as a lady's legs, contenting yourself with her head and her heart."' Ramsay of Ochtertyre, *Scotland and Scotsmen* (see Ch. 1, n. 3), vol. 1, p. 207.
119. 'Of Venery – of the Commerce of the 2 sexes', EUL La 620/12, in *Texas Studies* (see n. 92), p. 435.

120. Kames to Miss Gordon, Edinburgh, 17 June 1765, in Stewart of Afton Papers, BL Add. 40635, f. 61.
121. Kames to Miss Gordon, Edinburgh, 28 July 1766, in ibid., f. 69.
122. 'On Female Virtue, with Intellectual Accomplishments' in James Fordyce, *Sermons to Young Women* (Edinburgh, 1765), vol. 1, pp. 271–85.
123. Hutton, *Investigation* (see n. 114), vol. 3, p. 589.
124. 'On the influence of the Female Sex, especially the Younger Part' in Fordyce, *Sermons to Young Women*, vol. 1, p. 21.
125. 'Correspondence with Miss Rose of Kilravock', Savile Papers, quoted in H.W. Thompson, *A Scottish Man of Feeling* (Oxford, 1931), p. 145.
126. Alexander, *History of Women* (see n. 1), vol. 1, p. 2.
127. Ibid., vol. 2, p. 336.
128. Ibid., vol. 2, p. 323.
129. Francis Hutcheson, *A System of Moral Philosophy* (Glasgow and London, 1755), vol. 2, pp. 165–6.
130. 'No person who did not live at the time this cause was depending, can form any conception of the agitation, the anxiety, the polemical spirit, which it excited among the inhabitants of the metropolis, and, indeed, far and wide throughout the country. It was a constant subject of conversation in companies of every description. Families of all ranks ranged themselves on different sides of the contest. Members of the same family were often at variance with one another; and, as generally happens, the heat and violence of the opposite partizans were inflamed by the intricate and embarrassing circumstances which encumbered the question at issue. I shall never forget the keenness with which many ladies of my acquaintance entered into this controversy, which gave me a new and no depreciating view of feminine acuteness and eloquence.' Thomas Somerville, *My Own Life and Times, 1741–1814* (Edinburgh, 1861), p. 112.
131. 'You broke, I am told, your father's windows because they were not enough illuminate. Bravo, bravissimo!' Lord Marischal to Boswell, 26 August 1769, *Boswell in Search of Wife*, p. 191, n. 2.
132. 'Jeanie, without youth, beauty, genius, warm passions, or any other other novel-perfection, is here our object from beginning to end. This is "enlisting the affections in the cause of virtue" ten times more than ever Richardson did; for whose male and female pedants, all-excelling as they are, I never could care half so much as I found myself inclined to do for Jeanie before I finished the first volume.' Lady Louisa Stuart to Sir Walter Scott, in J.G. Lockhart, *Life of Sir Walter Scott* (New York and Boston, 1871), vol. 1, p. 317.
133. Edward Burt and his friends, while building roads in the north in the 1720s, were deeply impressed by the thrift of Scots gentlewomen: 'There is a Laird's Lady, about a Mile from one of the Highland

Garrisons, who is often seen from the Ramparts on *Sunday* Mornings coming barefoot to the Kirk with her Maid carrying her Stockings and Shoes after her. She stops at the Foot of a certain Rock, that serves her for a Seat, not far from the Hovel they call a Church, and there she puts them on, and in her Return to the Same Place, she prepares to go home barefoot as she came . . . What *English* Squire was ever blessed with such a Housewife!' *Letters from a Gentleman* (see Ch. 1, n. 33), vol. 2, p. 192.

Chapter 10: Earth to Earth

1. David Hume, *Enquiry Concerning Human Understanding*, p. 195.
2. To Boswell, London, 2 March 1784, in *Life of Johnson*, p. 1267.
3. In his petition to the Town Council to admit his son as conjoint professor, Monro *primus* said that the students were 'more than two hundred for many years past': Town Council Minutes, 19 June 1754, SLI/1/72.
4. Alexander Carlyle, *Autobiography* (see Ch. 1, n. 71), pp. 49–50. J.D. Comrie, in his *History of Scottish Medicine to 1860* (Edinburgh, 1927), gives the following numbers for students attending the anatomy class: 57 in 1720, 83 in 1730, 136 in 1740, 158 in 1750, and an average of 194 in the 1760s, 287 in the 1770s and 342 in the 1780s.
5. Dr William Buchan to Mr William Smellie, 20 August 1761, in *Memoirs of Smellie* (see Ch. 1, n. 13), vol. 1, p. 238.
6. James Gregory, *Memorial to the Managers of the Royal Infirmary by James Gregory, MD* (Edinburgh, 1800) and *Additional Memorial, etc* (Edinburgh, 1803).
7. *Lectures on the Elements of Chemistry, delivered in the University of Edinburgh, by the late Joseph Black, MD*, ed. J. Robison (Edinburgh, 1803), vol. 1, p. xxi.
8. 'I had mixed together some chalk and vitriolic [sulphuric] acid at the bottom of a cylindrical glass; the strong effervescence produced an air or vapor which, flowing out at the top of the glass, extinguished a candle that stood close to it.' To William Cullen, 3 January 1754, in John Thomson, *An Account of the Life, Lectures and Writings of William Cullen, MD* (Edinburgh and London, 1832), vol. 1, p. 50.
9. Sir William Ramsay, *The Life and Letters of Joseph Black, MD* (London, 1918).
10. *Lectures on the Elements*, vol. 1, p. xlvii.
11. 'Early Draft' in *Lectures on Jurisprudence* (see Ch. 8, n. 87), p. 570.
12. 'To a Louse, 1786'.
13. There were also twenty-nine redcoats and nine Highlanders brought in from Preston who died from their wounds, four casualties of the Castle

cannonade and one stabbing: Edinburgh Bills of Mortality, *Scots Magazine*, vol. 7 (1745), p. 622.

14. William Buchan, *Domestic Medicine; or the Family Physician* (Edinburgh, 1769), p. 110.

15. For example, in 1719 a surgeon-apothecary in the town of Elgin indented for such services as a phlebotomy (12s. Scots), vomitories (10s.), gilded pills (18s.) and antimony (12s.): 'Accompt Laird of Thundertown since Jan. 22, 1719 to Kenneth Mackenzie, Chyr Aporie in Elgin' in Dunbar, *Social Life* (see Ch. 9, n. 44), pp. 20–1.

16. *Memoirs of Sir John Clerk of Penicuik* (see Ch. 2, n. 7), p. xi.

17. Oliver Goldsmith to the Revd Contarine, 8 May 1753, printed in Forster, *Life of Goldsmith* (see Ch. 1, n. 15), vol. 1, p. 448.

18. Ibid., p. 452.

19. Scotus *pseud.*, 'Sketches of the Medical School of Scotland, No. 25: Dr Monro', *The Lancet* (1828–29), vol. 1, pp. 391ff.

20. *An Account of the Rise and Establishment of the Infirmary, or Hospital for Sick-Poor, erected at Edinburgh* (Edinburgh n.d., probably 1729), p. 2.

21. Ibid.

22. Ibid., p. 9.

23. Ibid., p. 6.

24. Ibid., p. 7.

25. Derived from the Edinburgh Bills of Mortality, *Scots Magazine*, vol. 7 (1745), p. 622. and vol. 51 (1789), p. 655.

26. Günter B. Risse, *Hospital Life in Enlightenment Scotland: Care and Teaching at the Royal Infirmary of Edinburgh* (Cambridge, 1986), pp. 46–7.

27. The figures given by Risse are £2,587 6s. in 1770 and £2,936 5s. in 1790: ibid., p. 36.

28. *Pharmacopoeia Collegii Regii Edinburgensis* (Edinburgh, 1699); and *Pharmacopoeia Edinburgensis: or the Dispensatory of the Royal College of Physicians in Edinburgh, translated and improved from the Fourth Edition of the Latin by Peter Shaw, MD* (Edinburgh, 1746).

29. 'Death and Dr Hornbrook: A True Story'

30. William Cullen, *Clinical Lectures, 1772–1773*, p. 616, Royal College of Physicians, Edinburgh.

31. J.W. Estes, MD, 'Drug Usage at the Infirmary: The Example of Dr Andrew Duncan, sr.': Appendix to Risse, *Hospital Life*.

32. William Cullen, *First Lines of the Practice of Physic, for the Use of Students in the University of Edinburgh* (Edinburgh, 1778), vol. 1. p. 2.

33. Ibid., vol. 2, p. 1.

34. *Memorial to the Managers of the Royal Infirmary by James Gregory, MD* (Edinburgh, 1800), p. 218.

35. Lecture of 21 April 1772, in Thomson, *Life of Cullen* (see n. 8), p. 108.
36. Ibid., p. 111.
37. Cullen, *First Lines*, vol. 1, p. 3.
38. Gregory, *Additional Memorial* (see n. 6), pp. 490–1.
39. *Edinburgh Records*, 1 July 1505.
40. Robert Chambers, *Domestic Annals of Scotland from the Reformation to the Rebellion of 1745* (Edinburgh and London, 1885), pp. 395–6.
41. J.D. Comrie, *History of Scottish Medicine* (London, 1932), vol. 1, pp. 258–9.
42. Alexander Carlyle, *Autobiography* (see Ch. 1, n. 71), p. 58.
43. *Scots Magazine*, vol. 4 (1742), pp. 140–1.
44. In *Tracts, 1731–47*, BL 1078.k.26, pp. 5–8. The writer exonerated the College, but blamed 'airy, flory Surgeon Lads' and the English – 'Surgeons from South that did attend the College/ Bought Corps and sent them South amongst ther Baggage'.
45. Dunlop, *Anent Old Edinburgh* (see Ch. 3, n. 11), p. 37.
46. Henry Cockburn, *Memorials of his own Time* (Edinburgh, 1910), p. 426.
47. Entries for 14 and 28 January, 1829 in *The Journal of Sir Walter Scott*, ed. W.E.K. Anderson (Oxford, 1972).
48. Buchan, *Domestic Medicine* (see n. 14), p. vii.
49. Andrew Wight, *Present State of Husbandry in Scotland, Extracted from Reports made to the Commissioners of the Annexed Estates, and Published by their Authority* (Edinburgh, 1778–1784), vol. 1, p. 379.
50. Ibid., vol. 3, p. 544.
51. Ibid., vol. 3, p. 374.
52. Ibid., vol. 3, p. 469.
53. *Dissertatio physico-medica inauguralis de sanguine et circulatione microcosmi* (Lugduni Batavorum [Leyden] 1749): BL T.52 (4).
54. John Playfair, 'A Biographical Account of James Hutton, MD, FRSEd' (n.p., n.d), p. 6.
55. 'Dr Hutton had formed this system or the principal parts of it more than 20 years ago and he has found reason to be more and more confirmed in it by his study of Fossils ever since that time': Black to Princess Dashkova, Edinburgh, 27 August 1787, printed in Sir William Ramsay, *The Life and Letters of Joseph Black, MD* (London, 1918), p. 124.
56. Martial, *Epigrammata*, Book 12, No. 57.
57. *Transactions of the Royal Society of Edinburgh*, vol. 2, 1790.
58. 'Le docteur James Hutton est peut-être le seul particulier à Edinburgh qui ait réuni dans un cabinet quelques mineraux et une suite nombreuse d'agathes, tirées particulièrement de l'Écosse; mais j'ai trouvé qu'il ne s'est pas assez attaché de recueillir les matrices diverses dans lesquelles elles sont renfermées, et qui servent au complément de l'histoire

naturelle de ces pierres. J'eus donc beaucoup plus de plaisir à m'entretenir avac cc savant modeste, que j'allai voir, que d'examiner sa collection . . .' *Voyage en Angleterre, en Ecosse, et aux îles Hébrides* (Paris, 1797), p. 266.

59. *Transactions*, vol. 1, 1788, p. 36.

60. *Account of Sir Isaac Newton's Discoveries* (London, 1750), pp. 409–10.

61. *Abstract of a Dissertation read in the Royal Society of Edinburgh upon the Seventh of March, and Fourth of April, 1785, concerning the System of the Earth, its Duration, and Stability* (reprinted, Edinburgh, 1997), p. 4.

62. Ibid., p. 28.

63. Smellie, in his *Philosophy of Natural History* of 1790, saw nature as a graduated scale in which sea mammals linked fishes to quadrupeds, and bats birds and mammals, with humanity at the head. 'Were there no other argument in favour of the UNITY of DEITY, this uniformity of design, this graduated concatenation of beings . . . seems to be perfectly irrefragable': *The Philosophy of Natural History* (Edinburgh, 1790), p. 526.

64. Hutton, *Abstract*, p. 30.

65. Ibid., p. 19.

66. For example, Playfair: 'Thus he arrived at the new sublime conclusion, which represents nature as having provided for a constant succession of land on the surface of the arth, according to a plan having no natural termination, but calculated to endure as long as those beneficent purposes, for which the whole is destined, shall continue to exist': Playfair, 'Biographical Account of Hutton', pp. 18–19.

67. Ramsay, *Black's Life and Letters*, p. 124.

68. 'Biographical Account of Hutton', p. 27.

69. Ibid., p. 34.

70. Ibid., pp. 53–4.

71. Ibid., p. 44.

72. James Hutton, *Theory of the Earth with Proofs and Illustrations* (Edinburgh, 1795), vol. 1, p. 3.

73. Ibid., vol. 2, p. 562.

74. Ibid., vol. 1, p. 200.

Chapter 11: The Man of Feeling

1. Walter Scott, *Lives of Novelists* (Paris, 1825), vol. 2, p. 155.

2. Ibid., p. 156.

3. 'We could suppose, a retired modest somewhat affected man with a white handkerchief and a sigh ready for every sentiment. No such thing.

H. M. is alert as a contracting tailor's needle in every sort of business, a politician and a sportsman.' Scott, *Journal*, 6 December 1825, p. 25. Cockburn called Mackenzie a 'hard-headed practical man, as full of worldly wisdom as most of his fictitious characters are devoid of it: Henry Cockburn, *Memorials of His Time* (Edinburgh, 1910), p. 257.

4. Adam Smith, *Lectures on Rhetoric and Belles Lettres* (see Ch. 5, n. 23), p. 112.

5. *The Works of Henry Mackenzie, Esq., with a critical dissertation on the tales of the author by John Galt, Esq.*, in H.W. Thompson, *A Scottish Man of Feeling* (Oxford, 1931), p. 364.

6. Cockburn, *Memorials of His Time*, pp. 34–5.

7. *Treatise of Human Nature* (see Ch. 3, n. 75), Bk 1, Pt 3, Sect. 7.

8. E.A. Baker, *The History of the English Novel* (London, 1934), vol. 5, p. 35.

9. *Letters of Hume* (see Ch. 3, n. 84), vol. 2, p. 314.

10. 'In contemplating Man, as at the head of those animals with which we are acquainted, a thought occurred, that no sentient being, whose mental powers were greatly superior, could possibly live and be happy in this world. If such a being really existed, his misery would be extreme. With senses more delicate and refined, with perceptions more acute and penetrating; with a taste so exquisite that the objects around him could by no means gratify it; obliged to feed on nourishment too gross for his frame; he must be born only to be miserable, and the continuation of his existence would be utterly impossible. Even in our present condition, the sameness and insipidity of objects and pursuits, the futility of pleasure, and the infinite sources of excruciating pain, are supported with great difficulty by cultivated and refined minds. Increase our sensibilities, continue the same objects and situation, and no man could bear to live.' William Smellie, *Philosophy of Natural History* (see Ch. 10, n. 63), p. 526.

11. In *Anecdotes and Egotisms* (see Ch. 1, n. 57), p. 184, Mackenzie says he first attended the College 'when I was but eleven years old'. He first appears in the Matriculation Album on 14 March 1758.

12. Ibid., p. 184.

13. *Scots Magazine*, vol. 26 (1764), p. 196.

14. *Anecdotes and Egotisms*, p. 186.

15. Ibid., p. 190.

16. H. Mackenzie to Miss Elizabeth Rose, 8 July 1769, in H.W. Thompson, *A Scottish Man of Feeling* (Oxford, 1931), p. 108.

17. To Miss Rose, 18 May 1771 and [?June, 1771], in ibid., pp. 111–12.

18. *The Rivals*, Act 1, Scene 2.

19. *Anecdotes and Egotisms*, p. 190.

20. H. Mackenzie to Miss Elizabeth Rose, 8 July 1769, in Thompson, *Scottish Man of Feeling*, p. 108.

21. I have counted the 'beamy moisture' in Edwards' eyes at the end of Chapter Ten towards the total.

22. *The Man of the World* (Edinburgh, 1773), vol. II, p. 190.

23. 'Danger of Regulating our conduct by the rules for romantic sentiment. Story of *Emilia*', *The Mirror: A Periodical Paper*, No. 101, 25 April 1780.

24. 'An Account of My Last Interview with David Hume, Esq.', Sunday, 7 July 1776, in *Boswell in Extremes* (see Ch. 3, n. 85), p. 12.

25. *Scott's Journal*, 10 July 1827.

26. 'Such, Echecrates, was the end of our companion, a man who, we may fairly state, was of all those we knew in our time the most virtuous, and on the whole the wisest and most just.' *Phaed.* 118. 'Upon the whole, I have always considered him, both in his lifetime and since his death, as approaching as nearly to the idea of a perfectly wise and virtuous man, as perhaps the nature of human frailty will permit.' *Correspondence of Smith* (see Ch. 5, n. 5), p. 221.

27. *The Mirror*, No. 42, Saturday, 19 June 1779.

28. Ramsay of Ochtertyre, *Scotland and Scotsmen* (see Ch. 1, n. 3), vol. 1, p. 205.

29. *The Mirror*, No. 44, Saturday, 26 June 1779.

30. Mackenzie, *Anecdotes and Egotisms* (see Ch. 1, n. 57), p. 171.

31. Scott, *Lives of the Novelists* (see n. 1), pp. 167–9.

32. To Dr John Moore, Mauchline, 2 August 1787, *The Letters of Robert Burns*, ed. J. de L. Ferguson (Oxford, 1931), vol. 1, p. 112.

33. To John Murdoch, Lochlea, 15 January 1783, in ibid., vol. 1, p. 14.

34. To William Burns, Irvine, 27 December 1781, in ibid., vol. 1, p. 5.

35. To John Murdoch, Lochlea, 15 January 1783, in ibid., vol. 1, p. 14.

36. *The Weekly Magazine, or Edinburgh Amusement*, 2 January 1772.

37. Matthew McDiarmid, ed., *The Poems of Robert Fergusson* (Edinburgh, 1954), vol. 1, p. 32.

38. John Buchan, *The Northern Muse* (London and Edinburgh, 1924), p. 227.

39. *The Weekly Magazine, or Edinburgh Amusement*, 2 January 1772.

40. 'The old City-Guard of Edinburgh', No. 170, *Kay's Original Portraits* (see Ch. 4, n. 102).

41. 'Some Aspects of Robert Burns' in *The Cornhill Magazine*, vol. 60, July–December 1879, p. 426.

42. To Dr John Moore, Mauchline, 2 August 1787, in *Letters of Burns*, vol. 1, p. 113.

43. 'Abstract of Expences, Anno 1750' in A.B. Grosart, *The works of Robert Fergusson; with life of the author and an essay on his genius and writings* (London, Edinburgh and Dublin, 1879), p. xxx.

44. His fellow student Charles Webster told Alexander Campbell that officiating at prayers one day as a precentor, Fergusson begged the congre-

gation to pray for the recovery of a fellow-student who was drunk. 'Had not Dr Wilkie (the ingenious author of *The Epigoniad*) stept in between him and the displeasure of the rest of the professors, it may easily have been conjectured what whould have been the consequences': Alexander Campbell, *An Introduction to the History of Poetry in Scotland* (Edinburgh, 1798), p. 291.

45. *The Ever Green, being a collection of Scots poems, wrote by the Ingenious before 1600*, ed. Allan Ramsay (Edinburgh, 1724), vol. 1, p. xi.

46. 'Caller Water' (21 January), 'Mutual Complaint of Plainstanes and Causey in their Mother-tongue' (4 March), 'The Rising of the Session' (18 March), 'The Farmer's Ingle' (13 May), 'The Ghaists: A Kirk-yard Eclogue' (27 May), 'On Seeing a Butterfly in the Street' (24 June), 'Hame Content' (8 July), 'Leith Races' (22 July), 'Ode to the Gowdspink' (12 August), 'To the Principal and Professors of the University of St Andrews, on their superb treat to Dr Samuel Johnson' (2 September), 'The Election' (16 September), 'Elegy on John Hogg, late Porter to the University of St Andrews' (23 September), 'The Sitting of the Session' (4 November), 'A Drink Eclogue' (11 November) and 'To my Auld Breeks' (25 November).

47. Grosart, *Works of Fergusson*, p. 126.

48. Alexander Peterkin, ed., *The Works of Robert Fergusson, To Which is Prefixed, a Sketch of the Author's Life* (London, 1807), pp. 50–1.

49. George Gleig, *Supplement to the Third Edition of the Encyclopaedia Britannica* (Edinburgh, 1801), vol. 1. p. 648.

50. McDiarmid, *Poems of Fergusson*, p. 76.

51. John Pinkerton, 'A List of All the Scottish Poets; with brief Remarks' in *Ancient Scottish Poems* (London, 1786), p. 140.

52. 'Reflections on genius unnoticed and unknown; anecdotes of Michael Bruce', *The Mirror*, No. 36, Saturday, 29 May 1779.

53. 'Hallow Fair'.

54. Mackenzie, *Anecdotes and Egotisms*, p. 150.

55. J.G. Lockhart, *Life of Robert Burns* (Edinburgh, 1828), p. 115.

56. Creech, *Letters to Sir John Sinclair* (see Ch. 1, n. 7), pp. 6–7. For comparison, Scotland's entire public debt at the Union of 1707 was just £200,000 sterling.

57. Ibid., p. 9.

58. Ibid., p. 8.

59. Ibid., p. 15.

60. *Provincial Antiquities* (see Ch. 1, n. 18), p. 77.

61. William Strahan (1715–85), bred as a printer at Edinburgh, moved to London and was taken on as partner by Andrew Millar. In 1767 Millar handed over his publishing business to Strahan and Thomas Cadell. John Murray, or McMurray, was born in Bailie Clerk's Land in 1737,

passed through the High School, the College and the Royal Navy, and took part in various Edinburgh literary controversies before setting up as a bookseller in London in 1768.

62. Creech, *Letters to Sir John Sinclair*, p. 11.
63. Ibid., p. 17.
64. Ibid., p. 11.
65. Forbes of Pitsligo, *Memoirs of a Banking-House* (see Ch. 1, n. 24), pp. 56–9.
66. Williamson surrendered his licence in 1793 for a pension of £25 a year: B. Auckland and G. Oxley, *Penny Posts of Scotland* (Edinburgh, 1978).
67. Creech, *Letters to Sir John Sinclair*, p. 18.
68. Ibid., pp. 38–9.
69. Ibid., p. 33.
70. Ibid., p. 35.
71. Ibid., p. 36.
72. Ibid., p. 37.
73. Ibid., p. 38.
74. Ibid., p. 32.
75. Ibid., pp. 42–3.
76. *Cornhill Magazine*, vol. 60, July–December 1879, p. 417.
77. To David Brice, Mossgiel, 12 June 1786, in *Letters of Burns*, vol. 1, p. 31.
78. To Dr John Moore, Mauchline, 2 August 1787, in ibid., vol. 1, p. 114.
79. To John Richmond, Old Romefoord, 30 July 1786, in ibid., vol. 1, p. 35.
80. Deed of Assignment, Mossgiel, 22 July 1786, in ibid., vol. 1, pp. 33–4.
81. To John Richmond, 30 July 1786, in ibid., vol. 1, p. 35.
82. Scott made the same point, but in comparison with Byron: '[Byron] wrote from impulse, never from effort, and therefore I have always reckoned Burns and Byron the most genuine poetical geniuses of my time and a half a century before me. We have however many men of high poetical talent, but none of that ever-gushing, and perennial fountain of natural water.' *Journal*, 9 February 1826.
83. 'My vanity was highly gratified by the reception I met with from the Publick; besides pocketing, all expences deducted, near twenty pounds': to Dr Moore, Mauchline, 2 August 1787, in *Letters of Burns*, vol. 1, p. 115.
84. Blacklock received a copy from the Revd George Lawrie, minister at Loudoun, Ayrshire and wrote back enthusiastically on 4 September 1786.
85. *Poems, Chiefly in the Scottish Dialect* (Kilmarnock, 1786).
86. *The Mirror*, No. 36, 29 May 1779.
87. Lockhart, *Life of Burns* (see n. 55), p. 92.
88. *Edinburgh Magazine, or Literary Miscellany*, 3 November 1786.
89. To Aiken, Edinburgh, 16 December 1786, in *Letters of Burns*, vol. 1, p. 58.

90. *The Lounger*, No. 97, 9 December 1786.
91. Ibid.
92. ibid.
93. To Gavin Hamilton, Edinburgh, 7 December 1786, in *Letters of Burns*, vol. 1, p. 55.
94. To John Ballantine, Edinburgh, 13 December 1786, in ibid., vol. 1, p. 56.
95. To John Ballantine, Edinburgh, 14 January 1787, in ibid., vol. 1, p. 67.
96. 'His person was strong and robust; his manners rustic, not clownish; a sort of dignified plainness and simplicity, which received part of its effect, perhaps, from one's knowledge of his extraordinary talents. His features are represented in Mr Nasmyth's picture, but to me it conveys the idea that they are diminished as if seen in perspective. I think his countenance was more massive than it looks in any of the portraits. I would have taken the poet, had I not known what he was, for a very sagacious country farmer of the old Scotch school – i.e., none of your modern agriculturists, who keep labourers for their drudgery, but the *douce gudeman* who held his own plough. There was a strong expression of sense and shrewdness in all his lineaments; the eye alone, I think, indicated, the poetical character and temperament. It was large, and of a dark cast, which glowed (I say literally *glowed*) when he spoke with feeling or interest. I never saw such another eye in a human head, though I have seen the most distinguished men in my time. His conversation expressed perfect self-confidence, without the slightest presumption. Among the men who were the most learned of their time and country, he expressed himself with perfect firmness, but without the least intrusive forwardness; and when he differed in opinion, he did not hesitate to express it firmly, yet at the same time with modesty. I do not remember any part of his conversation distinctly enough to be quoted, nor did I ever see him again, except in the street, where he did not recognise me, as I could not expect he should. He was much caressed in Edinburgh, but (considering what literary emoluments have been since his day) the efforts for his relief were extremely trifling. I remember on this occasion I mention, I thought Burns' acquaintance with English poetry was rather limited, and also, that having twenty times the abilities of Allan Ramsay and of [Robert] Ferguson, he talked of them with too much humility as his models; there was doubtless national predilection in his estimate . . . His dress corresponded with his manner. He was like a farmer dressed in his best to dine with the Laird . . . I never saw a man in company with his superiors in stations and information, more perfectly free from either the reality or the affectation of embarrassment.' Lockhart, *Life of Burns*, pp. 112–15.
97. To William Greenfield, *Letters of Burns*, vol. 1, p. 59.
98. Ibid.

99. 'Burns's Unpublished Common-place Book', pp. 5–6, in *Macmillan's Magazine*, vols 39–40 (1878–9) pp. 455–6.

100. To Gavin Hamilton, Edinburgh, 7 January 1787, in *Letters of Burns*, vol. 1, pp. 62–3.

101. *Memoirs of Smellie* (see Ch. 1, n. 13), vol. 2, pp. 350–1.

102. To the Honble the Bailies of the Canongate, Edinburgh, 6 February 1787, in *Letters of Burns*, vol. 1, p. 72.

103. To Mrs Dunlop, Edinburgh, 22 March 1787, in ibid., vol. 1, p. 82.

104. To the Earl of Buchan, Lawnmarket, 7 February 1787, in ibid., vol. 1, pp. 72–4.

105. To Mrs Dunlop, 22 March 1787, in ibid., vol. 1, pp. 80–1.

106. It appeared the following February.

107. To James Johnson, Edinburgh, 4 May 1787, in ibid., vol. 1, p. 89.

108. To James Smith, Mauchline, 11 June 1787, in ibid., vol. 1, p. 95.

109. Edinburgh, 28 September 1787, in ibid., vol. 1, pp. 126–7.

110. James Grierson, an avid collector of anecdotes of the poet, reported that 'A person met Burns coming up Leith walk brandishing a sapling & with much violence in his face & manner, said, Burns what is the matter? I am going to smash that S[hite] Creech.' 'Grierson MS, 1814–17' quoted in J. Mackay, *A Biography of Robert Burns* (Edinburgh, 1992), p. 396.

111. Lockhart, *Life of Burns*, p. 169.

112. To John Skinner, Edinburgh, 25 October 1787, in *Letters of Burns*, vol. 1, p. 134.

113. To Mrs Agnes McLehose, Edinburgh, 25 January 1788, in ibid., vol. 1, p. 173.

114. To Mrs Agnes McLehose, Edinburgh, 13 February 1788, in ibid., vol. 1, p. 185.

115. To Mrs Dunlop, Ellisland, 10 April 1790, in ibid., vol. 2, p. 20.

116. Ibid.

117. To Margaret Chalmers, Edinburgh, 24 March 1788, in ibid., vol. 1, p. 208.

118. To Capt. Richard Brown, Glasgow, 20 March 1788, in ibid., vol. 1, p. 212.

119. To Mrs Dunlop, Ellisland, 25 March 1789, in ibid., vol. 1, p. 317.

120. To James Smith, Mauchline, 28 April 1788, in ibid., vol. 1, pp. 218–19.

121. 'Account of the German Theatre', 21 April 1788, in *Transactions of the Royal Society of Edinburgh*, vol. 2 (1790), p. 167.

122. Lockhart, *Life of Burns*, p. 127.

Epilogue

1. Colin MacLaurin, *An Account of Sir Isaac Newton's Philosophical Discoveries* (Edinburgh, 1750), p. 9.

2. 'These men were all great peripatetics, and the Meadows was their academic grove. There has never in my time been any single place in or near Edinburgh, which has so distinctly been the resort at once of our philosophy and our fashion. Under these poor trees walked and talked, and meditated, all our literary and scientific and many of our legal worthies. I knew little then of the grounds of their reputation, but saw their outsides with unquestioning and traditionary reverence.' Henry Cockburn, *Memorials of His Time* (Edinburgh, 1910), pp. 51–2.

3. As far as I can tell, the word 'Enlightenment' is applied to Scotland and this era for the first time in William Robert Scott's *Francis Hutcheson: His Life, Teaching and Position in the History of Philosophy* (Cambridge, 1900), p. 257. John Buchan used the German term *Aufklärung* in 1908: *Eighteenth Century Byways* (see Prologue, n. 2), p. 157.

4. Montaigne, *Essais*, Bk 1, Ch. 19.

5. To Adam Smith, 26 August 1776, *Correspondence of Smith* (see Ch. 5, n. 5), p. 220.

6. Dugald Stewart, 'Account of Smith' in Smith, *Essays on Philosophical Subjects* (see Ch. 5, n. 3), p. 328n.

7. John Robison, *Lectures on the Elements of Chemistry, delivered in the University of Edinburgh by the late Joseph Black, MD* (Edinburgh, 1803), vol. 1, p. lxxiv.

8. John Millar, *Origin of Ranks* (see Ch. 9, n. 103), pp. 361–2. Ruling for Knight, the 82-year-old Lord Kames pronounced: 'We cannot enforce [the slavery laws of Jamaica], for we sit here to enforce right and not to enforce wrong.'

9. The first Catholic Relief Act (18 George III c. 60) was passed in 1778.

10. To Robert Graham of Fintry, Dumfries, 31 December 1792, in *Letters of Burns* (see Ch. 11, n. 32), vol. 2, p. 139.

11. Lord Craig to Dugald Stewart, quoting Lord Abercromby, 15 February 1794, in *Works of Dugald Stewart*, ed. Sir William Hamilton (Edinburgh and London, 1854–8), vol. 10, p. lxxi.

12. *Proofs of a Conspiracy against all the Religions and Governments of Europe, carried on in the secret meetings of Free Masons, Illuminati, and Reading Societies, Collected from Good Authorities* (Edinburgh, 1797), p. 431.

13. *Autobiography of Mrs Fletcher of Edinburgh* (Carlisle, 1874), p. 63.

14. Smellie to Patrick Clason, 27 June 1790, in *Memoirs of Smellie* (see Ch. 1, n. 13), vol. 1, p. 296.

15. 'Trial of Maurice Margarot', *Howell's State Trials* (London, 1817), vol. 23, p. 266.

16. 'If innocent of this atrocious sentiment, he was scandalously ill-used by his friends, by whom I repeatedly heard it ascribed to him at the time, and who, instead of denying it, spoke of it as a thing understood, and

rather admired it as worthy of the man and of the time': Henry Cockburn, *Memorials of His Time* (Edinburgh, 1910), p. 107.

17. H. Furber, *Henry Dundas, First Viscount Melville* (Oxford, 1931), p. 76.

18. Cockburn, *Memorials of His Time*, p. 377.

19. Henry Cockburn, *The Life of Lord Jeffrey* (Edinburgh, 1852), vol. 1, p. 157.

20. *Woodhouselee MS* (see Ch. 2, n. 41), p. 26. 'A poor Italian prince C. Stewart from Lochqwaber in the obscurest corner of Britain, with ane ill-armed mobb of Highlanders and a bankrupt Tweedall laird his secretary. And bagpyppes surprising Edinburgh o'rrunning Scotland at Cockeny, defeating a Royall armie . . . : *Woodhouselee MS*, pp. 88–9.

21. 'Life and Works of John Home' in Walter Scott, *Essays on Chivalry, Romance and the Drama* (London, 1888).

22. 'To a man like Scott, the different appearances of nature seemed each to contain its own legend ready made, which it was his to call forth: in such or such a place, only such or such events ought with propriety to happen; and in this spirit he made "The Lady of the Lake" for Ben Venue, "The Heart of Midlothian" for Edinburgh': Robert Louis Stevenson, *Edinburgh: Picturesque Notes* (Edinburgh, 1888), p. 59.

23. The use of post-horses in Scotland rose and continued to rise, the publisher Robert Cadell told Lockhart, 'the author's succeeding works keeping up the enthusiasm for our scenery which he had thus originally created': Lockhart, *Life of Scott* (see Ch. 9, n. 132), vol. 1, p. 199.

24. Ibid., vol. 1, p. 317.

25. Ibid., vol. 2, p. 409. 'Deportment', as a peculiar preoccupation of the men of the Regency and the 1820s, is a minor theme of Dickens's *Bleak House*.

Index